D0914027

DECEPTION OPERATIONS

Studies in the East–West Context

Also from Brassey's

BLUNDEN AND GREENE
Science and Mythology in the Making of Defence Policy

CHARTERS AND TUGWELL
Armies in Low-Intensity Conflict

GODSON
Comparing Foreign Intelligence

HIGGINS
Plotting Peace

JAMES
Imperial Rearguard

LAFFIN
The World in Conflict 1990

DECEPTION OPERATIONS

Studies in the East–West Context

Edited by

DAVID A. CHARTERS

and

MAURICE A. J. TUGWELL

BRASSEY'S (UK)

(a member of the Maxwell Pergamon Publishing Corporation plc)

LONDON · OXFORD · WASHINGTON · NEW YORK · BEIJING

FRANKFURT · SÃO PAULO · SYDNEY · TOKYO · TORONTO

UK (Editorial)	Brassey's (UK) Ltd., 24 Gray's Inn Road, London WC1X 8HR, England
(Orders, all except North America)	Brassey's (UK) Ltd., Headington Hill Hall, Oxford OX3 0BW, England
USA (Editorial)	Brassey's (US) Inc., 8000 Westpark Drive, Fourth Floor, McLean, Virginia 22102, USA
(Orders, North America)	Brassey's (US) Inc., Front and Brown Streets, Riverside, New Jersey 08075, Tel (toll free): 800 257 5755
PEOPLE'S REPUBLIC OF CHINA	Pergamon Press, Room 4037, Qianmen Hotel, Beijing, People's Republic of China
FEDERAL REPUBLIC OF GERMANY	Pergamon Press GmbH, Hammerweg 6, D-6242 Kronberg, Federal Republic of Germany
BRAZIL	Pergamon Editora Ltda, Rua Eça de Queiros, 346, CEP 04011, Paraiso, São Paulo, Brazil
AUSTRALIA	Brassey's Australia Pty Ltd., P.O. Box 544, Potts Point, N.S.W. 2011, Australia
JAPAN	Pergamon Press, 5th Floor, Matsuoka Central Building, 1-7-1 Nishishinjuku, Shinjuku-ku, Tokyo 160, Japan
CANADA	Pergamon Press Canada Ltd., Suite No. 271, 253 College Street, Toronto, Ontario, Canada M5T 1R5

First edition 1990

Library of Congress Cataloging in Publication Data
Deception operations : studies in the East–West context/edited by David A. Charters and Maurice A. J. Tugwell.—1st ed.
p. cm.
Bibliography. p. Includes index.
1. World politics—1945– 2. Disinformation.
I. Charters, David. II. Tugwell, Maurice.
D842.D43 1989 909.82—dc20 89-15774

British Library Cataloguing in Publication Data
Deception operations: studies in the East–West context
1. Western bloc countries. Foreign relations with Eastern Europe. History 2. Europe. Eastern Europe. Foreign relations with Western bloc countries. History
I. Charters, David A. II. Tugwell, Maurice A.J.
327′.09171′3

ISBN 0-08-036706-2

Printed in Great Britain by BPCC Wheatons Ltd., Exeter

Contents

Acknowledgements

The Editors wish to express their appreciation for the vital financial support provided by Canadian foundations and a number of generous individuals, without which the research and writing of this book would not have been possible. Funding for this project was provided by the Birks Family Foundation, the Carthy Foundation, the Richard and Jean Ivey Fund, the Jackman Foundation, the J W McConnell Foundation, the Molson Family Foundation, the Stephen B Roman Foundation, the Themadel Foundation, the R Howard Webster Foundation, and the W Garfield Weston Foundation. We also wish to acknowledge the key role in getting the project off the ground played by Mr Kitson Vincent, and the support he provided throughout.

The Editors also thank Dr Barton Whaley for permitting the use of his work as the theoretical cornerstone of the volume. We are also grateful for the contributions to various phases of the project by Kirsten Amundsen, Patrick Armstrong, Jerald A Combs, Dominick Graham, David Hamilton, Paul Hollander, Richard Immerman, Peter Kenez, Richard Morgan, Arie Ofri, David Stafford, and Wesley Wark. We are indebted to Deborah Stapleford, our secretary, for her indefatigable accuracy, patience and good humour in the word processing of this manuscript.

Finally, we owe a debt of thanks to the editorial staff of Brassey's (UK) Publishers, especially to Bryan Watkins and Jenny Shaw, for their faith in us and their enthusiasm for the book, which were essential in bringing it to completion.

List of Contributors

Hannes Adomeit is a specialist in Soviet foreign policy and military affairs. His present position is senior staff member at the *Stiftung Wissenschaft und Politik* near Munich. His publications include, with Robert Boardman, *Foreign Policy Making in Communist Countries* (1979), and *Soviet Risk-Taking and Crisis Behavior* (1982)

Christopher Andrew is Fellow and Senior Tutor of Corpus Christi College, Cambridge. From 1976 to 1985 he was editor of the *Historical journal*; he is now co-editor of the journal *Intelligence and National Security*. His most recent books are *The Missing Dimension: Governments and Intelligence Communities in the Twentieth Century*, with David Dilks (1984), and *Secret Service* (American edition) *Her Majesty's Secret Service: The Making of the British Intelligence Community* (1985).

Trevor Barnes is a producer on BBC Television's *Newsnight* Programme. He took a Double First in Law and History at Corpus Christi College, Cambridge before studying at Harvard as a Kennedy Scholar. His work on American and British intelligence has been published in *The Historical Journal* and newspapers around the world, including *The Boston Globe*, *The Sunday Times* and *The Observer*.

David Charters is Associate Professor of History and Director of the Centre for Conflict Studies at the Univerisity of New Brunswick. He is author of *The British Army and Jewish Insurgency in Palestine, 1945–47* (1989) and co-editor and contributing author of *Armies in Low-Intensity Conflict* (1989). He is also Executive Editor of the Centre's journal, *Conflict Quarterly*.

Michael Handel is Professor of National Security Affairs at the United States Army War College, Carlisle Barracks, having previously taught at Harvard and the Hebrew University. His publications include *The Diplomacy of Surprise: Hitler, Nixon, Sadat* (1981) and *Weak States in the International System* (1981). He has contributed to several volumes on deception in war, and is co-editor of the journal *Intelligence and National Security*.

Michael Hennessy is a doctoral candidate in history, and former Research Assistant at the Centre for Conflict Studies, University of New Brunswick. His MA thesis (UNB 1987) examined United States Marine Corps pacification operations in I Corps area of Vietnam.

Harold James is Assistant Professor of History at Princeton University and was formerly Fellow and Director of Studies in History at Peterhouse, Cambridge. He is the author of *The Reichsbank and Public Finance in Germany, 1924–1933: a Study of the Politics of Economics During the Great Depression* (1985) and *The German Slump: Politics and Economics 1924–1936* (1985), and of articles on German economics, politics and music.

Guenter Lewy, Professor Emeritus of Political Science, University of Massachusetts, is the author of *Religion and Revolution* (1974), *America in Vietnam* (1978), and *Peace and Revolution* (1988).

David Martin served as Executive Secretary of the Committee for a Fair Trial for Draja Mihailovich. From 1959 to 1970 he was foreign policy aide to the late Senator Thomas J Dodd of Connecticut and from 1971 to 1979 he served as Senior Analyst to the Senate Subcommittee on Internal Security. He is the author of *Patriot or Traitor: The Case of General Mihailovich* (1978) and numerous political articles.

Simon Ollivant is the author of two *Conflict Studies* for the Institute for the Study of Conflict, London, and has contributed to several research projects sponsored by the Centre for Conflict Studies. He received his Ph.D in History from St Andrews, Scotland.

Maurice Tugwell is Director of the Mackenzie Institute for the Study of Terrorism, Revolution and Propaganda, in Toronto, and served as Director of the Centre for Conflict Studies, University of New Brunswick from 1980 to 1986. He is author of *Peace With Freedom* (1988), co-editor and contributing author of *Armies in Low-Intensity Conflict* (1989). He has also written numerous studies on various aspects of low-intensity conflict, psychological warfare and airborne forces.

Jiri Valenta is Professor of Political Science, Director of Soviet, East European and Strategic Studies, Graduate School of International Studies, University of Miami. He is a Member of the Council on Foreign Relations. His most recent books include *Soviet Decision-making in National Security* (1984), co-edited with William Potter, and *Grenada and Soviet/Cuban Policy* (1986), co-edited with Herbert Ellison.

J A Emerson Vermaat is a specialist in international law and international affairs, and has written extensively on West European peace movements, terrorism, military issues, East-West relations, East German intelligence, international law and the ecumenical movement. He is also the author of a book on the World Council of Churches in International Affairs. He has been a journalist for Dutch radio and television and for secular and religious newspapers since 1973.

Philip Walters studied modern languages at Corpus Christi, Cambridge and went on to complete his PhD thesis in Early 20th Century Russian Religious Philosophy at the London School of Economics. He has visited the Soviet Union and Eastern Europe on several occasions, and spent a year at Moscow University from 1974 to 1975. During the subsequent four years, he was a Research Fellow in Russian Studies at Cambridge University, and since 1979 has been employed at Keston College, where he has been Research Director since the beginning of 1984.

Glossary

ABC	– American Broadcasting Corporation
AFL/CIO	– American Federation of Labor/Congress of Industrial Organisations
ANZUS	– Australia, New Zealand, United States (Defence alliance)
AWS	– Air Weather Service (United States Air Force)
BBC	– British Broadcasting Corporation
BLO	– British Liaison Officer
CAA	– Civil Aviation Authority (United Kingdom)
CBS	– Columbia Broadcasting System
CCG(BE)	– Control Commission for Germany (British Element)
CCI	– Citizens Commissional Inquiry
CIA	– Central Intelligence Agency
CISL	– *Confederazioni Italiani dei Sindicati Lavoratori* (Italian Confederation of Workers' Syndicates)
CPC	– Christian Peace Conference
CPN	– Communist Party of the Netherlands
CPSU	– Communist Party of the Soviet Union
DCI	– Director of Central Intelligence
DDRS	– Declassified Documents Reference System
DKP	– Deutsche Kommunistische Partei (West German Communist Party – since 1968)
DNVP	– *Deutsches National Volks Partei* (German National People's Party)
DRVN	– Democratic Republic of Vietnam (North Vietnam)
ECOSOC	– Economic and Social Council (United Nations)
END	– European Nuclear Disarmament

ExComm	– Executive Committee (of the US National Security Council)
FBI	– Federal Bureau of Investigation
FBIS	– Foreign Broadcast Information Service
FM	– Field Manual (US Army)
FO	– Foreign Office (UK)
FORD	– Foreign Office Research Department
Fr	– Father (religious title)
GRU	– *Glavnoye Razvedyvatel 'noye Upravleniye* (Main Intelligence Directorate of the General Staff) (USSR)
G-2	– Military operational intelligence branch (US Army)
ICAO	– International Civil Aviation Organisation
ICBM	– Intercontinental Ballistic Missile
ICFTU	– International Confederation of Free Trades Unions
ID	– International Department (of CPSU)
IDF	– Israeli Defence Forces
IIP	– International Institute for Peace
IKV	– Inter Church Peace Council (Netherlands)
INS	– Inertial Navigation System
IPPNW	– International Physicians for the Prevention of Nuclear War
IRBM	– Intermediate Range Ballistic Missile
IRD	– Information Research Department (UK)
ISC	– International Student Conference
IWA	– International Workers Aid
JCS	– Joint Chiefs of Staff (US)
JPRS	– Joint Publications Research Service (Part of FBIS)
J-2	– Joint Services intelligence staff
KAL	– Korean Air Lines
KGB	– *Komitet Gosudarstvennoy Bezopasnostiy* (Committee for State Security) (USSR)
KPD	– *Kommunistiche Partei Deutschslands* (German Communist Party to 1956)
LCS	– London Controlling Section
MACV	– Military Assistance Command Vietnam (US)
MI5	– British Security Service

MI6	– British Secret Intelligence Service (see SIS)
MP	– Member of Parliament
MRBM	– Medium Range Ballistic Missile
NACA	– National Advisory Committee on Aeronautics
NASA	– National Aeronautics and Space Administration
NATO	– North Atlantic Treaty Organization
NBC	– National Broadcasting Corporation
NCO	– non-commissioned officer
NKVD	– *Narodnyy Komissariat Vnutrennikh Del* (People's Commissariat of Internal Affairs) (USSR)
NLF	– National Liberation Front (Vietnam) (see also VC)
NSA[1]	– National Students Association (see chapter 12)
NSA[2]	– National Security Agency (see chapters 14,16)
NSAM	– National Security Action Memorandum
NSC	– National Security Council (US)
NSDAP	– *National Socialistische Deutsches Arbeiter Partei*
NVA	– North Vietnamese Army
OB	– Order of Battle
OPC	– Office of Policy Coordination (CIA)
Op Log	– Operations Log
OSS	– Office of Strategic Services (US)
PCI	– Italian Communist Party
PRG	– Provisional Revolutionary Government (of Vietnam)
PRO	– Public Records Office (UK)
PVO	– Voyska Protivovozdushnoy Oborony (National Air Defence Forces, USSR)
PWE	– Political Warfare Executive (UK)
RAF	– Royal Air Force
RFE	– Radio Free Europe
RIAS	– Radio in the American Sector
RL	– Radio Liberty
RM	– *Reichmarks*
SA	– *Sturmabteilung* (Storm Troopers)
SAM	– Surface to Air Missile
SD	– Self-Defence (units of Viet Cong)
SED	– Socialist Unity Party (East Germany)
SIS	– Secret Intelligence Service (UK); see also MI6
SNIE	– Special National Intelligence Estimate
SOE	– Special Operations Executive (UK)

SPD – Social Democratic Party (Germany)
SS – *Schutz Staffel* (Protection Detachment)
SSD – Secret Self-Defence (units of Viet Cong)

TCP – Technological Capabilities Panel (TCP)

UAR – United Arab Republic
UK – United Kingdom
UN – United Nations
UNEF – United Nations Emergency Force
UNESCO – United Nations Educational, Scientific and Cultural Organisation
USA – United States of America
USAF – United States Air Force
USIA – United States Information Agency
USMC – United States Marine Corps
USSR – Union of Soviet Socialist Republics

VC – Viet Cong; see also PLAF
VVAW – Vietnam Veterans Against the War
VWP – Vietnamese Workers Party

WAY – World Assembly of Youth
WCC – World Council of Churches
WFTU – World Federation of Trades Unions
WPC – World Peace Council

Introduction

If, like the truth, the lie had but one face, we would be on better terms. For we would accept as certain the opposite of what the liar would say. But the reverse of truth has a hundred thousand faces and an infinite field.

Montaigne, Essays[1]

On 2 May 1987, Dr Valentin Pokrovskiy, President of the Soviet Academy of Medical Sciences and former director of the Soviet Ministry of Health Institute of Epidemiology, was quoted on the origin of the AIDS virus. 'I don't think it came from medical experiments,' Pokrovskiy stated, 'I think it was caused naturally.'[2] In the same month the World Health Organisation reaffirmed that the virus was of 'natural origin'.[3]

These statements slowed, but did not halt, a Soviet deception operation designed to attribute blame for the AIDS epidemic to the United States, by spreading the false story that the virus had been brought into existence by genetic engineering experiments conducted at Fort Detrick, Maryland—allegedly to develop new biological weapons. The deception evidently began in July 1983 when a letter was published in the Indian daily newspaper *Patriot*.[4] Later, this was picked up by the Soviet weekly *Literaturnaya Gazeta*.[5]

To provide spurious proof, a retired East German biochemist conceived a bogus scientific foundation for the deception and succeeded in having his concoction spread by newspapers in Africa, South America and Europe.[6] On 26 October 1986, for example, the London *Sunday Express* ran an interview with the biochemist which included all the allegations concerning Fort Detrick, and between January and June 1987 and Soviet media relayed the story at least 32 times.[7]

On 25 August 1986, the *Wall Street Journal* ran a story that the United States was contemplating military or covert action against Libya, and that internal opposition to Libyan leader Qaddafi was growing. These reports were false, being the unintended domestic replay of a United States deception operation designed for overseas consumption. When, one month later, the deceptive nature of the story was revealed in the *Washington Post*,[8] the Reagan administration was criticised for deceiving the American as well as other publics.

According to journalistic sources, the deception operation had been designed to create alarm in Libyan ruling circles and possibly precipitate 'Qaddafi's overthrow by Libyans . . . a sequenced chain of real and illusory events.'[9] Apparently the operation involved diplomatic visits and hints relating to a future American attack, media disinformation alleging internal unrest in Libya, the placing of false evidence, such as abandoned rubber dinghies to suggest covert landings, and naval signals traffic designed to mislead Libyan intelligence staffs. Whatever success the deception may have achieved initially, it presumably fell apart when exposed in Washington.

These two examples demonstrate the continuing use of deception by both sides in the confrontation between Marxist–Leninist states and their ideological or tactical allies, and the Western democracies and their allies—deception in the East–West context. From its beginnings in 1917, this confrontation has been neither war in the full meaning of the word, nor peace as we in the West understand it. This book examines the calculated use of deception in this East–West rivalry. The method adopted is the case study.

The aim of the book is to improve knowledge in four principal areas: the circumstances in which deception has been used; its results; the proclivity of the two political systems to use deception in international relations under conditions short of war; and the systems' relative vulnerability to such deception. Obviously, these objectives can be accomplished only in a rather limited way. By its very nature, deception is difficult to research and document. The most successful examples presumably delude us and remain undetected. Many others so muddy the waters that the truth and falsehood remain indistinguishable.

Deception has been addressed by almost every philosopher and political or military commentator in history. In the pre-Christian era, Sun T'zu argued that 'all war is deception,' meaning, presumably, that all successful war involved its use.[10] The Indian minister Kautilya's *Arthasatra*—Essentials of Indian Statecraft—is full of advice on the subject.[11] Among the ancients, Plato alone saw the potential of building a political system on 'one noble lie' which would promote belief in the pre-ordained superior status of the ruling elite, thus ensuring the loyalty of the masses.[12] Two thousand years later, Machiavelli's writings showed that time had not eroded the power or importance of deception.[13]

Recently, there has been a revival of interest in the subject, particularly its use in war, at both the scholarly and popular levels. A select bibliography related to such writings is to be found at page 407. In this book we have sought to focus attention on deception as an extra-diplomatic tool in time of peace, or between allies in time of war.

If some of the best examples of deception in the East–West context go unrecognised or unproven, then at the other end of the scale there is a surfeit of material comprising the small change of the business. It is in this area of forgeries, planted newspaper stories and front operations that most research and writing has so far been concentrated. What usually emerges is a fascinating glimpse of a nefarious tradecraft, one that is often referred to as disinformation. One reads of action. Only rarely is there any explanation of the wider purpose behind the action, or how this fitted into diplomatic or other activity, or whether or not the operation succeeded. We present case studies here that do examine the wider settings and do assess the outcomes of the various deceptions uncovered.

Obviously, the selection of cases was limited by availability of documentation. The 'Trust' operation, for instance, would justify a separate chapter rather than passing mention, were the full facts available.[14] The editors have also striven to achieve a rough balance between Eastern- and Western-inspired deceptions, and to include cases representing the various phases of East–West relations since 1917. Authors were recruited from the ranks of scholars, journalists, and professional writers mainly for their knowledge of the periods and settings in which their studies arose. The evidence of 'operators' has proven invaluable as a source, but the authors themselves are students of deception, not practitioners.

It has not been one of our objectives to develop a new deception theory or to prove a particular hypothesis. Nevertheless, we need to know what we mean by deception and how it works. Theory has nowhere been more concisely analysed than in Barton Whaley's article 'Toward a General Theory of Deception,'[15] and, with Michael Handel's definition, this analysis has been accepted throughout as a working formula. Whaley contributed to the early research for this book and the principal components of his theory make up most of the remainder of this introduction. It is stressed that his theory covers far more ground than this digest, relating to deception in all circumstances—business or cards, love or war, sports or politics. We will also look briefly at some conclusions of other theorists, where these have relevance to one or more case studies.

* * *

Deception, as examined in this volume, is defined thus:

> a purposeful attempt by the deceiver to manipulate the perceptions of the target's decision-makers in order to gain a competitive advantage.[16]

This is Michael Handel's formulation, adopted for the purposes of our study. One point should be made, however, in qualification. Whereas in war the 'target's decision-makers' are almost invariably the political and military leaderships in the enemy camp, or their senior intelligence and operations staffs, in the setting of this study this may not necessarily be the case. When an electorate or faction has it in its power to influence or dictate policy, that group may become the 'target's decision-makers', even though it holds no official executive position.

According to Barton Whaley, deception is but one form of perception. It falls within a main form of perception called misperception. These terms and their relationships are illustrated by Whaley in this diagram:[17]

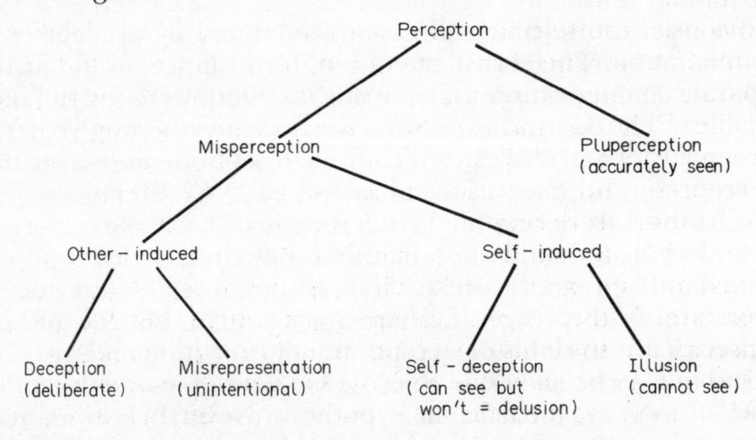

Whaley believes that all deceptions occur inside the brain of the person deceived and consequently in a narrow sense we are not deceived by others, only by ourselves. The deceiver's job is to *induce* deception by projecting a false picture of reality. To be deceived, the target or victim must both perceive the attempted portrayal and accept it in the terms intended. But this acknowledgement of the role of the victim in his own deceiving should not be carried to that common but mistaken conclusion that all deception is 'self-deception'. The latter is self-induced. Deception always involves a deceiver, the 'other inducer' of the diagram, and thus is a special case.

Fundamentally, deception is simply the distortion of perceived reality. The act of deceiving involves changing the pattern of distinguishing characteristics of the object or event detected by the victim's sensory system. Deception professes the false in the face of the real. Every deception operation is composed of two basic parts only: dissimulation and simulation. Dissimulation is hiding the real. It is covert. Its task is to conceal or at least obscure the truth. Operationally, dissimulation is done by hiding one or more of the characteristics that make up the distinctive pattern of a real thing. A military term for dissimulation is 'negative camouflage'; an intelligence term is 'cover'.

Simulation is showing the false. It is overt. Its task is to pretend, portray or profess an intended lie. Whaley explains that simulation is done by showing one or more characteristics that comprise the distinctive pattern of a false thing. Some soldiers call this 'positive camouflage'. Because both dissimulation and simulation are always present in any act of deception, secrecy by itself is not deception.

Whaley's theory sees ten steps in any deception process. First, the planner must know the wider goal: for instance, referring back to the cases at the start of this introduction, the isolation of the United States from her allies and the Third World, or the weakening and possible overthrow of Qaddafi. Second, he must decide how he wants his target audience to *react* to the deception, as opposed to what it should think: rage and disgust in the first example, fearful uncertainty in the other. The planner moves to Stage Three to decide what the audience is to *think* about the 'facts'—what it should perceive: that the United States military created and spread the AIDS virus, or that internal and external forces are on the verge of destroying Qaddafi. Fourth, he must decide what is to be hidden about those facts or impending events and what is to be shown in their stead: hiding the natural source of AIDS, showing the false; eclipsing the true situation inside Libya behind 'evidence' of forthcoming perils.

In the fifth stage, the planner analyses the pattern of the real thing to be hidden in order to identify distinguishing characteristics that must be deleted, and in Stage Six he considers persuasive characteristics or patterns that will create the false substitute. Dissimulation in both our examples was rather weak, with reliance on simulation—newspaper stories, 'medical evidence', abandoned boats, radio traffic.

Seventh, having designed a desired deception and method, the planner explores available assets for presenting the deception to its intended audience: secret services and their agents, propaganda or 'information' departments, diplomats, international news media. In Stage Eight, with the deception designed and the means in place, planning has ended and operations begin. The assets set about 'selling' the deception. In doing this, they must observe Stage Nine, which requires that the channels selected to pass the deception directly or indirectly to the target audience's brain or brains do in fact lead there. No amount of dissimulation and simulation will deceive someone who is looking in a different direction.

In the tenth and last stage, in order to succeed, the operation must attract the target's attention, hold interest, form the meaning intended by the planner in the audience's mind, without alerting that audience to the fact that it is being deceived. It should then achieve its goal as the audience accepts or 'buys' the false presentation, believing it to be true. Deception fails at this stage if the audience takes no notice, notices but judges the presentation irrelevant, misconstrues its intended meaning, or detects its method. The world must accept as true the story of AIDS being created at Fort Detrick; Qaddafi must be unnerved and his opponents encouraged.

In our two examples, weak dissimulation marred the Soviet operation when the international credibility of Soviet doctors seemingly took precedence over deception; United States simulation was undermined by domestic media exposés before the effects could be measured. Even where there are no such fumbles, success at Stage Ten is no guarantee that the deception will bring the desired operational outcome. Target reactions may not be as desired. Qaddafi, for instance, might have turned the American operation to his advantage by rallying his people against an evident external threat. Even when reactions follow the script, the operation may run out of control. Monitoring of the operation is therefore very important, to permit mid-course corrections or cancellation. Such monitoring is sometimes called 'playback'.

Two specialists in the field of deception studies, Donald Daniel and Katherine Herbig, see two variants of deception: 'ambiguity-increasing' and 'misleading' deception operations. The first clutters the target's sensors with a mass of contradictory indicators, Montaigne's 'hundred thousand faces,' ensuring that the level of ambiguity always remains high enough to protect the secret of the actual operation. The second misleads by reducing ambiguity. It builds up the credibility of one attractive but false hypothesis, thus causing the victim to concentrate his resources on the wrong contingency,[18] making him, in Whaley's formulation, 'quite certain, very decisive and wrong'.[19]

Michael Handel believes that deception in war must be considered a rational and necessary type of activity, even though its use does not guarantee success and can even prove counter-productive. He points to deception's potential value as a 'force multiplier' by magnifying the strength or power of the successful deceiver.[20]

When the 'target's decision-makers' comprise a mass audience, such as an electorate, deception channels may be expanded beyond the covert and specialist routes typically used in war to broader means of communication such as propaganda. According to a definition accepted by the Western Allies, the latter is any information, ideas, doctrines or special appeals disseminated to influence the opinion, emotions, attitudes or behaviour of any specified group in order to benefit the sponsor either directly or indirectly.[21] Propaganda is not intrinsically deceptive. It is 'loaded', meaning that it tends to select the facts it chooses to expose, and the interpretations it places upon them, to support preconceived bias. It may give a one-dimensional view of the world without actually telling lies. But when it is used as a channel for deception, propaganda dissimulates and simulates. Either it conceals the truth while propagating the false, or it conceals or falsifies the source from which the material emanates, or it does both.

The definitions applied to such falsifications have given rise to confusion because of rival classifications. Some writers refer to the truth as 'white' propaganda, descending through grey to black—blatant lies. This is not the usage adopted for this volume: there is no need to use coded language to describe a lie. In the accepted classification, a 'white' propaganda source is what it proclaims itself to be. TASS, the Soviet wire service, and the Voice of America are white propaganda sources. When the source is not identified—a voice on the radio without station identification, or a pamphlet published anonymously, then it merits the term 'grey'. A 'black' source pretends to be other than itself. A newspaper claiming independent editorial control would be black if it turned out that it was funded and directed by a foreign intelligence service. Radio messages claiming to come from an indigenous clandestine rebel radio station broadcasting from within the disputed territory would be black if it was discovered that they really originated in a neighbouring state, sponsored by that state's propaganda service.

* * *

The case studies that follow have been arranged in two groups: Eastern examples first, then Western. At the beginning of each section there is an editorial note about the cultural, historical and philosophical foundations that have influenced the Eastern and Western approaches to deception in time of peace. These notes also describe in outline the organisations that each system has developed for conducting deception operations, thus providing frameworks within which the case studies fit.

Endnotes

1. Michael Eyquem de Montaigne, 'Des Menteurs' [on Liars] *Essais* [1580], ed., Maurice Rat (Paris: Editions Garnier Frères, 1952), Book II, Chapter 9, quoted Sissel Bok, *Lying: Moral Choice in Public and Private Life* (New York: Pantheon, 1978), p. 3.
2. Quoted *Kepes* (Hungarian weekly magazine), 2 May 1987 in *Soviet Influence Activities: A Report on Active Measures and Propaganda 1986–1987* (Washington, DC: Department of State Publication 9627, 1987).
3. Proceedings of the 40th Meeting, World Health Assembly, 4–15 May 1987, cited *Soviet Influence Activities*, p. 42.
4. *Patriot* (New Delhi), 16 July 1983, quoted *Literaturnaya Gazeta* (Moscow), No 47, 19 November 1986, p. 9, cited *Soviet Influence Activities*, p. 34.
5. *Ibid.*
6. *Ibid.*, pp. 38–40. *Covert Action Information Bulletin*, No. 28 (Summer, 1987), pp. 36–7, 38–9, 40–41, 46–7, 61–2, gives extensive coverage to the charges of American genetic engineering experiments. While stopping short of endorsing them whole-heartedly, it provides enough planted axioms and 'corroborating evidence' to be suggestive in that regard.
7. For a detailed listing, see *Soviet Influence Activities*, cited.
8. *Washington Post*, 25 August 1986; 26 August and 2 October, 1986, and *New York Times*, 3 October 1986.
9. Quoted, Bob Woodward, *Veil: The Secret Wars of the CIA 1981–1987* (New York: Simon and Schuster, 1987), p. 471.
10. Samuel B Griffiths, trans., *Sun Tzu, The Art of War* (Oxford: Clarendon Press, 1963), p. 66.
11. T N Ramaswamy, *Essentials of Indian Statecraft: Kautilya's Arthasatra for Contemporary Readers* (London: Asia Publishing House, 1962).
12. See Bertrand Russell, *A History of Western Philosophy* (London: Allen and Unwin, 1946), pp. 130–31.
13. See Niccolò Machiavelli, *The Prince, the Discources and Other Writings*, ed., John Plamenatz (London: Fontana, 1972).
14. See 'Eastern Approaches', following this Introduction.
15. Barton Whaley (i), 'Toward a General Theory of Deception,' *The Journal of Strategic Studies* (London), Vol. V, No. 1 (March, 1982), pp. 179–93.
16. Michael T Handel, 'Intelligence and Deception,' revised edition in monograph form of article published in *The Journal of Strategic Studies*, Vol 5, No 1 (March 1982) pp. 122–54.
17. The original source of the diagram is Whaley (ii), *A Typology of Misperception* (draft, March, 1980) in which Whaley expresses thanks to Lewis Reich.
18. Donald C Daniel and Katherine L Herbig, 'Propositions on Military Deception,' in Daniel and Herbig, eds., *Strategic Military Deception* (New York: Pergamon, 1981), pp. 3–6.
19. Barton Whaley (iii), *Stratagem Deception and Surprise in War* (Cambridge, Mass.: unpublished thesis, MIT Center for International Studies, 1969), p. 135.
20. Handel, cited, p. 123.
21. North Atlantic Treaty Organization, *Glossary of Military Terms* (Mons: SHAPE, 1976), pp. 2–205.

Part One

Studies in Eastern Deception Operations

Eastern Approaches

Fine words are a mask to cover shady deeds.

Josef Stalin[1]

A reader coming uninitiated to the subject of Soviet deception could be forgiven for concluding that Western analysts had stumbled onto something new. Much of the Western writing on the subject is recent indeed, and a considerable proportion is devoted to contemporary cases, such as the AIDS deception mentioned in the Introduction, or charges of deception arising out of arms control initiatives.[2] A closer reading of history, however, demonstrates that, if Western interest in Soviet deception is a relatively recent phenomenon, Soviet practice is not. In fact, as Defence Intelligence Agency scholar John Dziak points out, deception has a long Russian history, and recent operations 'derive from a lengthy organisational milieu and operational tradition'.[3]

What follows will suggest not one tradition, but a blend of several. Russian history itself was one source. More firmly rooted in Byzantium than in Europe, by-passed by the Renaissance and Reformation, pre-1917 Russian society exhibited a medieval character in which the dominance of 'absolute truths' over empirical evidence was culturally and politically enforced. Russia was, after all, the land of the false front, the 'Potemkin Village'.[4] Marxism, with its claim to a monopoly of truth and emphasis on unity of theory and practice in the propagation of revolutionary ideas, provided yet another source. The Bolsheviks seized upon the power of propaganda as a political weapon with which to mobilise and 'correctly' educate the masses.

Finally, there is the tradition of clandestine warfare, which draws upon the experience both of the Bolsheviks themselves and of their Tsarist opponent, the Secret Police. If there is a special Soviet predilection to deception, then perhaps its genesis may be found in a fusing of these three traditions into an 'Operating code', which has put the lessons of experience to work for the protection and projection of the Soviet state.

From its inception, Marxism contained within itself many of the ideas, preferences and attitudes that would be beneficial for a modern propagandist or deception planner. The first, and perhaps most important, heritage from Marx to the Bolsheviks, was the conviction that they possessed the one and only true ideology. Marxism offered a belief system which claimed to provide the only 'correct' interpretation of the past and the only 'correct' path to the future; any other belief system was *ipso facto* 'incorrect', dangerous, and to be excluded from consideration.

However, the complexity of Marx's thought and analysis meant that only intellectuals could claim to understand him fully. Thus, the intellectual revolutionaries displayed a paternalistic attitude toward the common people who, the Bolsheviks argued, could not be expected to comprehend fully the subtleties of 'scientific socialism'.[5] In this regard Lenin wrote that, 'there can be no talk of an independent ideology formulated by the working masses themselves in the process of their movement'.[6] This conviction allowed Bolsheviks to dispense with pluralist notions according to which truth might emerge from a conflict of views and ideas. Lenin's rejection of the notion that the workers might arrive at socialist consciousness on their own, and his stress on the importance of organisation, implied that the workers would have to be led. He believed that the Bolsheviks' superior understanding of the working of history would enable them to interpret the immediate tasks of the proletariat and persuade them to carry out the revolution. The organising technique of intellectual revolutionaries was clearly going to be manipulative. So, in the process of creating Bolshevism, Lenin put emphasis precisely on those aspects of Marxism that were necessary for a good propagandist.

Nevertheless, Lenin's view accorded with the second component of Marxist thought seized upon by the Bolsheviks—the unity of revolutionary theory and practice. No distinction was drawn between spreading revolutionary ideas and organising revolution itself. 'Ideas are weapons,' he wrote. Propaganda, therefore, was deemed to be among the most important tasks of the revolutionary.[7]

This dual inheritance from Marxist thought determined the character of Bolshevik propaganda after 1917. No government had ever paid so much attention to influencing the minds of its subjects. Propaganda was so important that all Party members, organisations and sympathisers were expected to participate. Lenin's personal contribution to the process was his belief that the activists should conduct propaganda on several different planes simultaneously, in order to tailor the message to the appropriate level of sophistication of different audiences.

A commitment to the single absolute truth of Marxism brought an implicit intolerance of contrary views. The Bolshevik government introduced censorship. In the six to eight months following the October Revolution, all newspapers that did not identify themselves fully with the government's positions were closed down.[8] In their place arose what Paul Lendvai calls 'The Bureaucracy of Truth,'[9] a panoply of party-controlled media assets. From this time until the limited diversification under *glasnost*, the official organs maintained a monopoly of information. Not surprisingly, the most important publication was the Party newspaper, *Pravda (Truth)*.[10]

How deceptive was Soviet propaganda? In this respect the early history is contradictory. In Bolshevik vocabulary propaganda *per se* was a neutral activity; its tone depended on the truth of the ideology being propagated. The Bolsheviks came to power with the belief that their indoctrination work was bringing people the 'truth' and they needed no bourgeois obfuscations, deceptions and lies. It was this belief that allowed them to proclaim themselves proudly as propagandists. But they also experimented with black propaganda: that is, they printed forged declarations in the name of White Russian generals, the Bolshevik's rivals in the civil war, with the intention of embarrassing them.[11] Such tactical deception was not an essential or distinguishing feature of early Soviet propaganda. Once institutionalised and centralised, however, Soviet propaganda became deceptive as a matter of routine. This was not simply because it perpetrated lies, which it did, but because it made it impossible for Soviet citizens to discuss genuine political issues. In September, 1918, in an important article in *Pravda*, Lenin argued that there was no point in discussing politics, for the main political questions had already been resolved.[12] Instead, he wanted Soviet newspapers to describe life in individual factories and to 'unmask the class enemy', that is, those who did not work hard enough for the victory of the revolution.

It is hardly surprising then, that in a political environment in which there was no room for criticism, or for any heterogeneity of political views, the existing newspapers, indeed, the entire political language, acquired peculiar characteristics. For government-supported newspapers lying became easy, for there was no one to contradict. But more important, the very issue of what was a lie and what was truthful became almost meaningless. The newspapers lied in the sense that they did not reflect political reality; they simply did not discuss the genuine issues of the day.

It is not a mere figure of speech to say that the Soviet people came to speak a new language. This new political discourse distinguished between acceptable and unacceptable topics. Its language was full of set phrases, most of which, according to ordinary standards, were altogether meaningless. Slogans appeared on the walls such as 'Glory to work!' and 'Lenin is more alive than all the living!' Those who had come only recently to political participation never knew that political issues could be discussed in terms of alternatives. To them, the rules of political discourse in which they were being indoctrinated represented the only possible universe of such discourse. In 1928 a Soviet linguist, Selishchev, published a book on linguistic changes under the new regime. He pointed out that contemporary Russian usage contained a large number of words from the military vocabulary. Newspapers and orators spoke of struggle, war, *avant-garde*, muster, battalion, storming, besieging, 'war communism', workers' army, fronts, commanding heights, Leninist guards, and the 'Comintern as the General Staff of the world revolution,' and such like.[13]

The other influence which tilted Soviet propaganda in the direction of deception was the commitment to 'objective truth', that is, the truth that supports the historical process, which, of course, only the Party can interpret. In his book about Stalin, Alex de Jonge observes that, in place of evidence, the young communist idealist would feel that 'if the party decrees it, it must be necessary, if it is necessary, it is correct and, since it is correct, it must be true.'[14] As an example, he provides the story put about by the party that Trotsky was an agent of the Gestapo. It was widely understood that, subjectively speaking, Trotsky was innocent of the charge but that the party required him to be agent for reasons of historical necessity; so he could at one time be both subjectively innocent and objectively guilty.[15] Thus the line between Soviet propaganda and deception was blurred.

Soviet propaganda created *ersatz* institutions to discuss non-existent problems in a phoney language. The real problems were dealt with by an ever smaller political elite. Two of the means developed by this elite were the Cheka and the Comintern.

When the Bolsheviks came to power in November 1917, they had no plans for the establishment of a political police or for carrying out clandestine and deceptive operations against their enemies. It had always been assumed that the great majority of the people would enthusiastically welcome the revolution and therefore there would be no need for repression or deception. Further, they had always seen their revolution as part of a world-wide event. Thus, they were unprepared for the cruel reality in which they had to face a hostile majority of Russians and foreign-supported armies at the same time. In spite of their Marxism, which claimed to explain the course of history, the Bolsheviks were, of course, no better at predicting the future than anyone else.

So it was that circumstances forced the Bolsheviks, very shortly after taking power, to establish their own political police, the Cheka.[16] Any examination of the attitude of the Bolsheviks to the shadowy world of secret operations must keep in mind not so much their Marxist ideological background, as their underground experiences in fighting the Tsarist empire. Their long years of underground work deeply affected their behaviour. In this respect, their tutor was not Marx or any other socialist theorist, but some of the remarkably intelligent officers of the Tsarist political police, colloquially known as the *Okhrana*. It should be remembered that of all countries of the world, only Russia had such a police before 1917.[17]

The Tsarist regime was an inefficient autocracy. Officials, stupidly, did not distinguish between mere critics and determined enemies. Under the circumstances, political life necessitated deception. Revolutionaries, particularly Bolsheviks, became immensely skilful in communicating with an audience without arousing the suspicions of the censor. As the revolutionaries acquired strength and seemed increasingly dangerous, the regime tried infiltration in addition to repression. This policy turned out within a short time to be spectacularly successful. The underground movements were easily infiltrated and the *Okhrana* could buy the co-operation of some poor and ambitious workers cheaply. Many of the double agents showed extraordinary skill in dissimulation.

As a result, the history of the Russian Socialist parties is full of remarkable stories of betrayal. At one time, for example, four of the five members of the St. Petersburg Committee of the Bolsheviks were police agents.[18] The legendary Evno Azev, who murdered government ministers and members of the Imperial family; Bagrov, the assassin of Premier Stolypin; and Malinovskii, the head of the Bolshevik faction of Duma deputies; just to mention the most famous cases, all worked both for the revolutionary movement and for the police. Lenin was personally most deeply affected by finding out that Roman Malinovskii had betrayed the Bolshevik Party. For a long time he refused to believe that a person in whom he had placed so much faith could have been a traitor.[19]

This and similar experiences taught the Bolsheviks that in the world of politics things are not always what they seem. Of necessity, the party had to be conspiratorial in order to survive. Once the

Bolsheviks learned the skills of dissimulation and simulation, they never forgot them; deception became second nature. No political leadership of a European country had ever had so much experience in underground work, in carrying out deception, and in trying to ferret out deception. It is quite natural that after the Leninists established their own regime, and faced difficult times, confronting powerful enemies both inside and beyond the borders, they turned to those weapons that they knew from past experience.

* * *

The Soviets possessed three special assets in their deceptive endeavours. The first was secrecy, which has always been well-protected by the regime. Since deception consists of dissimulation (hiding the real) and simulation (projecting a false picture), a political system that virtually guarantees dissimulation as a constant asset operates at a considerable advantage.

The second asset has already been mentioned, a state security organisation (initially called the Cheka) steeped in a conspiratorial tradition. Finally, the victorious Russian communists encouraged a split in international socialism, forming from the radical wing, the Communist International, known as the Comintern. This provided the third important deception asset.

The link between state security and the Comintern was strong, but nevertheless deception operations by the Comintern network were often improvised locally in the target countries. These, conducted directly by state security (initially the Cheka and later, in the pre-Second World War period, the GPU, OGPU, GUGB and NKGB), were generally centrally conceived and controlled, as John Dziak has put it, under the auspices of senior party and state security leaders, beginning with Lenin and Cheka leader, Felix Dzerzhinskiy, and 'continuing with Stalin himself'.[20]

Although the details remain obscure, there is no doubt that early Soviet deceptions were as massive as they were masterly. Collectively, they can be viewed as survival exercises—operations to pre-empt possible threats to the young Bolshevik state. The main targets were the emigré White Russians in West Europe, the United States, and Japan, who were still actively plotting to overthrow the regime; the governments and intelligence agencies in the host countries, some of which might assist emigré operations; and, at a more abstract level, the very notion that resistance to Communism could ever succeed.

As explained earlier, dissimulation was made relatively easy by the high level of a secrecy surrounding the Soviet state, particularly its governing echelons. Simulation followed the advice about deception that Lenin is reported to have given Felix Dzerzhinskiy, head of Cheka: 'Tell them [the target audiences] what they want to hear.'[21] This advice runs like a golden thread through all the successful deceptions described in this book, whatever their origins. For the Soviets, it has often taken the form of simulating changes or re-packaging real changes taking place inside the Soviet Union. The direction and long-term implications of real or supposed changes are presented in such a way that the deceived target audience will act to the advantage of the regime.

Against the hostile emigrés, Dzerzhinskiy created a matrix of complex 'legends', cover stories founded on bogus internal opposition movements. The constant theme of the legends was that Communism need not be feared and that the regime was on the verge of collapse—precisely what the emigrés wanted to hear. Each legend acted as a channel for disinformation transmitted orally by agents posing as members of the Soviet branch of the supposed movement, or carried to the West in the form of fabricated documents. Historian Natalie Grant has described such operations in Europe, the Middle East, China and Japan. Some of these involved, in addition to the Comintern and state security, the People's Commissariat of Foreign Affairs and the Disinformation Bureau of the *Razvedur*, military intelligence.[22]

Operations 'Sindikat I' and 'Sindikat II' provide typical examples of this period's deceptions. The target of both was former Socialist Revolutionary Party terrorist and potent anti-Bolshevik chieftain, Boris Savinkov. According to the most reliable accounts, a Cheka agent called Eduard Opperput carried a suitcase full of forgeries to the West and succeeded in convincing Savinkov to trust him, the forged material, and the legend they supported. This legend simulated an important opposition movement, led by Opperput, in Byelorussia.

Opperput became Savinkov's close collaborator, gaining access to lists of genuine or potential anti-Bolsheviks inside Soviet territory. The liquidation of these people was the end of Sindikat I. Fake documents have retained their utility; Chapter 8, dealing with the death of Aldo Moro, describes a Soviet deception based on a forgery.

The legend underpinning Sindikat II lured Savinkov into Russia where he was later killed, although it is unclear whether he had already struck a deal with the regime before returning. What seems reasonably certain is that this second legend was the famous 'Trust'.[23]

Dziak has described the Trust as the prototypical strategic deception and provocation operation in the Soviet repertoire. Like several other legends, it took the form of a notional opposition within the Soviet Union. State security projected this false picture of reality towards several targets: the anti-Soviet emigration in the West; opponents inside the Soviet Union; and Western intelligence services and their governments. Planning began in 1921, three years before Savinkov left for Russia, and the operation ran until halted by Stalin in 1927, by which time it had achieved all its objectives.

Although based upon deception and provocation, the Trust operation also used the techniques of penetration, diversion, fabrication, influence, and what the Russians call *kombinatsiya*—the combining of operational undertakings in several times and places for maximum impact. The supposed opposition group was named the Monarchist Association of Central Russia. Its equally bogus cover title, also invented by state security, was the Moscow Municipal Credit Association—the Trust.

Through this complex legend, not only Savinkov, but the freelance agent Sidney Reilly, connected with British intelligence, was lured back. The degree of success can also be measured from a trip, arranged by state security (in their bogus capacity of Trust leaders), which was undertaken between the Autumn of 1925 and Spring 1926 by another prominent opposition emigré, V V Shulgin. This individual was duped into believing that he was making an 'underground' tour of Russian resistance forces. His report was first approved by the Trust (state security) and then published in Berlin. It announced that Communism was on the way out in Russia and that the Soviet leaders were really nationalist-monarchists of a new type. These welcome messages were topped with the operational element of the deception—that any anti-Soviet action by Western governments would be counterproductive.

The Trust overlapped the period of the New Economic Policy, which Dzerzhinskiy, through another of his portfolios, oversaw. Thus the deceptive messages were reinforced by apparent evidence of a return to a form of free market economy. When the New Economic Policy and the Trust were wound up in 1927, internal and external opposition movements had been eliminated or mortally wounded, Western intelligence agencies had been humiliated, and any lingering hopes among Western governments that the Soviet regime might be overthrown were shattered.

Under Stalin, state security turned its fury on the peasants, the Party, the Red Army, and itself. A major effort was made by the Moscow authorities to conceal the true horrors of the terror-famine of 1930–33 and the show trials.[24] The formula of telling audiences what they wanted to hear rescued the regime from the condemnation it deserved. For Bolshevism's opponents, the trick had been to deny communist power and to inflate opposition and change; for its sympathisers, the requirement was to play down embarrassing past and current events, redirecting attention to a radiant future.

In this, Communism has enjoyed a special advantage. Representing as it did the Utopian hopes of millions of socialists beyond the circles of committed communist party members, it could rely upon remarkable displays of amnesia, forgiveness, and plain refusal to believe information that contradicted the tenets of blind faith.[25] To tell this audience that mass murder had not been committed was to tell them what they wanted to hear; for that reason it would likely be believed. If empirical proof surfaced to shatter this delusion, it might be necessary to admit that some deaths had been 'historically inevitable' in the struggle to build the socialist future.[26]

So it was that American journalist Walter Duranty was able to misinform readers of the *New York Times* while covering up for Stalin's black deeds. So it was that such Western luminaries as George Bernard Shaw, Bertolt Brecht, Harold Laski and Henry Wallace deceived both themselves and their audiences with stories and opinions totally unfounded in fact.[27] Chapter 9, which deals with the 1983 shootdown of the Korean airliner, suggests that this syndrome was not merely a freak of history.

The Comintern played a major role in recruiting and sustaining the loyalties of these 'fellow-travellers'. From its very inception, the Comintern was a tool of Soviet policy. This organisation in 1920 obliged all participating communist parties to engage in illegal, underground work. It often happened that while one arm of the Soviet Government, the People's Commissariat of Foreign Affairs, negotiated with a capitalist country and maintained business-like relations with it, another arm, the Comintern, tried to foment social revolution. In order for the Soviet Government to maintain normal diplomatic relations, it tried to preserve the fiction that the Comintern was an independent agency.[28]

It quickly became evident, however, that the illusion of the Comintern as an independent organisation could not be sustained. The answer, then as later, was to mask the Comintern behind 'front' organisations. To the eyes of the Western publics in whose countries these fronts were formed, there was no easily visible link to foreign influence. Moreover, the communists who brought the fronts into existence were careful to persuade respected non-communists to assume key positions, reserving for themselves the low-profile secretarial appointments essential for effective control.

Such fronts usually addressed 'motherhood' issues that could be manipulated to Soviet advantage, such as opposition to Fascism. The first chapter in this volume describes a classic Comintern front operation of the 1930s, organised by the legendary Willi Münzenberg. In Chapter 2, the Comintern, aided by Soviet assets, is again in evidence, trying to ensure the success of Tito, its man in Yugoslavia. In May 1943 the Soviet Government formally disbanded the Comintern, presumably as a sop to their Western allies. It soon became evident, however, that its duties and assets had been transferred to other departments, particularly state security and the Ministry of Foreign Affairs. The 'two pronged' character of Soviet foreign policy, operating both through conventional diplomatic channels and unconventional party networks, was not abandoned. In Chapter 10, the Comintern's current successor, the International Department, is seen using the World Peace Council, both as a deceptive 'front' in its own right, and as the instigator of deceptive 'peace offensives'.

* * *

Soviet organisation for deception remained relatively centralised until the Khrushchev era. By the 1960s, however, deception had become, at one and the same time, bureaucratised and decentralised. According to several medium or high-level defectors, the *Komitet Gosudarstvennoy Bezopasnosti* (KGB), Committee for State Security—the Soviet intelligence and security service—was given pride of place in this new framework. At the end of 1958, Aleksandr Shelepin was appointed head of the KGB. Under him, the KGB was to return to an effective working partnership with the Party, particularly in foreign affairs, as had been the case before Stalin's excesses had strained the relationship. Apparently he was able to convince the Party leader, Nikita Khrushchev, that deception in the Lenin-Dzerzhinskiy tradition should be revived. One of Shelepin's earliest actions as KGB chief was to create a Department D (for disinformation) within the KGB's First Chief Directorate, and to place Colonel Ivan Agayants in charge. This officer had previously worked out of the Paris embassy, where his disinformation operations attracted approval from Moscow. At about the same time, the veteran Comintern controller, Boris Ponomarev, became the prime mover in the International Department of the Party's Central Committee. This department had assumed the work of the defunct Comintern and the short-lived Cominform as the second prong of the Party's foreign policy, the instrument of penetration and subversion working parallel to the more or less conventional diplomacy of the Ministry of Foreign Affairs.[29]

According to the same defectors, in 1959 Shelepin assembled

more than 2,000 senior KGB officers, the ministers of defence and internal affairs, and senior Central Committee members, to brief them on future deception policy within a long-term plan approved by Khrushchev. The security and intelligence services of the Warsaw Pact were to be mobilised to influence international relations by destabilising individual 'enemies', the United States, Britain, France, West Germany, Japan and other Western Alliance states, and by undermining their alliances.

These objectives were to be achieved by co-operative efforts on the part of the new disinformation department, relevant party and government departments at national and republic levels, and the security apparatus of all communist states. In addition the KGB would mobilise the Soviet intelligentsia for influence work against target countries.

Deceptions on the scale suggested by such reports are difficult to document. What follows in this volume should be considered only a partial survey. But it does provide an appreciation of the diversity of Soviet operations, the array of techniques, and the mixed record of results.

Chapters 8 and 9, examining the Aldo Moro and KAL 007 incidents, indicate that, as in the early years of the Revolution, the overt official media channels still play an important role in deceptive propaganda. At least, they reassure the faithful and present the 'correct' view to the home audience. At best, they can plant or replay ideas or 'facts' that may be useful to others engaged in disinformation.[30]

The usefulness of 'fronts', so clear from the Comintern period, is apparent again in Chapters 7 and 10. The Russian Orthodox Church, controlled and compromised by the Party's apparatus, is virtually a 'front' in its own right. The Church's role as a 'transmission belt' of deception is explained in Chapter 3.

During the Second World War, the Red Army developed doctrine and skills for strategic and tactical deception.[31] Chapter 5 indicates that these skills were not forgotten. The political and tactical deceptions in Soviet intervention operations demonstrated increasing subtlety and sophistication. By contrast, Khrushchev's own attempts at deception, described in Chapter 4, smack of the crudity and impatience that marked this leader's character.

Western analysts have devoted considerable resources to examination and discussion of the KGB's much-vaunted deception skills.[32] In Chapter 6, Agayant's creation, the disinformation department, was too successful by half in simulating an Israeli threat prior to the 1967 Middle East war; and the KGB achieved only a temporary and rather modest success with a forgery surfaced to support the overt deception described in Chapter 8.

Early in the 1970s, the disinformation department was elevated to the status of Service A of the First Chief Directorate, indicating a higher standing within the KGB hierarchy. In the same decade, Ponomarev's International Department seemed to act as the Politburo's co-ordinating body in international affairs, especially that prong of policy relying on the international communist movement, front organisations, propaganda and deception. The Department shared with Service A the manipulation of Soviet and foreign intellectuals and professionals in its various campaigns.[33] Under Mikhail Gorbachev, these organisations were strengthened by fresh blood, especially a more experienced Secretary for Propaganda,[34] but reform under *perestroika* has only begun recently. The new order is discussed briefly at the conclusion of Chapter 10.

Endnotes

1. Extracted from quote cited in Kerry M Kartchner, ' "A Mask to Cover Shady Deeds": Soviet Diplomatic Deception, 1917–1939,' in Brian D Dailey and Patrick J Parker, eds., *Soviet Strategic Deception* (Lexington, Mass and Stanford, California; Lexington Books/Hoover Institution Press, 1987), p. 147.
2. See for eg., Dailey and Parker, *op.cit*; Richard H Shultz and Roy Godson, *Dezinformatsia: Active Measures in Soviet Strategy* (New York and London: Pergamon-Brassey's, 1984); John Barron, *KGB Today: the Hidden Hand* (New York: Reader's Digest Press, 1983); United States Congress, House of Representatives, Hearing, Permanent Select Committee on Intelligence, *Soviet Covert Action (The Forgery Offensive)* (1980) and *Soviet Active Measures* (1982); Roger Beaumont, *Maskirovka: Soviet Camouflage, Concealment and Deception* Stratech Studies SS82-1 (College Station, Texas: Center for Strategic Technology, 1982); Joseph D Douglass Jr, 'Soviet Disinformation,' *Strategic Review*, Vol. 9, No. 1 (Winter 1981), pp. 16–25, and with Samuel T Cohen, 'Selective Targeting and Soviet Deception,' *Armed Forces Journal International*, (September, 1983), pp. 95–101; Edward Jay Epstein, 'Disinformation: Or, Why the CIA Cannot Verify an Arms Control Agreement,' *Commentary* (July, 1982), pp. 21–8; Michael Mihalka, 'Soviet Strategic Deception, 1955–1981,' *Journal of Strategic Studies*, Vol. 5, No. 1 (March, 1982), pp. 40–93; Jennie A Stevens and Henry S Marsh, 'Surprise and Deception in Soviet Military Thought,' *Military Review*, Vol. 62, Nos. 6–7 (June/July, 1982), pp. 2–11, 24–35; and Jiri Valenta, 'Soviet Use of Surprise and Deception,' *Survival*, Vol. 24, No. 2 (March–April, 1982), pp. 50–61.
3. John J Dziak, 'Soviet Deception: the Organizational and Operational Tradition,' in Dailey and Parker, p. 3.
4. Martin Ebon, *The Soviet Propaganda Machine* (New York: McGraw-Hill, 1987), pp. 6–7.
5. The single most important work for understanding Lenin's thought is *What is to be Done?* published in 1902. V I Lenin, *Polnow sobranie sochinenii*. 5th ed., (Moscow: Gosizdat, 1960), Vol. VI (hereafter Lenin, *PSS*). In *What is to be Done?*, p. 38, Lenin approvingly quoted the great German socialist leader, K Kautsky who wrote: 'Modern socialist consciousness can arise only on the basis of profound knowledge.'
6. Lenin, pp. 39–41.
7. Parts of this section dealing with early Soviet propaganda are based on an essay written for the editors by Professor Peter Kenez, author of *The Birth of the Propaganda State: Soviet Methods of Mass Mobilization, 1917–1929* (New York: Cambridge University Press, 1986). See also Gayle Durham Hannah, *Soviet Information Networks* (Washington, DC:

Center for Strategic and International Studies, 1977), p. 12. In *What is to be Done*, Lenin emphasised that to organise and to carry out propaganda were the opposite sides of the same coin.

8. John Keep, ed., *The Debate on Soviet Power. Minutes of the All-Russian Central Excecutive Committee of Soviets. Second Convocation. October 1917–January 1918* (Oxford: Clarendon Press, 1979), pp. 70–78.
9. Paul Lendvai, *The Bureaucracy of Truth: How Communist Governments Manage the News* (London: Burnett Books, 1981).
10. Ebon, pp. 29–49.
11. Peter Kenez, *Civil War in South Russian, 1919–1920* (Berkeley; University of California Press, 1977), p. 271.
12. Lenin, *PSS*, Vol. 37, pp. 89–91.
13. Ia M Selischev, *Iazyk revoliutsionnoi epokhi* (Moscow: Gosizdat, 1928), pp. 85–91. It was this militarized, but meaningless jargon that George Orwell reproduced with such frightening clarity in *Nineteen Eighty Four.*
14. Alex de Jonge, *Stalin and the Shaping of the Soviet Union* (London: William Collins, 1986), Fontana ed., p. 139.
15. *Ibid.*
16. On the establishment of the Cheka see Lennard D Gershon, *The Secret Police in Lenin's Russia* (Philadelphia: Temple University Press, 1976), pp. 15–24; George Leggett, *The Cheka: Lenin's Political Police* (Oxford: Clarendon Press, 1981), pp. 1–27; John J Dziak, *Chekisty: A History of the KGB* (Lexington, Massachusetts: D C Heath & Co., 1988), pp. 1–38.
17. Richard Pipes develops the implications of this fact in *Russia under the Old Regime* (New York: Scribner, 1974), pp. 290–318.
18. Leggett, p. xxiv.
19. Malinovskii's story is told in detail in Bertram Wolfe, *Three Who Made a Revolution.* Fourth ed. (New York: Delta, 1964), pp. 535–57.
20. Dziak, *Chekisty*, p. 40.
21. *Ibid.*, quoted p. 43.
22. Main sources on these operations are Dziak, *Chekisty*; Leggett; and Natalie Grant, *Deception, A Tool of Soviet Foreign Policy* (Washington, DC: Nathan Hale Institute, 1987).
23. On Sindikat I and II, and the Trust see Dziak, *Chekisty*, pp. 43–50; Grant, pp. 8–10; and Leggett, pp. 295–99.
24. See Robert Conquest (i), *The Great Terror*, rev. ed. (New York: Macmillan, 1973) and (ii), *The Harvest of Sorrow: Soviet Collectivization and the Terror Famine* (New York: Oxford University Press, 1986).
25. See Paul Hollander, *Political Pilgrims: Travels of Western Intellectuals to the Soviet Union, China and Cuba 1928–1978* (Oxford: Oxford University Press, 1981), Chapter Four, pp. 102–76.
26. In this respect, see Melvin J Lasky, 'The Cycles of Western Fantasy,' *Encounter*, February, 1988, pp. 3–16.
27. Hollander, pp. 119, 144–48, 156–60, 163–71; Lasky, p. 9.
28. Kartchner, in Dailey and Parker, p. 155–6. Comintern involvement in an abortive rising in Germany in 1921, and Soviet efforts to dissociate themselves from it, are described in Branko Lazich and Milorad Drachkovich, *Lenin and the Comintern* (Stanford: Hoover Institution, 1972), Vol. 1, pp. 471–527.
29. Main sources on reorganization are Dailey and Parker, Dziak, *Chekisty*, pp. 145–65; Ladislav Bittman, *The Deception Game: Czechoslovak Intelligence in Soviet Political Warfare* (Syracuse, NY: Syracuse University Research Corp., 1972), pp. 15–20; Anatoliy Golitsyn, *New Lies for Old* (London: Bodley Head, 1984), pp. 46–51; Jeffrey T Richelson, *Sword and Shield: The Soviet Intelligence and Security Apparatus* (Cambridge, Mass: Ballinger, 1986), p. 24. See also Jan Sejna, *We Will Bury You* (London: Sidgwick and Jackson, 1982). On recent organisation, see Shultz and Godson, pp. 14–33; and Richards J Heuer Jr, 'Soviet Organization and Doctrine for Deception', in Dailey and Parker, p. 27. The three defectors were Bittman, Golitsyn, and Sejna. (Specialists tend to be extremely sceptical about Golitsyn's analysis of events since his 1961 defection, which

most see as extravagant; his first hand accounts of events while he was still in KGB service are, however, regarded as sound).

30. The AIDS disinformation discussed in the Introduction was propagated through overt and covert channels. Shultz and Godson, pp. 51–110, examine the development of deceptive themes in overt Soviet propaganda from 1960 to 1980.
31. See Earl F. Ziemke, 'Stalingrad and Belorussia: Soviet Deception in World War II' in Donald C Daniel and Katherine L Herbig, eds., *Strategic Military Deception* (New York: Pergamon Press, 1981), pp. 243–76; and David M Glantz, 'The Red Mask: The Nature and Legacy of Soviet Military Deception in the Second World War', *Intelligence and National Security*, Vol. 2, No. 3 (July, 1987), pp. 175–259.
32. *Soviet Covert Action (the Forgery Offensive)* (1980) and *Soviet Active Measures* (1982); Shultz and Godson, pp. 19–20, 31–3, 133–5, 150, 172–84; 190; Ladislav Bittman, *The KGB and Soviet Disinformation: an Insider's View* (Washington, DC: Pergamon-Brassey's, 1986).
33. Shultz and Godson, pp. 19–25, 180–84.
34. Alexander Yakovlev, former Ambassador to Canada, who was later promoted to head the International Commission of the CPSU.

CHAPTER 1

Willi Münzenberg, the Reichstag Fire and the Conversion of Innocents

CHRISTOPHER M ANDREW AND HAROLD JAMES

In the mid-1930s it seemed to me and to many of my contemporaries that the Communist Party and Russia constituted the only firm bulwark against Fascism, since the Western democracies were taking an uncertain and compromising attitude towards Germany. I was persuaded by Guy Burgess that I could best serve the cause of anti-fascism by joining him in his work for the Russians.

Anthony Blunt[1]

The central delusion of the Cambridge moles, as of most Soviet sympathisers in the 1930s, was their belief that the Comintern and the Soviet Union represented 'the only firm bulwark against Fascism.' In reality, the anti-fascist rhetoric of Comintern and Soviet leaders was often belied by their actions. In the final phase of the Weimar Republic, the German Communist Party, in accordance with the Comintern principles of 1928, attacked not the Nazis but the Socialists, whom they termed 'social fascists'. Thus they may actually have assisted Hitler's rise to power. Before the Second World War, the Soviet Union had a machinery of totalitarian oppression even more highly developed than that of Nazi Germany. On 23 August 1939 the two dictators, Hitler and Stalin, concluded the Nazi–Soviet Pact which prepared the way for the invasion of Poland and division of eastern Europe into spheres of influence. After 1939 Stalin handed over German Communists to the Gestapo. Between 1939 and 1941 Stalin continued to supply Germany with vital raw materials and food-stuffs; and he refused to believe in the likelihood of a German invasion.

The belief of many left-wing intellectuals, at least until the Nazi–Soviet Pact, that Soviet Russia was 'the only firm bulwark against Fascism' derived from self-deception as well as from Soviet-inspired deception. Alienated by the injustices of their own society, they identified a Utopian image of a society free from those injustices, with the propaganda image of Stalin's Russia. They believed that Fascism was engulfing the democracies, and saw the cause of anti-fascism and the cause of the Soviet Union as one and the same. Anti-fascism thus became the most effective rallying cry of Soviet and Comintern propagandists.

The great virtuoso of the Comintern's anti-fascist crusade was the German Communist Willi Münzenberg, affectionately remembered by his comrade and 'life partner' Babette Gross as 'the patron saint of the fellow travellers'.[2] Münzenberg had caught Lenin's eye during the Swiss exile in the First World War when he became youth leader of the Zimmerwald Left. In April 1917 he was among the 'handful of faithfuls' who assembled to bid Lenin farewell as he boarded the sealed train taking him to Petrograd and revolution. When the Communist Youth International was inaugurated in Moscow in 1920, Münzenberg became its first president. But it was during the Russian famine of 1921 that Münzenberg emerged as the master propagandist of the Comintern. In August of that year, he founded the International Workers Aid (IWA) with headquarters in Berlin. Over the next two years, the IWA despatched to Russia fifty shiploads of supplies ranging from drugs to sewing machines. The shipments became the basis of a brilliant propaganda campaign. According to Babette Gross:

> His magic word was solidarity—at the beginning solidarity with the starving Russians, then with the proletariat of the whole world. By substituting solidarity for charity Münzenberg found the key to the heart of many intellectuals; they reacted spontaneously . . . When he spoke of the 'sacred enthusiasm for the proletarian duty to help and to assist' he touched on that almost exalted readiness for sacrifice that is found wherever there is faith.[3]

Each 'act of solidarity with the Russian people' forged an emotional bond between the giver and the idealised version of the Soviet worker–peasant state presented by Comintern propaganda.

The IWA became known in party slang as the 'Münzenberg Trust'. Arthur Koestler, who was sent to work for Münzenberg in 1933, found that he had acquired 'a greater measure of independence and freedom of action in the international field than any other Comintern leader. . . . Undisturbed by the stifling control of the party bureaucracy,' he was able to run campaigns 'in striking contrast to the pedantic, sectarian language of the official Party Press'.[4] The Münzenberg Trust quickly gained the support of a galaxy of 'uncommitted' writers, academics, and scientists. Käthe Kollwitz's poster 'Hunger', with its portrait of a large-eyed hungry child stretching out a hand for food, produced for Münzenberg in 1923, is one of the most potent and best remembered images of the century. In the course of the 1920s, the Münzenberg Trust established its own newspapers, publishing houses, book clubs, films, and theatrical productions. As far away as Japan, according to Koestler, the Trust controlled directly or indirectly 19 newspapers and magazines. Remarkably, Münzenberg even managed to make most of his ventures pay.[5]

The IWA was the progenitor of a series of what Münzenberg privately called 'Innocents' Clubs'[6] intended to 'organise the intellectuals' under covert Communist leadership in support of a variety of voguish causes. He had a friendly contempt for the 'innocent' intellectuals seduced by the lure of spiritual solidarity with the proletariat.[7] Though his main preoccupation remained propaganda, Münzenberg also used his Trust as a cover for intelligence work. This combination had an absolutely respectable justification in the works of Lenin. In an article in *Iskra* in 1901, Lenin had argued that the newspaper should not only be a propaganda instrument, but should also be the organisational centre of the party, and in particular that it should establish a network of agents.[8] Münzenberg proceeded to do just this. In 1927 he founded a League against Imperialism, swiftly denounced by the Socialist International as 'nothing but a communist manoeuvre'. And indeed, a police raid on its Berlin headquarters in 1931 found evidence that it received regular instructions from the Comintern, that it excluded non-communists on its executive from real power and that it had an intelligence network with agents in several countries.[9]

Usually, however, Münzenberg managed to keep both himself and the Comintern out of sight in the management of the 'Innocents' Clubs'. In August 1932 he organised a spectacular Congress in Amsterdam against Imperialist War, attended by 2,195 delegates representing 79 pacifist groups and 151 other organisations from 29 countries. The ostensible direction of the congress was entrusted to the celebrated French pacifist intellectuals Romain Rolland and Henri Barbusse, and messages of support were obtained from a galaxy of international celebrities, including Albert Einstein and Sigmund Freud.[10] The British Labour Party gave enthusiastic support to congress decisions. Its report 'United Front Against War' made no mention of Moscow, swallowing whole the claim that the Congress was the child of Rolland and Barbusse.

Anti-fascism gave Münzenberg the opportunity for his greatest propaganda coup. Early in 1933, shortly before Hitler became

German Chancellor on January 30, Münzenberg went to Moscow to discuss how to continue his propaganda if he were forced to leave Germany.[11] In view of his success in attracting the likes of Barbusse and Rolland, Paris seemed the best alternative base. On February 27, only six days before Reichstag elections were called by the Nazis, in an attempt to win an overall majority, a fire destroyed the plenary chamber of the Reichstag. Hitler and Hermann Göring, by now installed as Prussian Minister of the Interior, claimed that the fire was the work of communists, intended as the signal for a general communist rising against the German state. Immediately, warrants were issued for the arrest of leading communists. Münzenberg crossed the French border illegally, and came to Paris. There, despite the opposition of the Sûreté, he and his collaborators secured political asylum through the good offices of the radical deputy Gaston Bergery.[12]

The Reichstag fire was the work of a young Dutchman, Marinus van der Lubbe, who had once belonged to the Dutch Communist Party but had left in protest at its elitist Leninist character. He gave a coherent account of his activities which made sense to the police and fire officials: he had acted alone, was not in contact with any German communists, and his motive had been to inspire a popular uprising.[13] For nearly thirty years, these simple facts were submerged beneath two competing conspiracy theories. The first, put about by Hitler and Göring, that the fire was part of a communist plot, was used to legitimise the arrest of communists. The missing link, which was never satisfactorily established, was the identity of van der Lubbe's supposed German Communist Party (KPD) accomplices. Nevertheless, for those Germans who wished to believe in a communist threat, the Nazi's conspiracy theory provided the necessary evidence.

The rival theory, which will be described later in greater detail, was supplied by Münzenberg. It postulated a Nazi plot, as a provocation intended to justify an onslaught on the KPD. This theory survived the Nazi era and was accepted by most historians until 1962, when Fritz Tobias published his definitive account of the incident and the deception which has surrounded it. Tobias demonstrated with reasonable certainty that there was no conspiracy, whether Red or Brown, and that van der Lubbe's account was true.[14]

Once established in Paris, Münzenberg set to work to elaborate the theory of the Nazi plot. First, he created the World Committee for the Victims of German Fascism, ostensibly as a philanthropic organisation. Koestler writes that 'great care was taken that no Communist—except a few internationally known names, such as

Henri Barbusse and J B S Haldane—should be connected with the Committee.' The French section was led by a distinguished Hungarian emigré, Count Károlyi.[15] The international Chairman was a British Labour peer, Lord Marley. Einstein also agreed to join the Committee, and soon found himself described as 'President'.[16] In fact the Paris secretariat was, as Koestler later disclosed, 'a purely Communist caucus, headed by Münzenberg and controlled by the Comintern . . . Münzenberg himself worked in a large room in the World Committee's premises, but no outsider ever learned about this. It was as simple as that.'[17]

On 15 May 1933, an advertisement appeared in a Communist exile newspaper, *Der Gegen-Angriff* (published in Prague, Zürich and Paris) to announce the forthcoming publication of a *Brown Book on the Hitler Terror and the Burning of the Reichstag* (hereinafter referred to as the *Brown Book*).[18] This book, published anonymously from the Paris base in August 1933, was Münzenberg's most remarkable achievement. It was probably the single most effective piece of propaganda in the history of the Comintern. Koestler later made the (somewhat exaggerated) claim that it 'probably had the strongest political impact of any pamphlet since Tom Paine's *Common Sense* . . . It became the bible of the anti-fascist crusade.'[19] Within a few weeks, it was translated into at least 20 languages, including Greek, Hebrew, Japanese and Yiddish. 15,000 copies were printed for Britain, and around 135,000 were to be smuggled into Germany disguised as Goethe's *Hermann und Dorothea* or Schiller's *Wallenstein*, or hidden in commercial packages.[20]

The title page of the English edition said simply that the book was 'prepared by the World Committee for the Victims of German Fascism (PRESIDENT: EINSTEIN) with an Introduction by LORD MARLEY'. The formula surprised Einstein: 'My name appeared in the English and French editions as if I had written it. That is not true. I did not write a word of it.' But he added good-naturedly: 'The fact that I did not write it does not matter . . .'[21]

Lord Marley's Introduction, written from the 'House of Lords, London SW1,' gave the *Brown Book* an air of sober respectability and scrupulous accuracy:

> Many authentic documents have been placed at the disposal of the World Committee for the Victims of German Fascism: some by journalists, others by doctors and members of the legal profession, to whom special means of discovering the truth were available, but who did not dare and indeed were unable to publish their information in German. Other documents have been sent by the tortured and martyred victims themselves. For the greater part of the material the Committee has to thank its own reporters, who have been working in Germany at the risk of the lives.

> We have not used the most sensational of these documents. Every statement made in this book has been carefully verified and is typical of a number of similar cases. We would have been able to publish even worse individual cases, but we have not done this, just because they were individual cases. Not a single one of the cases published in this book is an exceptional case. Each case cited is typical of many others which are in our possession or in the hands of the National Committees.[22]

Perhaps Lord Marley believed all this. Like most successful deceptions, the *Brown Book* contained a substantial element of truth. But the facts and guesswork were mixed, as Koestler later acknowledged, with 'brazen bluff' devised by 'the Comintern's intelligence *apparat*'.[23] This cocktail was shaken during June and July by Münzenberg together with two other Germans, Otto Katz (who went under the pseudonym of André Simone and who, according to Koestler, did most of the writing) and Alexander Abusch who had briefly been the editor of *Rote Fahne*, the leading KPD newspaper, and of *Ruhr-Echo*.[24]

In addition to Katz and Abusch, the communist journalists Rudolf Feistmann, Albert Norden (who had been deputy editor of *Rote Fahne*) and Max Schroeder also worked on the book. Other Germans and foreigners supplied material. They included Alfred Kantorowicz, Koestler's predecessor as artistic correspondent of the *Vossische Zeitung* who had joined the KPD in the autumn of 1931 because he believed it to be 'the strongest force against Nazism,' and Bodo Uhse, a Nazi youth leader in the 1920s who had become disillusioned and joined the KPD. The communist Reichstag Deputy Wilhelm Pieck sent material from Germany; and Countess Katharina Károlyi visited Berlin to talk to the political underground.[25] Katz, Abusch, and other communists, tried hard to hide their political affiliations. After the publication of the book, a radical American lawyer visiting Paris found something fishy about the Committee and the *Brown Book*. 'I tried hard to find out who constituted the Committee and asked: "Who is the Committee?" Answer: "We." I made further inquiry: "Who are we?" Answer: "A group of people interested in defending these innocent men." "What group of people?" The answer came back "Our Committee".'[26]

The terror referred to in the book's title was documented in the second part of the book, partly from underground sources but also to a large extent from reports published in Germany by the legal but non-Nazi press.[27] There was little deceptive about this section. However, the opening part of the book contained the fraudulent thesis that the arson of the Reichstag on 27 February 1933 was not, as the Nazis had untruthfully claimed, a communist conspiracy, but was the work of members of the German (National Socialist) Government. According to the *Brown Book* account, Göring had first organised the fire and then used it to justify the introduction of emergency power provisions and to launch a terror in which 2,500 were arrested and many opponents of the regime were killed.

Specifically, the book claimed that the 24-year-old van der Lubbe, who had been arrested in the Reichstag building, was a vagrant with close links to the Dutch Fascist Party, whose homosexuality had brought him close to Ernst Röhm and his *Sturmabteilung* (literally 'Assault Squads') or SA—the 'Brown Shirts'. The *Brown Book* asserted that a group of SA men, led by Edmund Heines, but under the direction of Göring and following a plan devised by the Gauleiter of Berlin, Joseph Goebbels, had entered the Reichstag through an underground passage which connected with the official residence of the President (Speaker) of the Reichstag, Göring, had set fire to the building, and then left by the same route. They had been seen by Prussian police officers, but Göring, in his capacity as Prussian Minister of the Interior and Commander-in-Chief of the police, had silenced the officers concerned. The Berlin fire brigade had not been called out in full strength and had been late in answering the call because of directions from Göring.[28] After the fire, the firemen had allegedly found in the partly destroyed building enough incendiary material 'to fill a lorry'—which clearly could not have been carried by van der Lubbe alone and which had presumably been brought in by the SA. Finally, the House Inspector, a Nazi, had on the 27th, sent the Reichstag staff away early so that the SA could lay the incendiary material. Those who knew too much about the plot, such as Dr Georg Bell, who was supposed to have first brought van der Lubbe into contact with Röhm's homosexual circle in September 1931, and Dr Ernst Oberfohren, a prominent Nationalist (DNVP) politician, had been killed by the Nazis.[29]

The story was hastily drawn up, but it proved to be immensely attractive and convincing to the audiences targeted by Münzenberg. It offered an enormous amount of detail to support an instinctively appealing case. There was a large measure of consensus across the political spectrum outside Germany (and among anti-Nazis inside the Third Reich) that a large-scale conspiracy was required to produce a large-scale fire; and that the conspiracy was likely to have been a Nazi one. Münzenberg's book seemed to provide chapter and verse.

In fact, Münzenberg had few concrete facts, but he combined

them with guesswork and fabrication to produce a detailed argument. The argument rested upon four main supports: that the idea of a Nazi provocation fitted perfectly their general strategy for crushing communist opposition; that the Nazis themselves were uncovering 'evidence' that a conspiracy existed—evidence which could be re-interpreted to show the Nazis as the conspirators; that van der Lubbe was homosexually involved with SA members and worked for the Nazis; and, finally, that the Berlin fire brigade had been frustrated in their fire-fighting efforts. The first prop to the argument was a gift from the Gods, and needed no efforts on Münzenberg's part. The second was a gift from the Nazis, particularly Göring, who had not apparently foreseen how his conspiracy theory could be turned against him. The third posed a problem, and forged evidence was needed to make it plausible. The last was a bonus relying on distortions and gossip. Each of these elements of the *Brown Book* deception will be analysed in turn.

* * *

First, it was very easy to accept the general line that the Nazis rather than the Communists had been responsible for the fire. Any armchair detective works on the principle 'cui bono?', and it was clearly the Nazis who had drawn the most immediate benefit from the fire. At last they had a pretext to arrest their opponents and ban the Communist Party, the KPD, though it is not clear why the police raids on 24 February on the Karl–Liebknecht house, in which quantities of ammunition had been found, would not have sufficed for the purpose. The banning of the KPD, after the NSDAP and SPD the third largest German party, also ensured that the Nazis would have an absolute majority in the new parliament. Indeed Göring and Hitler were delighted with the fire. Hitler treated it as a sign from destiny; Göring said later that he would have had to set fire to the Reichstag himself if the communists had not been so obliging. The British, French and American ambassadors in Berlin all immediately thought that the fire was a Nazi provocation, one of those acts of lawless terror perpetrated by the SA in the early months after the Nazi seizure of power, though they were prepared to think that the Nazi leaders were not privy to the plan.[30]

Conservative non-Nazis were among the first to conclude that Göring was guilty. Gottfried Treviranus, who had been the main political adviser of Chancellor Brüning, describes how he was dining in Berlin with the Neuraths and the Rundstedts on the evening of 27th at the house of the Mayor. When he heard about the fire, Treviranus said to Frau Neurath that Göring was the only man he could think of who would do something like this.[31] A conservative publicist, Heinrich von Gleichen, announced in the second March issue of *Der Ring* 'Maybe members of the highest society are responsible.' According to Konrad Heiden, a visitor to the conservative *Nationalklub* on 28 February had said that van der Lubbe had been promised a prison sentence of only two years and a sum on release of 50,000 RM, but that he had lost his nerve, had attempted to flee, and had been involved in a struggle with the incendiaries of *Sturmabteilung* Detachment 17.[32]

In February and March there was a struggle going on between Hitler and the conservatives and, as the conservatives feared, after the 5 March elections the outlawing of the KPD gave the NSDAP, the Nazis, an absolute majority. Thus Hitler was no longer dependent on Nationalist parliamentary support. No wonder Berlin was buzzing with rumours about Nazi plots. The regime did little to discourage these rumours, for the story of the Reichstag fire and a Nazi conspiracy actually served its purpose in intimidating the conservatives. Hitler himself said: 'The old men of the DNVP, Hugenberg and friends, are upset and frightened. They believe that I have set light to the building myself. They take me for the devil in person. That's a good thing.'[33] In this atmosphere the idea of a Nazi plot seemed entirely credible.

Concerning the notion that the fire was the work of a conspiracy, the investigation by the Prussian police provided supporting material for the *Brown Book*'s case as well as their own. It showed, apparently conclusively, that a major conspiracy was required to set the Reichstag ablaze; but it failed to demonstrate at all convincingly that the communists were responsible. Göring was obsessed with the notion that there was a KPD plot and that there had been several incendiaries; there was thus a substantial amount of political pressure on policemen and firemen to provide evidence of a large conspiracy. Already, on the night of 27 February, before any evidence at all was available, Göring had shouted: 'There were 10 or even 20 men!'[34] The fireman who thought between six and eight persons would have been required to set fire at so many different places in the building helped to produce the story that there were *seven* conspirators.

On 1 March the official German press made the elementary error of trying to strengthen its case by revealing the existence of a secret undergound passage leading to the Reichstag.[35] The purpose was to explain how all but one of the arsonists had made their escape. But Münzenberg turned this 'evidence' on its head. Wilhelm Pieck, a leading figure in the underground KPD, supplied him with a plan of the area which clearly showed that the passage led to the residence of the Reichstag President.[36] The absurdity of the press statement was apparent: how could a band of communists have surfaced unnoticed in Göring's residence? But, if the incendiaries had been Nazi . . .?

Further evidence pointing to a communist conspiracy was turned round to point to a Nazi plot. It was said by firemen that the Reichstag carpets had been soaked with gasoline. Newspapers claimed that seven men would have been needed to carry in the incendiary material, and 10 to distribute it.[37] Since both communists and Nazis welcomed such statements, no one looked for more prosaic explanations, such as the liquid being water from firemen's hoses or the fire being started with blazing towels and curtains. At the time, no one knew about the physical obstacles which the tunnel's locked doors presented. In 1933 it was established that van der Lubbe *could*, as he demonstrated during the course of the investigation, have run so quickly that he *might* have lit all the incendiary sites. But the police, and then the prosecution at the subsequent court case, preferred to use evidence that pointed to a group of conspirators being responsible.

Next, it was necessary for Münzenberg to demonstrate van der Lubbe's Nazi connections. The authors of the *Brown Book* had no evidence, so they made it up. Their case rested on false claims of van der Lubbe's homosexuality: 'Enquiries into his life in Leyden have definitely established the fact that he was a homosexual.'[38] In fact there is much evidence that the Dutchman was *not* a homosexual,[39] though he had moved in vagrant circles where there was homosexuality. More importantly for the *Brown Book*'s deception, Röhm's sexual tastes were well known and had been widely discussed in late 1932 as a consequence of a trial in Munich.[40] The *Brown Book* authors chose Heines, a close friend of Röhm's, as the link between van der Lubbe, Nazi homosexuals, and the Nazi 'plot'. This invention, based on a fictitious account of a 'friend' of the (by now) dead Dr Bell,[41] not only supplied a reason why van der Lubbe should have become involved with Nazism, but added a piquant element of sensationalism to the story.

Georg Bell had been an international adventurer, who had been involved in the forgery of Russian *chervonets* banknotes and who had quarrelled with Röhm and the Nazi party. No one will ever know what he knew about van der Lubbe, and it is conceivable that he was responsible for a major part of the story. He was a fantasist who loved making up improbable stories, and he was in Switzerland in March 1933. It is just possible that he had met and talked to Münzenberg then. On the other hand, the *Brown Book*'s authors never produced a shred of evidence to back their story that Bell had met van der Lubbe, and that van der Lubbe had visited him in Munich and there been introduced to Röhm.[42] All attempts to make this crucial link failed, and the likelihood is that Münzenberg simply added his own fantasies to those of the dead fantasist.

The final element in the *Brown Book* deception rested on stories from the Berlin fire brigade, some from published sources, others from underground rumours. The fire brigade had been a bastion of the Republic. There had been a large amount of patronage from the SPD and the Catholic Centre party; and in consequence the tone in 1933 was distinctly hostile to Nazism. There were rumours that the firemen had been fighting with the SA at the scene of the blaze. A second story, which also appeared only later, was that policemen with pistols drawn had prevented firemen from going into the Reichstag cellars in order to look for other potential arson sites. Such stories were picked up by the SPD exile organisation and formed the basis of articles in *La République* on 21 April 1933 and in the *Saarbrückener Volksstimme* four days later. According to the articles, Walter Gempp, the director of the Berlin Fire Brigade, had complained that the fire engines had been summoned too late, that there 'were great masses of unused incendiary material lying about' and that Göring had forbidden Gempp to circulate a general alarm call.[43] In addition, some of the very early press reports gave very odd and confused accounts of when the fire brigade arrived: the morning editions on 28 February claimed that the fire brigade had been given a full alarm at 9 pm. Yet this was clearly untrue: at first only two individual sections arrived, and indeed the fire had not been reported (or even started) at 9 pm.[44] This confusion made it seem quite possible that the Berlin firemen's stories were indeed correct, and that there had been some kind of official cover-up. Then Gempp was suspended from office—though not as claimed by the *Brown Book* because of his role in the Reichstag fire or because of statements he had made, but because of his involvement in a corruption scandal. The news of Gempp's suspension had been widely publicised by the German press at the end of March: this suspension then gave the *Saarbrückener Volksstimme* and Münzenberg a name on which to hang the fire brigade material.

These four ingredients—the implicit credibility of rumours of a Nazi plot, Göring's insistence that a conspiracy existed and the material he assembled to prove this, the homosexual story, and the discontent in the Berlin fire brigade—were mixed in a bizarre but powerful cocktail. Next it was necessary to add a strong measure of ideological spirit. This, however, left an aftertaste which might have given the game away and revealed the nature of the cause being promoted in the *Brown Book*. The major problem of the book was, in fact, fitting the story to Comintern ideology. Since 1928 the Comintern had held that the 'stabilisation of capitalism' had come to an end, and that there would be a crisis which might produce similar, short-lived responses from fascism on the one hand and social democracy on the other. According to this thesis, fascism and social democracy, referred to by the communists as 'social fascism', were actually twins—futile attempts to perpetuate an unworkable economic system.

This doctrine was fully presented in the Introduction to the *Brown Book*, written by Rudolf Feistmann, which provided an allegedly historical account of the rise of Hitler. Here it would have been possible—even in 1933—to check the accuracy of the information, and to expose the crude manipulation of evidence in the name of ideology. The book described the public response to von Papen's 1932 government thus: 'Powerful anti-fascist demonstrations under the leadership of the Communist Party, which was carrying on the only serious extra-parliamentary fight against Fascism, were broken up. These reached their height in the Berlin traffic strike of November 1932, which demonstrated the helplessness of the Government in face of the determination of the workers.'[45] In fact, this strike against von Papen's wage-cuts was jointly conducted by communist union organisations (the 'Revolutionary Union Opposition') and by the Nazi Factory Cell Movement (NSBO). It certainly could not be described as a powerful anti-fascist demonstration.

Another point where the join between fact and ideology shows clearly and rather painfully is in the description of *why* the upper middle class officer Göring should have been the chief instigator of the fire plot.

> Goering represents the content of the policy of the National Socialists. National Socialism does not represent the workers or the employees or the middle class, but it represents the interests of the ruling class, of the *noble caste*. Power was put into the hands of the National Socialists in order that they should maintain the existing economic systems and protect it against the menacing forces of social revolution.[46]

Rigidly ideological interpretations of events are often so far removed from reality that they employ deception to bridge the gap. But to construct a credible bridge between the reality that is and the reality that theory dictates, the deceiver must sometimes surrender part of his ideological baggage. The *Brown Book* would have been better as an instrument of deception if it had shed more of its ideology, but was restricted by the Comintern's inflexibility. However, in the most important single piece of evidence used to defend the *Brown Book*'s argument—the 'Oberfohren memorandum'—Comintern ideology took second place to the needs of deception, and the inconvenient reality of the power struggle between the Nazis and German conservatives was openly acknowledged. The memorandum was a fake, the forger probably being the editor of the KPD newspaper, *Rote Fahne*—Albert Norden.[47]

The Oberfohren memorandum had been printed in part in the *Manchester Guardian* on 26 and 27 April 1933. Its alleged authorship was not revealed at the time: the document was simply presented as 'a serious attempt by one in touch with the Nationalist members of the Cabinet to give a balanced view of events'. In the *Brown Book* the piece was at last attributed—to Dr Ernst Oberfohren, who by then was dead.[48] Later, when it was asked whether Oberfohren, an educated man, would have used the rather crude and clumsy language of the report, a revised version stated that it was written not by Oberfohren personally, but by 'friends'. The memorandum consists of a reasonably intelligent but stylistically coarse attempt to guess what a German Nationalist might be thinking in Spring 1933, namely that an absolute Nazi majority was not in the conservative interest. This political analysis is coupled with details of how Goebbels planned to burn down the Reichstag; which SA men were involved in this conspiracy; and how the Nazis planned a *putsch* against the Conservatives. Thus the credibility gained from a frank statement of political realities was used to strengthen the plausibility of the deception which followed.[49]

But important though the memorandum was to the *Brown Book*'s case, it was flawed. Its surrender of the ideological dogma about the conservatives backing Hitler contradicted what was written in the book's Introduction: and in time several 'facts' relating to the fire were found to be untrue. The *Brown Book* and the memorandum claimed that the SA incendiary squad was led by Heines, but the later court case established that he had been in Gleiwitz on 27 February, a long way from Berlin. Schulz and Count Helldorf, the other SA men supposed to have been involved, could also establish alibis. Later versions of the conspiracy story simply changed the name and stated that the group had been led by Karl Ernst. Ernst too was dead by the time his name was mentioned: he had been killed on 30 June 1934, the 'Night of the Long Knives'. Once again, a forgery was used to support the new case: this time a letter from Ernst to Heines.[50] Since the Oberfohren document was the major piece of evidence for the *Brown Book*'s version of the Reichstag fire, doubt as to its authenticity was highly damaging.

The *Brown Book*, in its original form, was vulnerable for other reasons too. It tried to imply that Erik Hanussen, a Nazi occultist, was killed because he knew too much.[51] Yet what he had claimed to 'know' was that the fire had been the work of communists—hardly a reason why the Nazis should have shot him. The book also saw the fact that Hitler kept the period 25 to 27 February free of election speeches as evidence of a Nazi conspiracy.[52] But it did not explain why it would have been necessary for Hitler personally to be present in Berlin to oversee such a plot. Elsewhere, the *Brown Book* claimed that Göring and Goebbels were prime movers: there were in fact too many alleged leaders of the plot. The book also claimed that van der Lubbe spoke at a 'fascist meeting for fascists,'[53] but failed to offer any evidence. The *Brown Book* was quite convincing so long as the lies and forgeries that supported its argument went unchallenged. Its credibility began to fall apart as soon as facts and alibis emerged at van der Lubbe's trial.

* * *

Throughout the time between the fire on 27 February and the opening of the trial of van der Lubbe on 21 September 1933, the Nazis were working with just as much energy as Münzenberg to construct a conspiracy case, the only difference being that in their version it was communists who had planned and started the fire. The climax of the Nazi endeavour was the trial in Leipzig. The prosecution had little interest in establishing van der Lubbe's guilt: this was self-evident and unimportant. The purpose was to demonstrate that the accused could not have acted alone and must therefore have had accomplices. The next thing was to prove that these accomplices included the other men in the dock, four communists. So both the conspiracy theories would benefit from 'evidence' pointing to the involvement of accomplices and, in the struggle over which theory would prevail, everything would depend upon who these accomplices were. Few people had any interest during this polarised debate in weighing the evidence repeatedly offered by van der Lubbe, that he acted alone.

After the occupation by Nazis of the Karl–Liebknecht house, the German Communists, the KPD, had used the Reichstag as their main electoral office. For this reason, two KPD deputies, Torgler and Koenen, had been the last to leave the building on 27 February 1933. Koenen had fled abroad out of fear of being blamed for the fire: Torgler perhaps unwisely presented himself to the police in order to prove his innocence. Soon after, three Bulgarian communists were arrested, Georgi Dimitrov, Blagoi Popov, and Vassili Tanev. Originally the Bulgarians had been in Berlin in connection with the elections: notionally to 'observe' (though in fact only Dimitrov actually spoke German), but in practice to wait for the collapse of Nazism. In line with the Comintern's thesis, they believed that fascism was the final stage of capitalism, and that it would speedily collapse under its own contradictions, leaving the path clear for socialism. Their arrest was probably a case of mistaken identity: they were thought to be convicted Bulgarian arsonists. But then a web of false evidence was created to link them with van der Lubbe.

Some of the weaknesses of the case alleging communist involvement became apparent very quickly. Initially, Göring had claimed in a radio broadcast on 1 March that the arsonist was a Dutch communist who had just come back from Russia. Yet van der Lubbe was no longer a communist, and had never been to the Soviet Union. At the Leipzig trial, Göring's credibility was completely undermined. The restaurant waiter who, hoping for a 20,000 RM reward, said that he had seen van der Lubbe together with the Bulgarians (whom he described as Russians) turned out to have identified van der Lubbe on the basis of a photograph that had been released in the press. The Nazi Reichstag deputy who identified van der Lubbe as one of 'the criminals in Torgler's company' on the afternoon of 27 February had been allowed to see van der Lubbe under arrest before recognising him officially. A convicted thief named Lebermann alleged that in January 1932 Torgler tired to induce him to set fire to the Reichstag. Another witness named Kunzack, who had been convicted for indecency and theft, claimed to have assisted Torgler in experiments with explosives. An alcoholic then claimed to have met van der Lubbe in Konstanz in October 1932 (when the Dutchman was really in Leyden). Even the judge was embarrassed by this kind of evidence.

Such testimony made it easy for Dimitrov, who conducted his own defence, and for Torgler's counsel, Dr Sack, to demolish the prosecution case at Leipzig. On the other hand, Sack's argument, that van der Lubbe alone was responsible (an argument with which Dimitrov for obvious reasons never associated himself, as he was intent on proving the Nazi conspiracy) was unimpressive. The technical material assembled at Göring's urging in support of the conspiracy case did suggest that van der Lubbe could not have worked on his own. The notion of a conspiracy therefore triumphed. But the Leipzig trial could not prove the complicity of the four accused Communists, Torgler and the three Bulgarians, who were acquitted. When the case against the Communists collapsed, the alternative conspiracy theory gained strength. For now the only plausible conspiracy was that the Nazis had set the fire as a provocation aimed at their communist opponents.[54]

Although in this indirect way the Leipzig trial supported the *Brown Book*, in matters of detail it was very damaging. The book's depiction of van der Lubbe as a boastful, homosexual megalomanaic was demolished, and several of the weaknesses discussed earlier were exposed. It was clear to Münzenberg that while the outline of his conspiracy theory would hold up, the evidence supporting it would need to be improved. Before we examine the improved *Brown Book*, however, it may be useful to see what Münzenberg had been doing to undercut the legitimacy of the Leipzig trial.

* * *

The first stage in the defence and further elaboration of the *Brown Book*'s charges was prepared during a visit to Moscow in the Summer of 1933 when Münzenberg gained the approval of the Comintern leadership for the creation of an 'International Committee of Jurists', composed of sympathetic non-Communists who would consider with an apparent judicial impartiality the causes of the Reichstag fire—and then definitively pronounce the Nazis guilty.[55] Münzenberg and Katz then prepared in Paris for a 'Legal Inquiry into the Burning of the Reichstag' to be held in London a week before the German trial began in Leipzig. The Nazis, and German justice, would thus find themselves in the dock; the Inquiry became known as the 'Counter-Trial'.

Like the World Committee, the Jurists were a classic front organisation. The chairman was the prominent Labour MP and barrister, D N Pritt, KC, who was later to defend Stalin's show trials against the 'unscrupulous abuse' they received in England. He was eventually expelled from the Labour Party for supporting the Soviet invasion of Finland.[56] Pritt's colleagues on the International Committee were Arthur Garfield Hays, an American who was a champion of the Civil Rights Movement, and who was later to develop doubts about the London Inquiry;[57] Georg Branting, son of Sweden's first socialist Prime Minister; Maîtres Moro-Giafferi and Gaston Bergery from France; Valdemar Huidt from Denmark; Dr Betsy Bakker-Nort from the Netherlands; and Maître Pierre Vermeylen from Belgium.

The organisation of the London counter-trial was chiefly in the hands of Otto Katz, who appeared on MI5's black list as 'a red hot communist' and who, according to a Foreign Office memorandum, was admitted to Britain only 'as the result of intervention by Mr Arthur Henderson and other members of the Labour Party'.[58] Once in London, Katz stayed hidden behind the scene. As 'the invisible organiser of the Committee',[59] he knew that the Comintern's involvement had to be kept a close secret so as not to imperil the Inquiry's respectability. On 13 September, a reception was held for the jurists by Lord Marley and Sidney Bernstein in the prestigious surround of the Hotel Washington in Curzon Street.[60] The

Inquiry opened the next day in the Law Society's Court Room at Lincoln's Inn. This environment made the affair look like a British Crown Court. An opening address by the Labour lawyer Sir Stafford Cripps KC emphasised that 'none of the lawyers on the Commission belonged to the political party of the accused persons in Germany.'[61] Later Katz proudly claimed that the Committee acted as 'an unofficial tribunal whose mandate was conferred by the conscience of the world'.[62]

The proceedings were mixed with melodrama. Witnesses were brought in in disguise. The court doors were locked so that no one could leave during the testimony of sensitive witnesses. Pritt claimed that the National Government was trying to obstruct his 'trial', and that Sir John Simon, the British Foreign Secretary, was trying to make the Law Society revoke the temporary lease of its premises to the 'anti-fascists'.[63]

Despite all this drama, the investigations soon became rather boring. Public figures, such as H G Wells, who had initially been enthusiastic, quickly lost interest. The case for van der Lubbe's homosexuality looked inconclusive. 'The Counter-trial' merely concluded rather lamely: 'grave grounds existed for suspecting that the Reichstag was set on fire by, or on behalf of, leading personalities of the National Socialist Party.'[64]

This was hardly the pre-emptive indictment of the Leipzig trial that Münzenberg had been hoping for. So the task of repairing the case alleging a Nazi conspiracy and of fixing it in the public mind as a historical fact fell to a *Second Brown Book*. This appeared in the Spring of 1934 and offered a more solidly based case. The Leipzig trial had made a great deal more evidence available. Dimitrov had made a brilliant defence, and had successfully provoked Göring into losing his temper. The *Second Brown Book* made much out of the judge's attempt to cover up some of the key Nazi witnesses as their testimony fell to pieces, and also of Göring's hysterical outburst at Dimitrov: 'You wait till I get you out of the power of this court!'[65]

Obviously, the new book had to counter refutations of the old, a task which Münzenberg accomplished in part by pointing to the holes in Göring's evidence at Leipzig and in part by some rather improbable claims, such as arguing that Heines could have flown from Silesia to Berlin and back in next to no time, and that his alibi did not stand up.[66] Obviously incorrect *Brown Book* material had to be withdrawn, but in a way that left an element of suspicion that it might after all have been correct. For instance, in retracting the false statement that the Reichstag House Inspector had sent the Reichstag employees home early on the night of the fire, the *Second*

Brown Book nevertheless claimed that the original version was based on 'uncontradicted newspaper reports'.[67] The Oberfohren memorandum was still described as undoubtedly authentic,[68] but now a different account was given as to its provenance. According to the *Second Brown Book* 'the draft was prepared on Oberfohren's instructions by a journalist who frequently collaborated with him'.[69] Oberfohren was said to have made handwritten corrections himself, which established the authenticity; but in fact the draft was supposed to have been retyped before being submitted to the Commission. So no one ever saw Oberfohren's handwriting. The final proof of authenticity was that, after all, the memorandum had been printed in 'a most reliable newspaper'—the *Manchester Guardian*. The *Guardian*, of course, had only published the material because it had been the victim of a deception: Münzenberg's argument shows how one successful deception can become the foundation stone for another.[70]

A subtle escape clause was written into the *Second Brown Book* to pre-empt any future refutation of its case. Goebbels, the authors explained, had 'practically unlimited resources' for a world propaganda campaign involving a substantial amount of disinformation.[71] Past and future information contradicting the Münzenberg case could safely be disregarded by readers, because it would doubtless fall in this category. Finally, the *Second Brown Book* confused the picture further by highlighting odd items of information about the fire. There was a stranger supposedly seen entering the Reichstag; another young man in black boots and black coat (an SS officer?) was said to have reported the fire but had never been identified;[72] and incorrect but confusing stories were related about the fire brigade. Although these distractions added nothing to the logic of the case, they helped to strengthen the general belief in a conspiracy and thus divert attention from the sole supporter of van der Lubbe's version of events, Dr Sack. By 1934, after the Nazis had failed to implicate the communists, the idea that van der Lubbe had indeed worked alone was the only remaining alternative to Münzenberg's invention that the Nazis were to blame.

The *Second Brown Book* triumphed not merely over Nazi propaganda but also over the truth. As a result, the essential message of Münzenberg's deception was accepted until the 1960s. The early editions of Alan Bullock's biography of Hitler had no doubt: 'Göring and Goebbels were looking for some pretext to smash the Communist Party. After rejecting various plans—such as an attack on Hitler—they hit on the notion of setting fire to the Reichstag building.'[73] Because there was so much evidence of Nazi evil, the story fitted the popular conception precisely. Indeed, when Fritz Tobias released some of his findings in *Der Spiegel* ahead of the publication of his book, he discovered 'how tenaciously most people guard their familiar opinions'.[74]

In his introduction to the English language edition of Tobias's book, A J P Taylor reminds us that the 'essential thing is to acknowledge one's mistakes. On the Reichstag fire I was as wrong as everyone else; and I am grateful to Herr Tobias for putting me right.'[75] Not everyone was as 'grateful to Herr Tobias'. In the 1970s, Edouard Calic, a Croat emigré, was able to use further forged documents to convince an 'International Committee for Scientific Research on the Causes and Consequences of the Second World War' that the fire was indeed the work of a Nazi conspiracy.[76] The Committee received subsidies from the German *Auswärtiges Amt* and the *Bundespresseamt*; its work was defended by leading SPD politicians Willy Brandt, Egon Bahr and Herbert Wehner. Indeed, the Calic case shows how a thesis which had always been the domain of fantasists and forgers could still be made persuasive a generation later. Calic's forgeries included papers supposedly written by Richard Breiting, a 1930s right wing liberal and editor of *Leipziger Neueste Nachrichten*, but which were in bad German, including several phrases which were literal translations from Croatian. Other 'Breiting' papers were later proved to be false by Hans Mommsen. Yet another set of documents, the 'K' papers, which were supposed to have been given by a retired police Captain, Eugen von Kessel, to Breiting, and which implicated the Nazis, proved to be forgeries. Calic also claimed that while in Sachsenhausen concentration camp *General Eduard* Wagner—one of the 1944 plotters against Hitler—gave him documents proving the Nazi Reichstag conspiracy. But in fact General Wagner had been killed immediately after his arrest. So Calic later claimed that it was a wounded *Colonel Siegfried* Wagner who had told him that he had been sent to Sachsenhausen because of his knowledge of the 1933 fire.

Not surprisingly, the Münzenberg version continues to find equally enthusiastic advocates in the German Democratic Republic. A learned article published in 1980 by Klaus Sohl in the *Jahrbuch für Geschichte* acknowledged the drafting of the *Brown Book* by 'experienced cadres' of the KPD: 'The great significance of the *Brown Book* lies in its political and scientific approach to the events of 1933 from the standpoint of Marxism-Leninism.'[77] But Sohl was dismissive of all the other accounts which stressed the communist origins of the book. According to Sohl, Koestler could not have known about the writing of the book because he only arrived in

Paris during the Leipzig trial. Babette Gross's memoirs cannot be reliable since they were written from 'the standpoint of a renegade' who 'had returned to her bourgeois origins'. Sohl also sought to demonstrate that the *Brown Book* was not based, as Koestler had claimed, on forgery. But here Sohl relied upon Calic's bogus Breiting materials for vital evidence to support his case.[78]

The longevity of the Münzenberg account resulted from a brilliant publicity campaign begun in the 1930s. The recipe was simple. Begin with an inherently attractive hypothesis—Nazi conspiracy. Garnish with details, more or less accurate, based on the testimony of dead men (Ernst, Oberfohren, Bell) who could not protest. Finally, forge the documentation.

* * *

What was the ideological point of the London counter-trial and the vigorous defence of the original *Brown Book* case? The approach of the Comintern was gradually shifting in 1933 and 1934. It had become clear that the Nazi dictatorship was unlikely to collapse easily and quickly 'under its own contradictions'. Rather a broad alliance would therefore be needed in the struggle against Nazism. The London trial was 'intended to enlighten world opinion'[79] and demonstrate that the Hitler state was persecuting not only communists. Nazism would not tolerate opposition from any quarter, and the regime's opponents were imprisoned and killed without trial. For Münzenberg, the revelation of atrocities was the key step in the creation of an anti-Hitler coalition. This campaign had to depart from the assumption of the original *Brown Book* that the only real opposition came from the Communist movement; and it represented the beginning of the move from the 1928 Comintern line towards the strategy of the Popular Front finally officially adopted in 1935 at the Seventh Congress. For the Comintern it had originally been vital to establish that Nazism was the final stage of capitalism. Now there was the simpler propaganda task of convincing as many people as possible of the evil irrationality of Nazism (not its capitalistic rationality). That was, of course, not a difficult argument to present. Nazism really was profoundly evil. But it was important to present the argument in such a way that the corollary followed that alliance with the communists represented the only conceivable form of anti-fascism.

Münzenberg had always, in fact, been an enthusiastic advocate of a Popular Front strategy, even in 1931 and 1932 when this involved opposition to the Soviet demand for a tactical alliance with the Nazis against Chancellor Brüning and the SPD. But in public Münzenberg had had to swallow the orthodox Soviet line of Ernst Thälmann and the party leadership, and he had obediently attacked the SPD. In 1932 he had denounced Trotsky for proposing an 'action front' composed of social democrats and communists.[80] The Reichstag fire gave Münzenberg the opportunity to present a new and more congenial case.

Like Münzenberg's earlier 'Innocents' Clubs', the Reichstag fire campaign was designed to serve the purposes of the Comintern and Soviet intelligence *apparat* as well as to win a propaganda victory. Perhaps his primary aim was to conquer public opinion; but he also hoped to lure some British intellectuals into a *secret* war against Fascism under Communist direction. Preparations for a recruiting drive among British intellectuals began at the same time as preparations of the counter-trial. The bait was anti-fascism: the example allegedly being set by German workers in forming secret *Fünfergruppen* (groups of five) to launch a counter-attack against Nazism.

The phrase 'ring [or group] of five' was to haunt the British Security Service—MI5. It was later used by Soviet defectors and in intercepted Soviet communications to describe a dangerous network of Soviet moles at work in Britian. MI5, however, failed to discover either the origin of the phrase or its role in the recruitment of the Cambridge moles.[81] The first group of five had been formed in 1869 by the student revolutionary Sergei Nechaev, the model for Dostoyevsky's Peter Verkhovensky in *The Devils*. Though Dostoyevsky depicts a psychopath, the conspirators of Narodnaya Volya regarded Nechaev as a revolutionary visionary.

The KPD revived 'groups of five' in the tense political atmosphere of the last years of Weimar. In the summer of 1932 the KPD began planning to replace its existing semi-open cells of 10 to 30 members by secret 'groups of five' (not all with exactly five members) in the event of a Nazi seizure of power. Only the leader of each group was to learn the identity and addresses of the other members; and he alone was to make contact with the next level in the party hierarchy.[82] The problem was that von Papen's *putsch* against the Prussian state on 20 July, 1932, six months before Hitler became Chancellor, had shown the impotence of the left. Koestler knew that the KPD was now a 'castrated giant'.[83] After 1933, many former communists joined the Nazi party. The bulk of communist resistance in the Third Reich, such as it was, occurred in a rather chaotic and ill-organised way among the badly paid construction workers of Hitler's labour army.[84]

That so many communists were joining the Nazis was highly embarrassing to Moscow and the Comintern. There was, however, an attractive way of explaining it away. The party was going secret, and moreover was infiltrating the Nazi movement! By the summer of 1933 the Comintern claimed that the groups of five had created 'a new subterranean revolutionary Germany . . . dogging Hitler's every footstep'.[85] The chief propagandist of the groups of five was a Soviet journalist Semyon Nicolayevich Rostovsky, who had established himself in London under the alias of Ernst Henri. In August and September 1933 he wrote three articles for *The New Statesman*, which revealed publicly for the first time the existence of the groups of five. The articles bore the title 'The Revolutionary Movement in Nazi Germany'. They argued that Hitlerism was the result of a conspiracy of monopoly capitalists ('Thyssen's plot'). The resistance was organised clandestinely in 'revolutionary groups of five' under communist leadership. These were supported by many former Social Democrats, republicans, Catholics and liberals who allegedly ignored the corrupt and bankrupt old Social Democratic leadership, now in exile in Prague—though in fact the Prague Sopade had a broader range of contacts and influence within Germany in the mid-1930s than Henri admitted. According to Henri, the 'groups of five', which were also to provide a basis for the organisation of espionage later in the 1930s in western countries, had spread through 'practically the whole of German industry'. They were ignorant of each other's existence, and the whole organisation was held together by a very few co-ordinators, termed revolutionary 'workshop inspectors'. Their activities were diverse: to print clandestine leaflets, co-ordinate demonstrations, supply information on the Hitler Terror, but also to infiltrate the Nazi labour organisation and paralyse the system from within.

Henri's theories showed how infiltration and espionage followed from anti-fascism. Anti-fascism required secret organisation, because the secret organisations of Nazism were so powerful and widespread that they formed a covert 'fascist international'. On the other hand, anti-fascists belong only in the ranks of a revolutionary movement, and only that (communist-led) revolutionary movement was genuinely anti-fascist. In 1934 Henri included these arguments in a book, *Hitler over Europe?*[86] This book, said *The Times*, will 'make the democrat's flesh positively creep'.[87] Indeed, the conclusion was both frightening in its implications and ruthlessly candid in describing the leading role of the party:

> This power will and can only have the form of an anti-fascist world alliance. It will cover all genuinely anti-fascist forces—those that are already anti-fascist *and those that must be made so* (emphasis added): the proletariat, the lower middle class, the small and intermediate farmers, the progressive intellectuals, the Jews and Catholics who are prepared to struggle, the former front-line soldiers, the active pacifists, the emancipated women, the

> youth threatened with annihilation, the oppressed colonial and semi-colonial peoples; all individual persons and groups, finally, that have recognised the import of Fascism and are willing to fight. They will have their battalions in every country—just like Fascism. They will be under the leadership of that column having the strongest fighting organisation, the longest experience, the most energetic will to fight, and the clearest political vision; the working class.[88]

The implication of Henri's message was that anti-fascist British intellectuals, if their anti-fascism amounted to anything more than mere words, and if they were to display their 'solidarity' (a key Münzenberg phrase) with the oppressed workers suffering in Germany under Nazi rule, should join in the *secret* war against Fascism. To Guy Burgess, in particular, that was a heady message. According to one of those who knew him, Burgess set out to form in Cambridge his own 'light blue ring of five'.[89] Henri's book was reviewed in *The New Statesman* by Brian Howard, one of the closest of Burgess's friends, and, like Burgess, an Old Etonian marxist. Howard held, like Henri, that Nazi expansionism was 'point for point identical with the map of the expansion of German heavy industry', and urged a European Anti-Fascist Alliance to avert 'a second and far more appalling world war'. Anti-fascist intellectuals should also be anti-capitalist; and they should 'band themselves together without delay'.[90]

Countess Károlyi later recalled what an impact she had on Cambridge undergraduates when Münzenberg sent her to collect funds for the counter-trial and the Dimitrov defence in Leipzig. The collection turned into a tremendous propaganda coup:

> I remember my trip to Cambridge in the rickety car of a young communist undergraduate who, on the way, explained to me dolefully that it was imperative though most regrettable, that the beautiful ancient universities of Oxford and Cambridge should be razed to the ground when the Proletarian Dictatorship was proclaimed. For centuries, he said, they had been the symbols of bourgeois privilege. He seemed suspicious of my genuine revolutionary spirit when I expressed my doubts as to the necessity for demolition. In Cambridge we drove to one of the colleges, where white-flannelled undergraduates were playing tennis on perfectly kept green courts. We were received most enthusiastically. It was odd to see students of such a famous university, obviously upper-class, with well-bred accents, speak about Soviet Russia as the land of promise.[91]

The success of the propaganda campaigns of Münzenberg, Katz and Henri owed much to their own charismatic personalities. The Comintern had not yet succumbed to the all-Russian bureaucratised blight that in the later 1930s deadened its appeal. Both Münzenberg and Henri were unconventional figures who stood apart from the party orthodoxy and its narrow doctrinaires. Koestler considered Münzenberg a 'genius' whose 'person emanated such authority that I have seen Socialist cabinet ministers, hard-boiled bankers and Austrian dukes behave like schoolboys in his presence . . . His collaborators were devoted to him, the girl secretaries worshipped him.'[92] Though not the equal of Münzenberg, Otto Katz too was a charmer. During the 1920s he built up a wide range of contacts in the theatre and film industry. According to Babette Gross, 'in Hollywood he charmed German emigré actors, directors and writers. Katz had an extraordinary fascination for women, a quality which greatly helped him in organising committees and campaigns'.[93] Ernst Henri shocked conventional Soviet diplomats and journalists by his disdain for Socialist Realist aesthetics, by his taste for well-cut English suits and 'decadent' modern art, and by his unconventional manners. Edith Cobbett, who worked with Henri on the *Soviet News* and *Soviet Weekly* at the end of the Second World War without realising his intelligence role, remembers him as 'really a charismatic personality' who could always 'turn things into a great joke'.[94]

Though these men were ideally suited for luring 'innocents' into Comintern front organisations and for persuading a minority of enthusiasts to enter the intelligence *apparat*, their unconventional charm brought them into trouble with the Stalinist secret police, the NKVD. After the NKVD had tried unsuccessfully to lure Münzenberg back to Moscow in 1937, he was expelled from the Communist Party for alleged right wing bourgeois deviationism. Henceforth, the great virtuoso was kept under surveillance by the same intelligence *apparat* that he had previously assisted and lived in fear of his life. In 1940 he died in mysterious circumstances, probably murdered by the NKVD.[95] Katz tried to save himself by breaking off all contact with Münzenberg but, in 1952, his collaboration with Münzenberg on 'Trotskyist lines' was used against him in the Slánský show trial in Prague. After denouncing Münzenberg for working with 'hostile bourgeois and capitalist elements' Katz denounced himself. He told his judges: 'I regard myself as a criminal, I am a Jew. I stand before the court a traitor and a spy.' He asked for, and was given, the death sentence.[96]

Henri fared less badly than Münzenberg and Katz. Though probably not a formal control, he remained in contact with Burgess and other British moles, helping them to overcome their doubts as they found themselves gradually transformed from anti-fascist conspirators into fully-fledged Soviet agents. Early in 1942, probably at the instigation of Guy Burgess, then an influential BBC producer in the talks department, Henri broadcast news from the

Eastern Front of special significance for the moles. The Soviet Union, he declared, had 'an intelligence service which is among the best in the world'.[97] Henri's unorthodox flair was tolerated by the NKVD as long as he encouraged fellow travellers and reassured the Cambridge moles. But after the defection in 1951 of Burgess and Maclean he suddenly became expendable, was recalled to Moscow and sent to the *Gulag*. Unlike Münzenberg and Katz, Henri survived the Stalinist era to emerge from the labour camp in 1953; he renewed contact with Burgess and Maclean and eventually re-established himself as a Soviet propagandist and intelligence officer.

Dimitrov, whose success at the Leipzig trial in attacking Nazi distortions while planting some of his own must have impressed Moscow, went on to become head of the Comintern.[98] After the war he was to become the first communist Minister-President of Bulgaria. From 1948 to 1950, Alexander Abusch was a member of the Communist Party Executive in the German Democratic Republic. In 1950 he was dismissed during the process of Stalinisation, but he was later rehabilitated and in 1957 became a member of the Central Committee and in 1958 Minister of Culture.[99]

There remains a perhaps appropriate historical irony about the careers of Münzenberg, Katz and Henri. All three fell victim to the Stalinist oppression whose reality their propaganda had so brilliantly disguised under the banner of the anti-fascist crusade. In the 1930s they stirred a powerful and mendacious cocktail. When they drank it themselves, it proved to be poison.

Endnotes

1. Statement by Anthony Blunt at a press conference on 20 November 1979.
2. Babette Gross (transl. M Jackson), *Willi Münzenberg: a Political Biography* (Michigan: Michigan State University Press, 1974), p. 217.
3. *Ibid.*
4. Arthur Koestler, *The Invisible Writing* (London: Hutchinson, 1969), p 251. (There is no connection between the Münzenberg Trust and the Cheka 'Trust' operation mentioned in Eastern Approaches.
5. Gross, p. 120; Koestler, p. 253.
6. R N Carew Hunt, 'Willi Münzenberg', David Footman, ed., *International Communism*, St Antony's Papers IX (Oxford: Chatto and Windus, 1960), p. 87.
7. Gross, p. 219.
8. 'With the aid of the newspaper, and through it, a permanent organization will naturally take shape that will engage, not only in local activities, but in regular general work and will train its members to follow political events carefully, appraise their significance and their effect on the various strata of the population, and develop effective means for the revolutionary party to influence those events.' 'Where to Begin', *Iskra* No. 4, 1901, in V I Lenin, *Collected Works*, V (Moscow: Foreign Languages Publishing House, 1961).
9. Carew Hunt, p. 77.
10. Gross, pp. 223–7.
11. *Ibid.*, p. 240.

12. *Ibid.*
13. van der Lubbe's statement to police, 3 March 1933, cited Fritz Tobias, *The Reichstag Fire: Legend and Truth* (London: Secker and Warburg, 1963), pp. 34–6; (originally published in German at Rastatt, Baden, in 1962).
14. Tobias, *op. cit.*
15. Koestler, p. 243.
16. Gross, p. 241–2.
17. Koestler, pp. 242–3.
18. Klaus Sohl, '*Entstehung und Verbreitung des Braunbuchs über Reichstagsbrand und Hitlerterror 1933/1934*,' in *Jahrbuch für Geschichte*, 21, 1980, p. 294.
19. Koestler, p. 243.
20. Sohl, pp. 307, 323–7.
21. Ronald W Clark, *Einstein: The Life and Times* (London: Hodder and Stoughton, 1973), p. 463.
22. *The Brown Book of the Hitler Terror and the Burning of the Reichstag* prepared by the World Committee for the Victims of German Fascism (President: Einstein), (London: Victor Gollancz, 1933), Introduction by Lord Marley, p. 9.
23. Koestler, p. 243.
24. See Alexander Abusch, *Nachwort* (Epilogue) to reprint of *Braunbuch* (Frankfurt/Main, Röderberg–Verlag, 1973).
25. Sohl, pp. 395, 304.
26. Garfield Hays, cited Tobias, p. 123.
27. See for instance *Vossische Zeitung* 106, 3 March 1933, detailing the deaths of a worker, Busch, and a communist Landtag deputy, Gerdes, and 140 arrests.
28. *Brown Book*, p. 111.
29. *Ibid.*, p. 51.
30. PRO, FO 371/16718C 1974, 2 March 1933, Rumbold Cable.
31. Gottfried Treviranus *Das Ende van Weimar: Heinrich Brüning und seine Zeit* (Düsseldorf: Econ–Verlag 1968), pp. 361–2.
32. Jacques Delarue, *Geschichte der Gestapo* (Düsseldorf: Droste, 1964), p. 51; Konrad Heiden, *Das Zeitalter der Verantwortungslosigkeit: Eine Biographie* (Zürich: Europa Verlag, 1936), p. 412.
33. Hermann Rauschning, *Gespräche mit Hitler* (New York: Europa Verlag, 1940), p. 78.
34. Tobias, p. 90.
35. *Vossische Zeitung*, 101, 1 March 1933.
36. Abusch, *op. cit.*
37. *Vossische Zeitung*, 100, 28 February 1933 and 102, 1 March 1933.
38. *Brown Book*, p. 55.
39. Tobias, pp. 72–5.
40. *Frankfurther Zeitung*, 6 October 1932.
41. *Brown Book*, p. 55.
42. *Ibid.*, pp. 55–6.
43. *Ibid.*, p. 121.
44. *Vossische Zeitung* 99, 28 February 1933.
45. *Brown Book*, p. 41.
46. *Ibid.*, p. 142.
47. Tobias, p. 221.
48. *Manchester Guardian*, 26 April 1933; *Brown Book*, pp. 81–2.
49. F A Voight, a *Manchester Guardian* star correspondent based in Paris, considered the reporting on Germany by his paper to be inadequate. So he built his own Paris network to collect news from Germany and was the immediate and natural victim of forgery. See David Ayerst, *Guardian: Biography of a Newspaper* (London: Collins, 1971), pp. 507–18.
50. Tobias, pp. 249–57.
51. *Brown Book*, p. 125.
52. *Ibid.*, p. 85.
53. *Ibid.*, p. 59.
54. The trial is described in Tobias, pp. 179–284.
55. Gross, p. 251.

56. See D N Pritt, *From Right to Left* (London: Lawrence and Wishart, 1965), especially chapters V, VII, XI and XII.
57. See A Garfield Hays, *City Lawyer* (New York: Simon and Schuster, 1942).
58. PRO, FO 371/16755, Minute of 14 December 1933.
59. Koestler, p. 244.
60. *The Times* (London), 14 September 1933.
61. *The Times*, 15 September 1933.
62. Koestler, p. 244.
63. P Stojanoff, *Reichstagbrand: Die Prozesse in London und Leipzig* (Vienna: Europa Verlag, 1966), p. 183.
64. *The Times*, 21 September 1933, p. 12.
65. Tobias, p. 228.
66. *The Second Brown Book of the Hitler Terror* (London: Bodley Head, 1934), pp. 212–13 (herafter, *Second Brown Book*).
67. *Second Brown Book*, p. 298.
68. *Ibid.*, p. 294.
69. *Ibid.*
70. See note 46: Voight later told his editor 'I have long known both these persons [Otto Katz and Willi Münzenberg] as being quite unscrupulous.'
71. *Second Brown Book*, p. 77
72. *Ibid.*, pp. 301, 306.
73. Alan Bullock, *Hitler: A Study in Tyranny* (London: Odhams Press, 1952), p. 237.
74. Tobias, author's preface, p. 18.
75. A J P Taylor, Introduction to Tobias, p. 16.
76. On Calic and his Committee, see *Die Zeit* Dossier September–October, 1979; also E Calic, ed., *Der Reichstagbrand: Eine wissenschaftliche Dokumentation* Band II (Munich: K G Saur Verlag, 1978): the Committee included historians as distinguished as Karl Dietrich Bracher and Golo Mann.
77. Sohl, pp. 297–8.
78. Sohl, especially pp. 293, 295–6.
79. Stojanoff, p. 181.
80. Gross, pp. 214–15; Istvan Deak, *Weimar Germany's Left-Wing Intellectuals* (Berkeley: University of California Press, 1968), p. 283.
81. See Christopher M Andrew, 'F H Hinsley and the Cambridge Moles: Two Patterns of Intelligence Recruitment' in Richard Langhorne, ed., *Diplomacy and Intelligence during the Second World War: Essays in Honour of F H Hinsley* (Cambridge: Cambridge University Press, 1985), p. 25.
82. Koestler, pp. 28–9.
83. *Ibid.*, p. 30.
84. See Michael Kater, *The Nazi Party: A Social Profile of Members and Leaders 1919–1945*, (Oxford: Blackwell, 1982), pp. 81–3.
85. Ernst Henri (i), 'The Revolutionary Movement in Nazi Germany,' *New Statesman and Nation*, 5 August 1933, p. 153.
86. Ernst Henri (ii), *Hitler over Europe* (London: Dent, 1934).
87. *The Times*, cited in 1939 edition of *Hitler over Europe?*, p. 292.
88. Henri (ii), pp. 299–300.
89. Andrew, p. 26.
90. *New Statesman and Nation*, 7 April 1934.
91. Catherine Karolyi, *A Life Together* (London: Allen and Unwin, 1961), pp. 298–9.
92. Koestler, pp. 244, 250; Gross, p. 229.
93. Gross, p. 311; Also Koestler, pp. 255–6.
94. C M Andrew interview with Edith Cobbett, 18 March 1981.
95. Gross, pp. 304 ff.
96. Gross, pp. 310–13.
97. Andrew, p. 29.
98. At the Leipzig trial Dimitrov argued that the Communist Party never used terrorist tactics, that the Bulgarian Party had not been responsible for the blowing up of Sofia Cathedral in 1924 in which 150 had died, and after which many communists were

arrested. In his concluding speech he said, 'That act of provocation was actually organized by the Bulgarian police.' But 15 years later, at the Bulgarian Communist Party's fifth congress, Dimitrov admitted that the party executive, suffering ultra-left deviation, had been driven to 'desperate acts . . . culminating in the attempt at the Sofia cathedral'. (Koestler, p. 247; also Georgi Dimitrov, *Selected Speeches and Articles* (London: Lawrence & Wishart, 1951), p. 23 and pp. 202–3.)

99. Herman Weber, *Die Wandlung des deutschen Kommunismus*, Bd II (Frankfurt, Europäische Verlagsanstalt, 1969), pp. 58–9.

CHAPTER 2

James Klugmann, SOE-Cairo, and the Mihailovich Deception

DAVID MARTIN

The fact is, I am sure, that SOE-Cairo . . . do not want us to come to a satisfactory arrangement with Mihailovich. We have been on the verge of doing so many times, but on each occasion a spanner has been thrown in to prevent us . . .[1]

In the early morning of 27 March 1941, a number of Yugoslav officers overthrew the government of Prince Paul and installed the youthful King Peter on the throne. The ousted government had just concluded a pact with Germany that harnessed the Yugoslav economy to the Nazi New Order in Europe. The revolution of 27 March was primarily the work of Serbian officers, implementing the will of the Serbian people. Hitler never forgave the Serbs for their temerity in blocking the pact.

On 6 April 1941, the Nazi Army, without troubling to declare war, bombed Belgrade and invaded Yugoslavia. They threw into the attack 33 Axis divisions, against which the ill-prepared, ill-armed, and ill-deployed Yugoslav Army could do very little. The battle for Yugoslavia was over in 12 days and on 4 May 1941, Hitler proclaimed to the world that the Yugoslav state no longer existed.[2]

On 10 May, an unknown Yugoslav army officer, Colonel Draja Mihailovich, hoisted the Yugoslav flag on the mountain of Ravna Gora and proclaimed the continuation of the war against Germany. During 1941 and 1942 and the first half of 1943, Britain gave its full political support to Mihailovich as the leader of the national resistance movement in Yugoslavia. But by the end of 1943 he had been abandoned and his place as the officially accepted resistance leader had been taken by Josip Broz Tito, the Yugoslav Communist leader. Some months earlier, by October 1943, Tito had become the monopolistic beneficiary of the greatly augmented Allied support which had become logistically possible after the collapse of Italy.

This switch in Allied policy has inspired one of the great debates of the Second World War. The debate is all the greater because, while the majority (but not all) of the British and American officers who were attached to Tito defend the switch in Allied policy, the 27 British and the seven American officers who were attached to the Mihailovich forces at different times and different places are convinced that the Allies made a serious strategic mistake, that a grave injustice was done, and that much of the intelligence on which the switch was based was false or exaggerated.

It is not the author's intention to argue in this chapter the merits of the switch in Allied policy; this the writer has done elsewhere.[3] This analysis is concerned with the campaign of deception, targeted at the policy elites in London and at the Allied publics, which destroyed faith in Mihailovich and his national resistance movement, thus making the swing to Tito and his communists inevitable. It is also concerned with the roles in this deception played by SOE-Cairo and especially by one of its officers, James Klugmann.

* * *

On 22 June 1941, Hitler staked the future of Nazi Germany on the rapid conquest of Soviet Russia. The surprise attack threw Stalin and his Party apparatus into confusion and dismay. Messages went out from Georgi Dimitrov, Secretary General of the Comintern, to communists everywhere to rally to the defence of socialism. Parties in the non-communist world that had been following orders to sabotage the Allied war effort were abruptly ordered to change sides. Communists were encouraged to make tactical alliances with 'bourgeois' forces opposing Fascism—a return to the popular front of the 1930s.

The Secretary General of the Yugoslav Communist party was Josip Broz Tito, a professional revolutionary and former Comintern agent who had somehow survived Stalin's terror which liquidated so many of the central European communist political leaders. Tito did nothing when the Nazis invaded Yugoslavia, but the moment they marched on Moscow he reacted instinctively. That same afternoon he drafted a proclamation: 'A fateful hour has struck! . . . The precious blood of the heroic peoples is being shed. This is also our struggle which we are obliged to support with all our strength . . .'[4] This battle cry was followed by instructions from the Comintern exhorting the Yugoslav comrades to put the anti-fascist struggle first, socialist revolution a poor second, an order which Tito ignored.[5] On 1 July, a further Moscow directive ordered: 'Without wasting a moment, organise partisan detachments and start a partisan war behind the enemy's lines.'[6]

Thenceforth there would be two guerrilla armies fighting the Germans in occupied Yugoslavia. On the one hand, there was the Mihailovich movement—the Chetniks in popular parlance—which was basically traditionalist and nationalist in orientation. On the other hand, there was the Tito movement—the Partisans—which at a rank and file level consisted for the most part of non-communist peasants but which, at the top, was tightly controlled by the communists from the beginning.

During the Summer and Autumn of 1941, both the Chetniks and Partisans engaged in widespread actions against the German occupation forces. The Mihailovich forces even used captured or salvaged tanks.[7] But the Nazi policy of ruthless reprisals against civilians, applied in Yugoslavia with unprecedented ferocity, forced Mihailovich to adopt a strategy more in line with the majority of resistance movements in Nazi-occupied Europe. This was by no means a passive policy, as is proved by the combat record and the official German record of reprisals, but it did attempt to protect the people on whose behalf the Chetniks were fighting. Additionally, as was the case with the resistance movements in France and other Western European countries, the Chetnik leader relied upon perfecting his organisation against the day, in this case the day when Italy would collapse and the Chetniks would seize their arms to drive out the Germans.[8] Some tactical agreements with the Italians, not directly involving Mihailovich, flowed from this policy in 1942 and 1943. These agreements involved accommodations between enemies rather than collaboration between ideological allies and were designed primarily to protect the peasantry from the horrors of liberation by Tito's Partisans.

Mihailovich's strategy was in contrast with Tito's, which 'stood for all-out immediate resistance, regardless of cost, in response to Comintern directives and the need to relieve enemy pressure on the eastern front'.[9] Milovan Djilas, Tito's chief lieutenant during the war, has admitted frankly that the communists' goal was, from the beginning, the establishment of a communist dictatorship and that it used the resistance movement as an instrument to achieve this objective. Said Djilas: 'The military operations which we communists launched were motivated by our revolutionary ideology . . . A revolution was not feasible without a simultaneous struggle against the occupation forces.'[10]

The 'Chetnik' situation was more complicated than that of the Partisans, who were organised exclusively under Tito. The term 'Chetnik' was traditionally applied to all Serbian guerrillas. It is to be noted that Mihailovich himself consistently referred to his forces as the Yugoslav Army in the Homeland, but in popular parlance they were generally referred to as 'Chetniks'. While most of the Second World War Chetnik units bore allegiance to Mihailovich, others were little more than bandits and armed gangs. Some Chetniks, in particular the collaborationist Chetniks of Kosta Pecanac, fought many bitter battles with the Mihailovich Chetniks. To the extent that this situation created confusion, it lent itself to undermining the reputation of Mihailovich and his national resistance movement.

The followers of Tito and the followers of Mihailovich initially collaborated, but this fragile political alliance disintegrated when the Communists insisted on keeping their own military command and on establishing miniature Soviet Republics wherever they seized power—including mass executions of prominent citizens. Djilas described how 'communist sons confirmed their devotion by killing their own fathers, and there was dancing and singing around the bodies.'[11]

During the closing months of 1941, Draja Mihailovich became the toast of the Allied press. Nothing was said about the Partisans for three reasons: first, what little was known about them had been treated with reserve; second, and closely linked, the British and American Governments were disposed to be more sympathetic to traditionalists, such as the Mihailovich movement, than to revolutionary Communists; third, the Soviets said nothing about Tito. With access to the British press through the Yugoslav government-in-exile and through sympathetic British government departments, it was natural that Mihailovich should have become internationally known as the leader of the Yugoslav resistance.

Western ignorance of the existence of the Tito movement was in a very large measure due to the fact that Stalin at that time, in the interests of the popular front against Hitler and of aid from the allies, was deliberately muting all information that suggested revolutionary intentions by the Soviet Union for the post-war period.[12] To this end, Stalin engaged in strategic deception. Stephen de Mowbray has listed some of the major stratagems.[13]

In a speech of 6 November 1941, Stalin denied that the Soviets had expansionist ambitions towards non-communist states or intentions of interfering in their internal affairs. Non-interference was reaffirmed in a twenty-year Anglo-Soviet treaty signed on 26 May 1942. Stalin's commitment to the Atlantic Charter and the United Nations Pact was followed by a speech of 6 November, 1942, promising 'liberation' of countries overrun by Hitler's troops and the 'restoration of their democratic liberties'. The deception operation reached its climax in May 1943, when the Comintern was formally abolished by Stalin. Thenceforth its activities would be carried out by covert methods.

The historian Stephen Clissold who, during his SOE days, was sympathetic to the Partisans, shares de Mowbray's view that the abolition of the Comintern was a tactical move designed to allay Allied fears that the Russians, through the control they exercised over local Communist Parties, were intervening in the internal affairs of the occupied nations by subversion. In practice, these parties, including Tito's, continued as before to report to Moscow through Dimitrov and his Comintern assistants, who remained in their old appointments but without the Comintern title.[14]

The Soviet Government's overt policy in Yugoslavia was guided by the same caution about offending Churchill by suggesting that the Soviets harboured revolutionary intentions. According to Edvard Kardelj, who in later years was Prime Minister of Yugoslavia under Tito, Stalin exerted pressure on the Partisans 'to reach an understanding with the Chetniks at all costs and set up a joint army under the command of Mihailovich'.[15] But while the deception implying moderation was to continue under different formulations, the Soviets changed their real policy in Yugoslavia in mid-1942. In one way or another, the canny Stalin seems to have decided that the situation was developing in a manner which made it safe to break with Mihailovich without endangering his relationship with Churchill.

The events marking this shift in Soviet policy were as follows. On 18 July 1942, the editorial columns of *World News and Views*, the international communist organ published in London, proclaimed: 'The forces of Mihailovich, the patriot leader, are growing stronger and bolder, until they are raiding over the Italian border toward the great port of Trieste.' Until approximately the same date, *The Daily Worker*, organ of the American Communist Party, was also supporting Mihailovich as the leader of the Yugoslav resistance forces.

Four days later, on 22 July, the BBC monitoring service recorded a broadcast emanating from an apparently clandestine transmitter, calling itself Radio Free Yugoslavia. The broadcast said that it was untrue that the Chetniks had been fighting against the Axis; that, in fact, Mihailovich and his chief lieutenants had been collaborating with the enemy; and that all the fighting against the occupying forces during the previous year had been done by the Partisans. The site of Radio Free Yugoslavia was subsequently identified as within the USSR; thus it was a 'black' propaganda station with a disguised point of origin.[16] On 28 July 1942, the

British Communist newspaper, *Daily Worker* carried an article stating that Mihailovich had been labelled a Fascist and a collaborator by Radio Free Yugoslavia.

On 3 and 5 August 1942, the Yugoslav government-in-exile protested against the attacks on Mihailovich, first to the Soviet Government and then to the United States State Department. On 5 August the Soviet Foreign Ministry formalised its charges of collaboration in a detailed memorandum to the Yugoslav Legation in Kuibyshev—the Soviet wartime diplomatic capital; and on 7 August, Soviet Ambassador Maisky presented a copy of the memorandum to the British Foreign Office. This was the first time the Soviet Government had attacked Mihailovich. On 20 August 1942, Foreign Secretary Anthony Eden replied to Maisky that Britain believed that the charges were not based on accurate and objective intelligence. Eden proposed a discussion of the matter, but there is no record of a Soviet reply to the proposal.[17]

Having reversed itself in a manner which must have completely satisfied Edvard Kardelj and his comrades, the Soviet Government from that point on embarked on a merciless campaign of character assassination and disinformation directed against Mihailovich. In retrospect, it is possible to identify three main themes in the Soviet-Yugoslav campaign: first, that all the fighting against the Germans was being done by Partisans; second, that the Chetniks, and Mihailovich personally, were collaborating with the Nazis; and third, that the dispute between Chetniks and Partisans arose only out of Mihailovich's intransigence. These themes were soon to be echoed within a small but influential section of the British war fighting establishment.

* * *

Guy Burgess, Donald Maclean and Kim Philby were the first three members of the 1930s Cambridge conspiracy—Burgess's 'light blue ring of five'—to be unmasked as Soviet agents.[18] In 1978, a fourth member was revealed—Sir Anthony Blunt. This led to speculation about the identity of the fifth member, with fingers pointed at James Klugmann. Klugmann, who died in 1977, was a prominent member of the British and European Communist youth movements in the 1930s, a member of the Executive Committee of the British Communist Party in the post-war period, and editor of its theoretical and historical organ, *Marxism Today*. At his death, the British Communist newspaper *Morning Star*[19] (successor to the *Daily Worker*) published an obituary photograph showing a genial Klugmann, with bald pate and curly fringe—a kind of middle-aged cherub—conversing warmly with young communists. It was hardly the kind of image that would bolster the impression of a master agent. The impression conveyed by the photograph is further reinforced by descriptions of Klugmann during his student days. Michael Straight, a scion of a wealthy and respected American family, who was part of the Klugmann group at Cambridge, remembered him as a stooped, bespectacled, warm-hearted and compassionate intellectual whose commitment to Communism left him with no time for such minor preoccupations as taking a bath or cleaning his fingernails.[20] Straight is also convinced that Klugmann served as principal adviser to Anthony Blunt on potential student recruits, because Blunt did not know the students personally, whereas Klugmann knew them well.[21]

On 26 February 1975, Lord Clifford, speaking on a motion in the House of Lords that dealt with 'subversive and extremist elements', named Klugmann as 'an example of an intelligent, highly educated, and dedicated agent of a foreign power.'[22] In 1981 in a popular account of British spies in Soviet pay, Chapman Pincher referred to Klugmann as 'a sinister figure' and alleged that Klugmann had recruited John Cairncross, a confessed agent, for the Soviet apparatus. Pincher also suggested that Klugmann had been instrumental in recruiting Bernard Floud, a Labour MP and alleged Soviet agent who committed suicide in 1967 while he was being interrogated by British intelligence.[23]

According to the London *Observer*, Anthony Blunt confessed to having invited Stuart Hampshire to a dinner in Paris hosted by 'James Klugmann, the Russian spy', to vet Hampshire as a possible recruit.[24] Blunt's story was that they were simply considering Hampshire as a potential recruit for the British Communist Party and that the entire incident had taken place in 1934 before he started recruiting for Soviet intelligence. However, according to the *Observer*, MI5 had firm information that the dinner took place in 1938, at a time when Blunt was very busy indeed as a recruiter of Soviet agents.

Klugmann was also the ideological mentor of Donald Maclean and a frequent companion of his, first at Gresham's, a boys' public school, and later at Cambridge.[25] At Cambridge, Klugmann was, by common consent, the most brilliant member of the pro-Communist group which spawned Kim Philby, Donald Maclean, Guy Burgess, Anthony Blunt, John Cairncross and, for extra measure, the convicted nuclear spy, Allan Nunn-May.

Even Klugmann's very close friend of the period, Michael Straight, found it difficult—and finally impossible—to resist the circumstantial evidence that Klugmann was already a Soviet agent before the Second World War. In retrospect, Straight was convinced that the purpose of a meeting he had in 1935 with Guy

Burgess, Anthony Blunt, and James Klugmann was to recruit him—Straight—into the Soviet apparatus. Apparently still hurt by the thought, he asks: 'And James, whom I loved, did he know what was going on that evening? And was he a part of the snare?' Straight relates that he found it 'difficult to accept that Klugmann was a Soviet agent.'[26] But after discussing the matter with fellow students who had been communists at Cambridge, he came to the conclusion that the charges were essentially correct, and that Klugmann, 'in his gentle way' would have justified his deception as a 'historical necessity.'[27]

In one important respect Klugmann differed from the better known Cambridge conspirators: he did not renounce his Communism and seek to use establishment credentials to burrow into government service. Klugmann, as an avowed communist, walked head erect into one of the most sensitive departments of British intelligence—the Special Operations Executive (SOE). Whether or not he worked under orders from Soviet intelligence, as did his Cambridge colleagues, is a matter which must be left open for the time being.

* * *

SOE has been described by its historian, M R D Foot, 'as a small, tough British fighting service. It was formed in deadly secrecy in July 1940 to tackle one of the nastiest regimes even of this century, Hitler's German empire . . . It worked with the forces of resistance: classically, the resource of the weak against the strong.'[28] Foot has described how the British 'old boy' system gained Klugmann entry to SOE. Terence Airey, a capable and ambitious regular officer, was working in a Cairo office of Military Intelligence. One day he recognised the NCO who brought him his tea as the 'cleverest boy who had ever been (after his time) at his school—Gresham's'.[29] Airey arranged for this clever boy, who of course was Klugmann, to be transferred to SOE-Cairo and work on the Balkans desk, not serving tea, but as an embryo staff officer. Security checks revealed nothing detrimental, due to the chance destruction of a mass of files including Klugmann's by a German incendiary bomb.[30] When Private Klugmann, Pioneer Corps, joined the Yugoslav section, SOE-Cairo on 9 February 1942, Russia had been in the war for eight months. British assistance to Yugoslav guerrillas had hardly begun but, emotionally, politically and in propaganda, the British public was behind Mihailovich. About Tito and his Partisans, only the initial reports of Radio Free Yugoslavia were available to SOE, and this not until mid-1942.

An able linguist, Klugmann far outshone the other members of SOE-Cairo in his knowledge of Serbo-Croat. The fact that he was one of the founding staff of the Yugoslav section, moreover, stood him in good stead in making his expertise indispensable to his superior officers. His initial duties are recorded as 'looking after and keeping records of agents.'[31] In 1943, he was in charge of Intelligence, and later of Coordination. A junior intelligence officer who served under (the then, Captain) Klugmann beginning September 1943 wrote to the author that, to the best of his knowledge, Klugmann 'would have been the first to see decoded messages from the field, and would almost certainly have had control of distribution (within prescribed guidelines). I am almost certain he would have had delegated authority to discriminate what went into the OP Log.'[32]

Was Klugmann a Soviet agent? The author has discussed this matter with many veterans of SOE. One school of thought has it that Klugmann, because he was publicly known as a communist, would probably not have invited investigation by engaging in covert activities under Soviet command. As Michael Foot put the matter, while sceptics can point out that no connection between Klugmann and the Comintern has been proved, 'any Bolshevik as bright as Klugmann knew where his party duty lay'.[33] Other SOE veterans hold that, since the Soviets were playing for important strategic stakes in Yugoslavia, they were not likely to leave the direction of communist policy to a random and undirected association with a sympathetic ideologue like Klugmann—and, as brash as it might appear at first glance, Klugmann was able to disarm suspicion by publicly expressing his sympathy with Communism. Yet they also point to Klugmann's alleged recruiting activities for Soviet intelligence: if these reports are true, the reasoning goes, Klugmann must himself have been an agent.

Klugmann's responsibilities grew steadily. On 16 June 1942 he was granted an emergency commission as 2nd Lieutenant in the General List. On 22 May 1943 he was promoted Acting Captain, at which date his designation was GSOIII(I) (General Staff Officer, Grade III, Intelligence) in the Yugoslav section, which was known by the code name B.1. In September 1943 his designation became GSOIII (Briefing) and on December 15 1943 he became GSOII (Co-ordination). On 15 May 1944, by which time he had moved forward with some of the staff from Cairo to Bari, he was promoted Acting Major as second-in-command of the Yugoslav section.[34] But this formal listing of Klugmann's record of promotions tells only a very small part of the story. Because of his brilliance and the assiduousness with which he applied himself to his work, and also because of the sympathy of his superior, Major Basil Davidson,[35] Klugmann came to exercise an influence altogether disproportionate to his rank.

Basil Davidson was a formidable character in his own right. A pre-war journalist, he had been recruited early into SOE, whereupon he was sent on a mission to Hungary. He escaped, through Yugoslavia, just as the Germans were invading. His next posting was a covert one in Istanbul, after which he was sent to SOE-Cairo as head of the Yugoslav desk. He arrived in late August 1942, by which time Klugmann was well established as a commissioned officer, a veteran of the section who knew it all.

Time and again, Davidson quotes Klugmann as a fount of oracular wisdom:

> He could talk with brilliance on almost any subject, but what he really liked to talk about was politics . . . The fact is that politics moved in at this period, around the last months of 1942, and unavoidably; or, as James said, reality came home to roost. *It could even be called the Klugmann period and it changed a great deal.*[36]

These careful words reveal the immense impact that Klugmann's Communism was soon to have on SOE-Cairo's attitudes to Yugoslav resistance, attitudes which were to affect the course of Yugoslav history.

Basil Davidson was to devote his life as an historian to the glorification of Utopian causes.[37] In his book about the Second World War, he devoted several pages to an explanation of why 'large and serious resistance came and could only come under left wing leadership and inspiration.'[38] For Davidson the Yugoslav communists under Tito represented the forces of history, while Mihailovich's Chetniks were 'beaten generals' and 'ruling classes [who] had collapsed in defeat.'[39]

When a young officer named Peter Kemp reported for briefing by Davidson in preparation for being parachuted into Yugoslavia,[40] Davidson, as Kemp wrote later,

> could not conceal his antipathy to the Chetniks. He wanted me to sign a declaration that I had been subjected in London to indoctrination on behalf of Mihajlović; I refused, but the incident warned me of the sort of feeling that was to embitter relations between British officers in the field as well as at headquarters.[41]

Klugmann, under Davidson's orders, was used extensively to brief Allied officers about to be dropped into Yugoslavia. In the Summer of 1942, to prepare for the possibility that they might want to establish contact with Tito's Partisans, the SOE had recruited some 30 Canadian Croats who were known to be members of the Communist Party—apparently on the theory that the use of liaison officers who were identified communists would inspire trust on the part of the Partisans. Before the first group was dropped in to Tito in late April 1943, Klugmann had charge of preparing them for their assignment. According to Davidson, Klugmann told his Canadian-Yugoslav wards: 'You've got to see that this war has become more than a war *against* something, against Fascism. It's become a war *for* something, for something much bigger. For national liberation, people's liberation, colonial liberation.'[42]

Any liability that Klugmann suffered from the fact that he was known to be a prominent communist was offset by an amiability that he used to good advantage to promote his objectives. Professor Hugh Seton-Watson, a respected authority on Central European affairs, who once shared an office with Klugmann at SOE-Cairo, has recalled their ambiguous relationship. On the one hand, Seton-Watson never doubted that if ordered by the Communist Party to kill him, Klugmann would not have hesitated. On the other hand, he remembered his colleague 'as a brilliant man who was once my friend, and as a person of selfless, almost saint-like, character.'[43]

Klugmann's web of influence ultimately embraced Brigadier Keble, SOE-Cairo's Chief-of-Staff. On one occasion, according to Davidson, Keble pushed Klugmann into a lavatory, warning that security was after him but that he, Keble, would protect him.[44]

Peter Boughey, who was in charge of the Yugoslav section of SOE-London, resigned in frustration and anger when it became apparent during the summer of 1943 that control of the situation had passed into the hands of SOE-Cairo. This was perhaps unavoidable in view of the fact that all communications with Yugoslavia had to pass through the hands of SOE-Cairo, and that Cairo controlled the greatly increased provision of munitions and supplies to that country. M R D Foot confirms this transfer of control to SOE-Cairo, adding the point that the judgement of country section staffs would as a rule be deferred to by their seniors, 'who were both too remote and too busy to wish—or to need—to involve themselves in detail'.[45] The views of the head of SOE-Cairo on resistance operations in the Balkans would, of course, have weighed heavily with the Middle East Defence Committee, if only because the Committee had so few sources of information on this subject.

Intelligence about events inside Yugoslavia came from two principal sources: first, from SOE-Cairo's British liaison officers (BLOs) in the field, whose reports passed over the Yugoslav desk in Cairo; and, second, from intercepted enemy communications. Davidson and his section, including Klugmann, were the 'gate' through which the first source of intelligence passed. German signals traffic was intercepted and, to the degree that was possible, deciphered in Cairo and the United Kingdom. Cairo could break various low and medium-grade ciphers.[46] Only at Bletchley, in England, was it possible to decipher the highly secret material encoded by the German's supposedly invulnerable *Enigma* system. Copies of the Enigma decrypts, suitably modified and heavily encrypted in the Ultra system, were passed on to military officers who might have a need to know.

Access to this decrypt material was severely restricted. But Keble, by a staff accident, retained throughout his stewardship of SOE-Cairo access to this intelligence information because he had earlier been cleared to see it. Keble was a thrusting, ambitious Chief-of-Staff, anxious to promote the reputations of SOE-Cairo and Brigadier Keble. His boss was a civilian, Lord Glenconner, whose busy schedule in and out of the politico-military power centres of Cairo divorced him from day-to-day SOE policy.[47]

Davidson describes the struggle inside SOE-Cairo to convert that office from support for Mihailovich to support for Tito: this, obviously, was the necessary prerequisite for a wider conversion. The greatest coup was convincing Keble that Tito was the man for an ambitious brigadier to back. Davidson's account suggests that Keble simply wanted another mission to swell his empire—a somewhat unlikely story, given the enormous size of SOE-Cairo's existing responsibilities.[48]

All deception requires that certain truths be hidden and that the desired illusion be perceived in their place. Emphasis and filtering can be as effective in some instances as invention. If two grey objects were displayed, one in murky darkness and the other in brilliant light, the first might be perceived as black and the second as white. The contrast would tend to reinforce the illusion. The stories circulated by senior staff officers of SOE-Cairo, that Mihailovich's forces were doing nothing to harm the Germans but instead were collaborating with the Italians or with the Germans and waging civil war against the Partisans, relied upon the suppression of reports from British Liaison Officers in the field and the amplification of other reports supporting such charges.

SOE's first British Liaison Officer in Yugoslavia was Captain D T Hudson, who was put ashore by submarine in September 1941. He was sent to make contact with the only resistance force known about in the West—Mihailovich's Chetniks. By chance, however, the first armed men he encountered were from Tito's Partisans and he was taken to meet Tito before proceeding to Mihailovich's headquarters. According to the official historians, Hudson 'gave the first of many warnings that Tito's Partisans suspected that Mihailovich's Chetniks were collaborating with Nedic's government in Serbia and with other pro-Axis elements against them'.[49] In fact, Hudson understood and did not disapprove of Mihailovich's policy of infiltrating the Nedic forces, noting at the same time that this might give the impression of collaboration. His report was used by SOE-Cairo to give a completely different impression. Hudson later told the author that when, after the events, he was shown a sheaf of reports that SOE-Cairo had sent to London, based on his reports from the field but in fact heavily slanted by means of selection, he was so angry that he tore up what had been handed to him.[50] The clear deduction is that the Yugoslav desk at SOE-Cairo was selecting the material that was to be seen by London in a manner that displayed Mihailovich as black and Tito as white.

* * *

The 'Cairo partisans'—that is Klugmann and Davidson and probably Keble and possibly William Deakin—now set to work to convince the British Foreign Office and the Prime Minister, Winston Churchill, that Mihailovich was responsible for the civil war and was collaborating, while Tito was blameless in both respects and was a resistance leader of rare martial virtues.

SOE-Cairo's campaign opened when Churchill visited Cairo in late January 1943. The office prepared a briefing paper which claimed that, of the 40 Axis divisions in Yugoslavia, Mihailovich's forces were holding down only nine, while Tito's Partisans were engaging 31, but the Partisans, it said, could not be of real military value to the Allies unless supplied independently of the programme aiding the Chetniks. Mihailovich was damned with faint praise, in the sense that his accommodations with Italians were understandable in a man whose real concern was 'the future political struggle with other Yugoslav groups'. Playing upon Churchill's fears of being squeezed out by more powerful allies, the paper also advised early liaison with Tito's forces to forestall possible Soviet or American influence in the region.[51] SOE-Cairo did not at this stage advocate a break with Mihailovich. This suggestion came in March 1943, after the idea of backing both groups had been digested in London.[52]

This enormously inflated estimate of Axis strength in Yugoslavia was a product of an inclination to build up the image of Tito as a redoubtable fighter in the Allied interest. There were no liaison officers with Tito's forces and Colonel S W Bailey, who had joined Hudson on Christmas Day 1942 as senior BLO with Mihailovich, had certainly not provided the force estimates. To a considerable extent the paper was a creative work, presented as fact on the basis of the Yugoslav Section's near monopoly of intelligence. This monopoly had been strengthened by Keble's questionable release of local decrypt signals to the Section. These, however, were confined to 'German medium grade and low grade hand ciphers, chiefly those used by the Abwehr and Police'. Military Intelligence in London, which had access to the Ultra decrypts, estimated that there were about 17 Axis divisions in Partisan areas and 14 in Mihailovich's at the beginning of 1943. It was, however, SOE-Cairo's figures that lodged in Churchill's consciousness.[53] By bypassing intermediaries, the briefing succeeded in raising some serious questions in the Prime Minister's mind.

Davidson attributes the authorship of the briefing paper[54] to William Deakin. A young Oxford history don who was now a captain in SOE, Deakin had helped Churchill with his research on *The Life of Marlborough* and his word, therefore, carried special weight with the Prime Minister. Although it seems reasonable to assume that Deakin, recently arrived from New York, used the expertise of James Klugmann in preparing the briefing, he has assured the author that Klugmann was not involved. Davidson was elated by the impact of the Churchill briefing. He relates that news of the event 'exploded round the corridors of Rustem Building with the scattering effect of shrapnel. The Cairo "partisans" had won.'[55]

Indeed, two months after the briefing, Churchill wrote to Professor Slobodan Jovanovic, the Yugoslav Prime Minister-in-Exile, saying that 'His Majesty's Government are becoming seriously disturbed at recent developments in Yugoslavia.'[56] It was ironic that Churchill's expressed concern over the Chetnik-Italian accommodations came one month after German Military Intelligence for Eastern Europe warned about the Michailovich movement, and at the very moment when Tito's envoys had agreed to collaborate with the Germans in resisting a British invasion. General Reinhard Gehlen's Top Secret memorandum to the German General Staff was dated February 9 1943, and read as follows:

> Among the various resistance movements which increasingly cause trouble in the area of the former Yugoslav state, the movement of General Mihailovich remains in the first place with regard to leadership, armament, organization, and activity . . . The followers of D M come from all classes of the population and at present comprise about 80 per cent of the Serbian people. Hoping for the liberation from the 'alien yoke' and for a better new order, and an economical and social new balance, their number is continuously increasing.[57]

Six weeks after the Gehlen memorandum was written, in March 1943, Tito sent a high-level delegation to the Headquarters of the German Commander-in-Chief at Sarajevo. The delegation consisted of Milovan Djilas, General Koca Popovich, Dr. Vladimir Velebit, three Partisan officials who ranked next only to Tito himself.[58] The ostensible purpose of the meeting was to arrange for a prisoner exchange. The three Partisan leaders were subsequently flown to Zagreb by a special German military plane. There the discussions were continued. Walter Roberts, who discovered this interesting documentation in German military archives, summarised it as follows:

> The Partisan delegation stressed that the Partisans saw no reason for fighting the German Army—they added that they fought against the Germans only in self-defence—but wished solely to fight the Chetniks . . . that they would fight the British should the latter land in Yugoslavia . . . [and that,] inasmuch as they wanted to concentrate on fighting the Chetniks, they wished to suggest respective territories of interest.[59]

The agreement was finally vetoed by Ribbentrop at the end of March, despite the fact that Kasche, the German minister in Zagreb, argued passionately that the agreement was to the Germans' advantage, and that, *in all negotiations with the Partisans to date*, 'the reliability of Tito's promises' had been 'confirmed'.[60]

Unlike the Gehlen memorandum, which remained secret until Walter Roberts unearthed it in the 1970s, rumours of the talks quickly reached Mihailovich, though the scope of the agenda was unknown to him at the time. Within 10 days of the meeting, Colonel Bailey signalled SOE-Cairo with the news.[61] However, this news of Partisan-German collaboration never seems to have reached London. According to both Peter Boughey of SOE-London and Douglas Howard, who was in charge of Yugoslav affairs for the Foreign Office, London did not receive the many references to Partisan initiatives in waging war against the Chetniks which figured in the dispatches of the BLOs with Mihailovich. Nor did they receive more than a token of the numerous despatches dealing with anti-Axis actions by Chetniks, sent out by the large corps of British and American liaison officers scattered across the country.[62] This must be why there is no reference to them in Foot's history of SOE or in the official history of British Intelligence in the Second World War by F H Hinsley *et al.*

Because the January briefing of Churchill had committed SOE-Cairo to portraying Tito as white and Mihailovich as dark grey, they could hardly allow news of the Tito-German negotiations to contradict their argument and destroy their credibility. The fact is that SOE-Cairo was potentially compromised.

The deception concerning Mihailovich was doubtless carried to London on as broad a front as the Soviets could orchestrate. This front embraced diplomatic action, covert measures, and propaganda—direct and indirect. The diplomatic and propaganda channels were activated, as described earlier, in the Summer of 1942. No doubt 'moles' in other government service pushed the themes of Chetnik collaboration and inaction as opportunity permitted, and the wider community of communists or Soviet-sympathisers would have been quick to support Tito's cause. But in spite of these influences, certain policy elites in London, particularly in SOE headquarters and the Foreign Office, remained committed to Mihailovich, or at least unpersuaded by the propaganda against him, until roughly the early Autumn of 1943.

Intercepted German reports sometimes spoke of Chetnik collaboration with the Italians, but these messages just as often presented the Chetniks as the greater threat to German interests. The one constant source of disinformation about Mihailovich was to be SOE-Cairo. The most knowledgeable sectors of the wartime bureaucracy in London detected a bias, and they discounted some of what was received from this quarter. But governments tend to place trust in their own covert services, and no one suspected that the bias could be more than skewed professional judgement and that it might contain deliberate disinformation.[63]

While Keble's use of decrypt material was of only tangential importance to the Churchill deception, his access to the source of such material became the nub of a cover story designed to explain to London why SOE-Cairo was so persistently anti-Mihailovich. This second deception has apparently also confused historians and effectively concealed the role of Klugmann and others in manipulating government perceptions and policy. The cover-up relied on the restricted access to decrypt enemy signals traffic, a restriction which persisted for 30 years after the events and is still in force for much of the material. It said in effect that the decisive evidence of Chetnik collaboration and Partisan prowess came less from agents in the field than from the secret intercepts. Davidson implies as much in his book: '. . . much must be forgiven to the man [Keble] who got the intercepts and used them'.[64] The effect of this argument, which for so long rested on hidden and therefore unchallengeable evidence, was to undercut the assertions by other historians who suggested that SOE-Cairo had been distorting reports from the field.

Bickham Sweet-Escott, who worked in SOE-London, has

explained how this cover-up deception was innocently conveyed to London by Lord Glenconner, who visited the UK in June 1943. Presumably briefed by Keble and by the Yugoslav section, Glenconner told SOE-London:

> that for some time past he and his people had been allowed access to the information I have mentioned [intercepts]. They accordingly knew a great deal more than we did . . .[65]

As SOE-London was not privy to decrypt material, this disclosure made the subordinate headquarters, SOE-Cairo, master. But when in 1984 F H Hinsley's Official History related the contents of this intercepted material, it turned out that late in 1942 the Germans 'remained determined to eliminate Mihailovich's organisation' and in March 1943 they instructed their own forces that 'no Chetnik formations whose leaders were proved to be in touch with Mihailovich are to be spared.' In the same month, an *Enigma* signal referred to 'the menacing consequences' of the Italian wish to use Chetnik forces against the Partisans and ordered the German supreme command to treat the Chetniks as enemies.

It is true that the Hinsley material did document, or appeared to document, accommodations with the Germans by certain Chetnik commanders; but what the Hinsley history failed to record is how these accommodations looked in practice and how enduring they were. At the highest levels, the signals demonstrated that the Germans were set on Mihailovich's destruction. Hinsley does refer to a decrypt message after the British had decided to abandon the Chetniks that apparently implicated Mihailovich in negotiations with the SS in Zagreb in December 1943.[66] However, in response to a letter from the author, he said that it had not been his intention to imply that Mihailovich was personally involved or that anything came of the meeting. The relevant point is that, during the period when British policy over Yugoslavia was under review, decrypted enemy signals traffic was not an important factor.

SOE-Cairo's cover-up was effective for so many years because the truth was hidden by authority. The illusion presented, of surreptitious access to secret material which explained their anti-Mihailovich stance, could only be challenged by historians, not fully discredited, because access to most decrypt material is still denied.

* * *

On 28 May, 1943 SOE-Cairo sent Colonel Bailey, at Mihailovich's headquarters, a lengthy message which, in retrospect, marked the formal turning point in SOE's relations with the Mihailovich movement. The message, which bore the signature of General Sir Henry Maitland Wilson, Allied Commander-in-Chief, Middle East, was ostensibly designed to prevent hostilities between Chetniks and Partisans by arranging for delineation of their respective territories. Actually, it asked, in insulting terms, that Mihailovich pull back all his forces to a small area in eastern Serbia, leaving all the rest of Yugoslavia to Tito. The message said:

> General Mihailovich does not represent a fighting force of any importance west of Kapaonik. His units in Montenegro, Herzegovina, and Bosnia are already annihilated or else in close cooperation with the Axis; it is also difficult to say that his units exist in Croatia, Slovenia, and Slavonia . . . The Partisans represent a good and effective fighting force in all parts, whereas only the quislings represent General Mihailovich . . . You will advise General Mihailovich that he immediately go to Kapaonik with all his faithful officers and men; if necessary, he is to break through by armed force . . . In the future the Supreme Command will consider the districts under his command and influence to be bordered on the West by the fighting elements already existing on the right bank of the Ibar River and toward the south to Skoplje.[67]

Colonel Bailey had no alternative but to show the text of the directive to Mihailovich, although it is obvious that he had the most serious misgivings about its wisdom. Mihailovich's reaction, as could have been predicted, was immediate and explosive. His reply to the directive said, among other things:

> The statement . . . that my forces are not of any importance west of Kapaonik, is absolutely without foundation. It is not true that the units of the Yugoslav Army in Montenegro, Herzegovina, and Bosnia are annihilated, and there is even less truth in the same statement that they are in close cooperation with the Axis. It is not true that it cannot be said that no Yugoslav Army units exist in Slovenia and Slavonia. Quite on the contrary, the Yugoslav army exists in all parts, and it will prove its existence to the entire world. *I will not tolerate such insults any longer* . . .[68]

It developed that the message to Colonel Bailey had been sent off without the knowledge or approval of the Foreign Office. The Foreign Office, indeed, strongly disapproved of the message and ordered it rescinded.

Who actually wrote the 28 May directive? Colonel Bailey told the author categorically that it had been Basil Davidson. Davidson has denied authorship, suggesting that Glenconner was responsible. It is however inconceivable that Glenconner could have penned such a message without help from a Yugoslav specialist – which points directly at Klugmann. Commenting on Mihailovich's reaction to the 28 May directive, Douglas Howard, head of the Foreign Office Southern Department, wrote to his superiors on 15 June, 1943:

> In one way or another, SOE have excelled themselves in the handling of this question.
>
> First, Lord Glenconner goes and sends Bailey a telegram off his own bat and completely at variance with our own policy . . . we have suggested to the chiefs of staff that the disastrous telegram should be rescinded . . . even if it is withdrawn, so much bad blood will have been spilled that it may be very difficult to come to terms.[69]

In one sense, SOE-Cairo had overplayed their hand, as the indignant reaction from London indicated. But within Klugmann's notion of historical determinism, Mihailovich's 'I will not tolerate such insults any longer' must have been reassuring. Just as Churchill had been made to doubt Mihailovich's good faith, so the latter had been made to question Britain's. Politics were pushing history in the right direction. SOE-Cairo's message can be seen as a classic provocation.

* * *

The supply operation in support of the Mihailovich forces in 1943 was conducted in a manner that compels one to ask whether it formed part of an organised campaign of provocation by Klugmann and company. It is almost inconceivable that the many gaffes in this operation should have been simply the random outcome of departmental inefficiency. Inevitably, this was a situation that increased mistrust. Some of the messages dealing with supplies from the BLO's with Mihailovich's forces were personally addressed to Klugmann.[70]

Colonel Bailey reported that 'during the 10 weeks following my arrival at Mihailovich's headquarters, we received only two sorties. The few tons of supplies dropped included 130 million occupation lire, overprinted in bright red "Ethiopia", and several hundred boxes of tropical anti-snakebite serum.'[71]

Brigadier Armstrong, who was dropped in as the Head of the British Mission to Mihailovich on 28 September, 1943, brought with him a personal letter from General Sir Henry Maitland Wilson to Mihailovich promising stepped-up deliveries of military equipment, contingent on the activity of the Mihailovich forces. In fact, Armstrong subsequently reported that, despite considerable Chetnik activity, he had received one drop consisting primarily of used office lighting equipment and another shipment of rubber boots (useless in the mountains of Yugoslavia), and several drops consisting of rusted and useless military equipment. He felt strongly that it was criminal that RAF fliers were being asked to risk their lives to deliver such rubbish to the Mihailovich forces.[72]

The supply of equipment illustrated the way in which the British Mission to Mihailovich was virtually ignored, even though, in addition to the Headquarters mission, it included 17 sub-missions by the end of summer 1943. In a wire dated 4 September, 1943, Colonel Bailey said that in the eight months that he had been in Yugoslavia he had received 'only two relevant telegrams' from his government, 'one in May and one in August. I consider neither adequate . . . Though nominally in command of all sub-missions to Mihailovich's forces, I was, until August, in practically complete ignorance of sub-missions' locations, plans, and activities.'[73] SOE-Cairo was evidently starving liaison officers with Mihailovich's forces of information, encouragement and support, while denying them the ability to communicate with each other, and filtering the flow of intelligence from Yugoslavia. Cairo prohibited direct communications between Bailey and his missions, insisting that each deal directly and exclusively with SOE-Cairo. In effect, this made communications between the various Mihailovich sub-missions virtually impossible.

A third provocation deprived the Chetniks of recognition for their actions against the Axis forces. This operation was additionally aimed at British and Allied public opinion. It aimed to discredit Mihailovich as a traitor and consign him to the memory hole of history, and to promote Tito in his place as the great Yugoslav resistance leader. This campaign, of course, had begun on the international scene as soon as Radio Free Yugoslavia came on the air. But it became really effective only when the British Broadcasting Corporation, which had previously backed Mihailovich, changed sides in the first half of 1943.

Further research is needed to establish definitively what forces were at work to bring about so dramatic a change. Clearly, the passage of doctored intelligence from SOE-Cairo to London, in which can be seen the hand of Klugmann, was a major factor. But, from Hinsley's account, it is also clear that neither SOE-London nor the Foreign Office accepted SOE-Cairo's reports uncritically, so it is probable that there were other factors. Whatever its causes, the consistent misreporting by the BBC changed public perceptions in Britain and even influenced reluctant high-ranking officials.

In July 1943, the German occupation authorities in Yugoslavia had published a poster bearing drawings and descriptions of Mihailovich and Tito and offering 100,000 gold marks for the head of either. The BBC was slow to mention this, but when it did it reported only the offer for Tito. Even when Kenneth Pickthorne MP, and Yugoslav officials in London protested, the BBC made no correction.[74]

Repeatedly, especially during the months of September and October 1943, the Chetniks engaged in military actions against the Axis forces—only to hear the BBC report that the centres in question had been captured by the Partisans. On 11 September, Chetnik forces attacked a strong German garrison at Prijepolje, killing more than 200 of the enemy and putting the rest to flight. This was credited to the Partisans. 48 hours later, the Chetnik forces moved on the town of Berane, where they took the surrender of the 6,000 Italians of the Venezia division (under orders from SOE-Cairo, Bailey brought the Venezia division under Allied command, thus depriving the Chetniks of the possibility of disarming it). This, too, was credited to the Partisans. On 12 September, a force of Chetniks, accompanied by Lieutenant Colonel Duane Hudson, succeeded in negotiating the surrender of 1,800 Italians at Priboj. This was credited to the Partisans. On 5 October, after blowing up four smaller bridges at Mokra Gora, Chetnik forces stormed the town of Visegrad, killing over 200 of the enemy and then, with the aid of British sappers blew up the Visegrad bridge, a 450-foot steel span. Both the bridges of Mokra Gora and the bridge at Visegrad were credited to the Partisans. On 14–15 October, the same Chetnik forces, pushing north in the direction of Sarajevo, took the town of Rogatica, again killing in excess of 200 of the enemy. Once more, the Partisans were credited with the attack. In the month of November, the Chetniks attacked substantial Bulgarian forces at Kalna and Nova Varos. These actions, too, were credited to the Partisans by the BBC.[75]

The persistent misreporting by the BBC was the subject of repeated protests by the British Mission attached to Mihailovich and finally of an omnibus protest which Brigadier Armstrong fired off to SOE-Cairo on November 18, 1943:

> If you want to get the best out of MIHAILOVICH you must give him fairer press and broadcasts. BAILEY was with MIHAILOVICH Chetnik forces when [they] took PRIBOJ and PRIJEPOLJE and BERANE. I saw capture VISEGRAD, destruction bridges, and know OSTOJIC took ROGATICA. MIHAILOVICH never credited with any [of] these, although reported to you. On the other hand, when PARTISANS drove his forces out, PARTISANS credited on BBC [with capture of these places from enemy].[76]

None of these protests was acknowledged by SOE-Cairo.

As for the BBC, it could point out that its broadcasts faithfully mirrored the communiqués put out by Tito's forces. By filtering out the reports from the field sent by British Liaison Officers who had been with Mihailovich for many months, SOE-Cairo ensured that the Partisan version of the war went unchallenged.

Throughout the Summer and Autumn of 1943 the propaganda campaign continued, aided presumably by Soviet agents and sympathisers in various departments. In the previous chapter, Christopher Andrew and Harold James referred to Guy Burgess at this period as 'an influential BBC producer in the talks department;'[77] Chapman Pincher and Stephen de Mowbray spoke of H Peter Smollett, head of the Soviet Relations Department of the Ministry of Information, as a Soviet agent. Sir George Rendel, British Ambassador to the exile Yugoslav Government, described the Political Warfare Executive, the propaganda arm, as 'anti-Mihailovich and pro-Partisan' and labelled the BBC's Balkans editor as a 'Leftist'.[78]

Churchill provided the decisive factor. In a personal minute to the Foreign Secretary, dated 28 July, 1943, Churchill 'openly and directly intervened, by insisting that Fitzroy Maclean should head the proposed high-level mission to Tito, and overruling all opposition . . .'[79] The special status conferred upon Maclean as Churchill's personal representative to Tito provides confirmation that Churchill had indeed been impressed by the January Cairo briefing, and that he was already halfway persuaded to back Tito.

* * *

Tito was never shy of proclaiming his Communism: it was the Comintern which kept prodding him to present a less ideological profile, for fear of frightening off British support.[80] Under this pressure, Tito did agree to modify the form and scope of his National Liberation Committee of Yugoslavia, which he had planned to use as a provisional government as early as November 1942. According to Clissold, the Comintern urged Tito in the following terms:

> 'You must not fail', the Comintern urged, 'to give the Committee an all-national Yugoslav and all-Party anti-fascist character, both in its composition and its programme of work . . . At the present stage, do not raise the question of the abolition of the monarchy. Do not put forward any slogan of a republic. The question of the regime in Yugoslavia, as you understand it, will come up for settlement after the German-Italian coalition has been smashed . . .[81]

This form of deception was to become an almost standard feature of communist-backed 'wars of national liberation' in the post-war era. In the case of Yugoslavia, Churchill, on the one hand, never doubted Tito's communist commitment. However, as difficult as it may be to believe, the British leader did seem to entertain hope that Yugoslav nationalism would prove stronger than Marxist-Leninist internationalism, and that by dealing fairly with Tito, he, Churchill, could persuade him to accept King Peter in some capacity and create a pluralist future for the country.[82] The Soviet technique of appearing to react reasonably to the British viewpoint at the diplomatic level while backing Tito and defaming Mihailovich in propaganda and by covert means, seems to have worked.

The Foreign Office, during much of the period in question, leaned strongly toward continued support of Mihailovich. For example, as late as 9 September, 1943, Douglas Howard wrote the scathing memorandum quoted at the start of this chapter in which he condemned SOE-Cairo for consistently souring relations between the British and Mihailovich.[83] Sir Orme Sargent, the Deputy Under Secretary for Foreign Affairs, minuted his agreement with this evaluation.

In late September 1943, however, Brigadier Maclean and his team parachuted into Tito's headquarters, assessed the situation as seen through Partisan eyes, apparently accepting without question all that was said about Mihailovich's 'treachery', and reported back accordingly. When, in early November, this report reached London, the debate was over. On 22 November, 1943, Douglas Howard was taking a line completely at odds with his minute of two and a half months earlier:

> Since [Mihailovich] is . . . doing nothing from a military point of view to justify our continued assistance, we should consider cutting off supplies while maintaining our political and moral support (eg., propaganda).[84]

The military support which had always been promised to the Chetniks but which, because of the logistical limitations which existed prior to mid-1943, had never been delivered, began to pour into Tito's hands. The assessment that Tito's forces were the strongest became a self-fulfilling prophecy.

Once leaders have made up their minds, they tend to close them to evidence that might contradict their decision. In November 1943, when arrangements were being made to bring Maclean and a Partisan delegation to Cairo to meet Churchill, Sir Orme Sargent wired Sir Ralph Stevenson, the British Ambassador to the Yugoslav Government in Cairo:

> We think it would be useful if Brigadier Armstrong and Colonel Bailey were both summoned to Egypt to arrive at the same time as Brigadier Maclean and the Partisan delegation and to be available for consultation during negotiations beween Commander-in-Chief and delegation.[85]

This proposal, which was repeated in the exceptionally impressive 7,000-word summary which Bailey and Armstrong sent to Cairo on 24 November,[86] was ignored.

But if minds were closed to any reassessment of Mihailovich, Churchill's fondest hopes that Tito would seek reconciliation with Yugoslav nationalists were soon to be demolished. At the end of November, Churchill records, Tito summoned a political congress in Bosnia, 'and not only set up a Provisional Government, "with sole authority to represent the Yugoslav nation", but also formally deprived the Royal Yugoslav Government in Cairo of all its rights'.[87] Churchill was dismayed but retained the hope that something could be done to persuade Tito to unite his countrymen. The decision was therefore made to persuade King Peter to dismiss Mihailovich as Minister of War and to withdraw the British missions to the Chetniks. In effect, this meant leaving the nationalists to the fate they feared: 'We will be persecuted and killed, and the outside world will know nothing of it.'[88]

Had Mihailovich been the 'traitor' that communist and later British propaganda painted him, the defection of his allies would presumably have brought out the worst in him. Despite ceaseless provocations, however, he remained remarkably steadfast in his attitudes. One of the most eloquent witnesses against the charge of collaboration was Dr Walter T Carpenter who in mid-August 1944 was dropped in to Mihailovich to take care of the sick and wounded among the American airmen undergoing evacuation. Dr Carpenter reported finding a Chetnik hospital, staffed by competent physicians, but without anaesthetics or analgesics or other medicines or proper surgical instruments—without bandages or even a bar of soap. This was the situation in Serbia. In the areas peripheral to Serbia proper, where the Chetniks were being hard-pressed by both the Germans and Partisans, some of his subordinate officers did enter into accommodations—as distinct from collaboration—with the enemy. But these arrangements were fragile and limited.

SOE's final provocation came at the beginning of December. Mihailovich was presented with an absurd 'ultimatum' that certain well-defended bridges were to be blown up by the Chetniks as a condition of continuing support. The deadline set was 29 December. Mihailovich initially replied that he would need some heavy equipment in order to assault the bridges. Brigadier Armstrong backed him on this request—which was promptly denied by Cairo. Despite the fact that he had never refused the ultimatum, on 13 December, 1943, more than two weeks before the ultimatum expired, Cairo ordered all 17 British sub-missions to leave their positions and escape to the Partisans.[89] This was in most cases physically impossible, and Armstrong challenged the decision in a series of signals, one of which read:

> I wish to point out in strongest terms very serious mistakes you are making in breaking [with] Mihailovich . . . what am I to say to Mihailovich, please? I ordered my officers to run away? It is perfectly maddening . . . Probably Mihailovich has heard about Paddle [code name of British officer who obeyed Cairo's message and sought to evacuate himself via the Partisans] . . . that was why in such a temper when he saw me on 17th. What do you imagine is now going to be attitude of Jugs to all my officers who remain their posts implicit trust put?[90]

Of course there was no satisfactory answer. Subsequently, Armstrong and his 17 sub-missions, totalling 27 officers and approximately 80 other personnel, were evacuated by air on 31 May, 1944, with the full co-operation of the Mihailovich forces.

When the time came for the British officers to leave, Mihailovich comported himself with dignity that greatly impressed all the BLOs. According to the account of Major Jasper Rootham:

> He said that whatever anybody might say or think, he and those who fought with him regarded themselves as the friends and allies of the Western Democracies, and for that reason, if for no other, he regarded it as his duty by all means in his power to ensure that . . . our evacuation was successfully carried out.[91]

When the BLOs reached Bari, Italy, Major Jasper Rootham and Major Archie Jack decided to visit the SOE-Yugoslav Section, which had moved there from Cairo. On the wall was a map on which coloured pins marked the supposed positions held by Partisans and Chetniks, Germans and other Axis troops. All over Serbia, there were Partisan pins, with here and there an isolated Chetnik pin. The officers could scarcely believe their eyes or control their indignation. For a moment they were stunned. Eventually Jasper Rootham explained to the SOE duty officer that he had been in Serbia for the better part of a year, that he had come out only 'the day before yesterday' and that it was the observation of the British Mission that the Mihailovich forces were very strong throughout Serbia and that the Partisans were limited to a few isolated pockets of strength.

The operations officer replied that Rootham was misinformed. 'Are you calling me a liar?' Rootham shot back. 'I said you are misinformed,' said the young officer. At this point Rootham could no longer contain his anger. He walked up to the map and swept his sleeve over the mendacious pins, wiping 'the whole bloody lot' off the map of Serbia. After this incident, no one who had served as a liaison officer with the Mihailovich forces was allowed in the operations room.[92]

During 1944, the Chetniks managed to rescue some 500 American airmen who had been forced down in Yugoslavia. With the co-operation of the United States Army Air Force, these rescued men were evacuated to Italy in a series of dramatic airlifts.[93] Between 26 August and 31 October, 1944, Lieutenant Colonel Robert H McDowell with an intelligence team from the Office of Strategic Services (OSS), the United States equivalent of SOE, visited Yugoslavia to contact Chetnik resistance forces (which he referred to as 'Nationalists') and assess their status and significance. His report was not declassified until 1984.

No doubt McDowell came to identify with the Nationalist cause in the same way that Deakin and Maclean did with the communists. His report should be read with this in mind. However, McDowell's qualifications ought to be recognised. Born of missionary parents in the Middle East, he learned Arabic, Turkish, Syriac and other languages. During the First World War he worked for British intelligence with Armenian guerrillas fighting the Turks. Later, the British sent him to work with anti-Bolshevik guerrillas in the Caucasus. He then became Professor of History at the University of Michigan, where he specialised in Balkan history. Considered a liberal by his colleagues, on his assignment to the Yugoslav section of OSS his instinctive sympathies were with Tito's Partisans. By no means did he approach his mission to Mihailovich with a preconceived favourable bias.

McDowell reported that the Yugoslav Nationalist Movement was devoted to the liberation of the country from the Germans and to saving it from the threat of communist rule. It was equally determined to defend the nation against the old parties and leaders who had controlled the country during many of the previous 25 years. The Movement sought free elections. Its strongest supporters were peasants, intellectuals, and students, all determined to achieve a legal revolution and a more realistic democracy.[94]

McDowell concluded that hatred of Communism by the masses derived not only from the still basically conservative and individualistic character of Yugoslavs, but also from the record of communist collaboration with Axis and Quisling groups, attacks upon Chetniks while the latter were engaging the Germans, Partisan failure to attack Axis forces, falsification of Partisan communiqués, and atrocities against Nationalist civilians, including women and priests.[95]

McDowell was on hand when Mihailovich called for a national

uprising against the Germans on 1 September, 1944, and he and the members of his mission saw the Mihailovich forces go into action against the Germans at a number of points. Among other things, he related that when he interviewed Partisan prisoners taken by the Chetniks, not a single one of them said that they had fought against the Germans. They had fought only against the Chetniks.

In terms of eyewitness testimony by Allied LOs, the McDowell documentation was more extensive than anything produced by Tito's Partisans in their case against Mihailovich. Rather sadly, in the light of developments, McDowell reported that, despite the support given by Britain to the Partisans, he was astonished at the strength of pro-British feeling among Nationalists, and the hopes still entertained that 'Britain and the Serb Nationalists would eventually resume their traditional friendly relations'.[96]

When the war was over and Tito had revealed his true colours, Churchill wrote ruefully to King Peter: 'I cannot conceal from your Majesty that recent developments in Yugoslavia have disappointed my fondest hopes.'[97]

Colonel S W Bailey, Chief of the British Mission to Mihailovich, wrote to *The Times* on August 6, 1971: 'I do believe that everything must now be done to give Mihailovich his rightful, honourable place in history. He and his Chetniks did more than is generally appreciated for the Allied cause.' It should be noted that Colonel Bailey, an engineer, had spent much of his adult life in Yugoslavia and was completely fluent in Serbo-Croat.

Colonel Hudson, who had spent two and a half years with the Chetniks, and had gone into Yugoslavia with a fluent command of Serbian, commented bitterly:

> How the hell would the English have behaved if they had got a knife in their throats? How different would their behaviour have been from that of the Chetniks under occupation in Yugoslavia?[98]

On May 31, 1978, Sir Douglas Howard, wartime head of the Foreign Office Southern Department, wrote to the author: 'If, of course, as now seems more than possible, our information re Mihailovich's activities was incomplete, to say the least, then the whole affair changes colour . . . If the information contained in your work had been available to us at the time, our whole policy might and probably would, have been very different.'

* * *

The deception operation to discredit Mihailovich relied on a mixture of dissimulation to hide the truth and simulation to create a false picture. Dissimulation took the forms of SOE-Cairo's selective and distorted passage of intelligence, the BBC's denial of Chetnik successes after mid-1943, and Tito's skill at concealing his own dealings with the Germans. Simulation created the myths of Mihailovich's unique treachery and collaboration, of his sole responsibility for the fighting between the two guerrilla armies, and the notion that all the fighting against the Germans was being done by Tito. Hints relating to decrypt material provided excellent cover. The transmission belts worked through Moscow and the Comintern network and, with greatest effect, through SOE-Cairo where James Klugmann was nudging history in the direction indicated by Marx. Everywhere sympathy for the Soviet Union as an ally, strengthened by Soviet strategic deception, disarmed suspicion.[99]

As to Klugmann's precise status, the jury remains out. No one doubts that, as a dedicated Communist, he acted in the Soviet interest. One of the principal items of evidence was the fact that he was the only officer in SOE-Cairo who had been there from the beginning and who was on hand during all the provocations that took place from May 1943 until the abandonment of Mihailovich. SOE-Cairo was a very small bureaucracy totalling at any one time no more than five or six officers. The author is satisfied on the basis of interviews and letters that during the months of September and November of 1943, three of these officers (Dennis Ionides, William Crawshay, David Erskine) were not pro-Tito. This narrows the list of suspects to Klugmann and a handful of others. It cannot be emphasised too much, however, that SOE-Cairo had a rather quick personnel turnover and Klugmann was the only officer who was there from the begining to the end.

What about the argument that Klugmann, as an unconcealed Communist, would not have tempted the Fates by serving as a Soviet agent? A contemporary case may provide the answer. Burgess seems to have used his drunken, homosexual misbehaviour as cover against being taken for anything so serious as a Soviet agent. So, perhaps, if Klugmann was indeed the fifth member of the Cambridge ring of five, he disarmed suspicion by his open and conspicuous Marxism.

Either way, he helped to change the course of Yugoslav history.

Endnotes

1. PRO FO371/37590. Memorandum by Douglas Howard, Head of the Southern Department, Foreign Office, 9 September 1943.
2. See, for instance, Stephen Clissold (i), ed., *A Short History of Yugoslavia* (Cambridge, Cambridge University Press, 1966), pp. 201–10.

3. David Martin, *Patriot or Traitor: The Case of General Mihailovich* (Stanford, California: Hoover Institution Press, 1978).
4. See Stephen Clissold (ii), ed., *Yugoslavia and the Soviet Union 1939–73: A Documentary Survey* (London: Oxford University Press, 1975), p. 127. 'CPY CC manifesto on the German invasion of the USSR, June, 1941' (Item 19).
5. Clissold (ii), p. 128, Item 20.
6. Clissold (ii), p. 129, Item 23.
7. Colonel D T Hudson (i), interview with the Author, London, November, 1977.
8. F H Hinsley, with E E Thomas, C F G Ransom, R C Knight, *British Intelligence in the Second World War*, Vol. III, Part I (London: HM Stationery Office, 1984) p. 140–1.
9. Clissold (ii), p. 14.
10. Milovan Djilas, *Wartime* (New York: Harcourt Brace Jovanovich, 1977), p. 244.
11. Djilas, p. 149.
12. Concerning popular front appearances in Yugoslavia, see Clissold (ii), pp. 145–52, 'Policy on United Front tactics . . .'
13. Stephen de Mowbray, 'Soviet Deception and the Onset of the Cold War,' *Encounter*, July-August 1984, pp. 16-24.
14. Clissold (ii), p. 28.
15. Edvard Kardelj, 'The Struggle for Recognition of the National Liberation Movement of Yugoslavia,' *Macedonian Review*, Vol. 11, No. 2, 1981, p. 190.
16. Hinsley, p. 140; Clissold (i), p. 223 places the radio in or near Tiflis.
17. Walter R Roberts, *Tito, Mihailovich and the Allies, 1941/45* (New Jersey: Rutgers University Press, 1973), p. 62–63; see also Clissold (ii), pp. 18–19.
18. See Chapter 1.
19. *Morning Star* (London), 16 September 1977.
20. Michael Straight, interviewed by Author, Bethesda, Maryland, 10 December 1984.
21. *Ibid.*; for additional references to Klugmann's activities, see Richard Deacon, *Biography of Sir Maurice Oldfield* (London: Macdonald, 1985), p. 37 and Leopold Trepper, *The Great Game* (London: Michael Joseph, 1978), p. 63.
22. *Hansard (Lords)*, 26 February, 1985, p. 892.
23. Chapman Pincher, *Their Trade is Treachery* (London: Sidgwick and Jackson, 1981), pp. 127–36.
24. *Observer* (London), 29 July, 1984, p. 1.
25. Andrew Boyle, *The Climate of Treason* (London: Hodder & Stoughton, 1980), pp. 52–4.
26. Michael Straight, *After Long Silence* (New York, London: W W Norton, 1983), p. 72–3.
27. *Ibid.*
28. M R D Foot, *SOE: An Outline History of the Special Operations Executive, 1940–46* (London: BBC, 1984), p. 9.
29. Foot, p. 46.
30. *Ibid.*, a photocopy is said to have been made of files prior to the fire, but in several cases, including Klugmann's, the copies were illegible.
31. For this information about Klugmann and other facts about the personnel of SOE-Cairo, the Author is indebted to Mr C M Woods, SOE Adviser at the Foreign and Commonwealth Office, London.
32. David Erskine, *Letter* to the Author, dated 20 November 1984, quoted with permission.
33. Foot, p. 46.
34. Woods, *op. cit.* and *letter*, 9 November 1984.
35. Foot, p. 46; When the Author sought to interview Basil Davidson in October 1984, he received a point blank refusal.
36. Basil Davidson (i), *Scenes from the Anti-Nazi War* (New York and London: Monthly Review Press, 1980), pp. 84, 86 (emphasis added).
37. See, for instance, Basil Davidson (ii), *The Liberation of Guiné: Aspects of an African Revolution* (Harmondsworth: Penguin, 1971); (iii), *Black Star: A View of the Life and Times of Kwame Nkrumah* (London: Allen Lane, 1973); (iv) *Southern Africa: The New Politics of Revolution* (with Joe Slovo and Anthony R Wilkinson) (Harmondsworth: Penguin, 1976).
38. Davidson (i), p. 93.
39. *Ibid.*

40. Peter Kemp was redirected to operations in Albania. Here, too, communists in SOE, notably John Eyre, were manipulating policy by various means, assisting Enver Hoxha to power. See Peter Kemp, *No Colours or Crest* (London: Cassell, 1958), pp. 74, 242–3; David Smiley, *Albanian Assignment* (London: Chatto & Windus, 1984), pp. 152–3; see also Nicholas Bethell, *The Great Betrayal* (London: Hodder and Stoughton, 1984) pp. 16, 21.
41. Kemp, p. 74.
42. Quoted Davidson (i), p. 100; see also David Stafford, *Camp X* (New York: Dodd, Mead and Co., 1987), pp. 170–86.
43. Hugh Seton-Watson, 'Reflections of a Learner,' *Government and Opposition* (London), Autumn 1980, p. 520.
44. Davidson (i), p. 123; powerful though Klugmann's position in SOE-Cairo was throughout the critical period, from mid-july, 1943 onwards he became even more influential. It was at this period that Davidson began preparing himself for his next asignment in occupied Central Europe, one that took him from SOE-Cairo permanently in mid-August. Thereafter, there ws a succession of short-term appointees to the position of section head—four by December, 1943—each of whom must have relied heavily on the continuity man—Klugmann. (Woods, *op. cit.*).
45. Interview, Author with Mr Peter Boughey, formerly Head of the Yugoslav Section, SOE-London, 25 November, 1977, London; Foot, p. 44, 232.
46. Hinsley, p. 141.
47. Concerning Keble, see Foot, pp. 28–45; Davidson, pp. 114–15; concerning Glenconner, see Bickham Sweet-Escott, *Baker Street Irregular* (London: Methuen, 1965), p. 170.
48. Keble died soon after these events. Between 1942 and his death, some of the actions attributed to him were so bizarre that observers have questioned his stability. See Davidson (i), pp. 129–35; Foot, p. 41; Martin, pp. 124–6.
49. Hinsley, p. 138.
50. Interview with D T Hudson (ii), London, 27 October 1976.
51. Hinsley, pp. 141–2; Davidson (i), p. 119. (On the same page Davidson identifies the briefing paper as PRO document FO 371/37579).
52. Hinsley, p. 143.
53. Hinsley, pp. 141, 142. (Churchill used a slightly enhanced version of SOE-Cairo's figures, not Military Intelligence's, in a letter to President Roosevelt).
54. Davidson, pp. 117–18.
55. Davidson, p. 119; for another report on the briefing see J R M Butler ed., History of the Second World War, UK Military Series, Michael Howard, *Grand Strategy*, Vol. IV (London: HMSO, 1972), pp. 389–90.
56. Quoted Clissold (ii), p. 20.
57. Hermann Neubacher, *Sonderauftrag Südost, 1940–1945* (Göttingen: Musterschmidt-Verlag, 1956), p. 170; Walter Warlimont, *Inside Hitler's Headquarters, 1939–45* (New York: Frederick A Praeger, 1964), p. 469; Records of Headquarters, T-77-781-5507576. (Gehlen was later to be the architect of the West German intelligence system, and carved an envied international reputation).
58. Martin, p. 44.
59. Roberts, *op cit.*, p. 108; for more extensive documentation of these talks, see Stevan K. Pavlowitch, 'Dedijer as a Historian of the Yugoslav Civil War,' *Survey*, Autumn, 1984, p. 101. See also *SOE, The Sword and the Shield—Yugoslavia*, BBC television programme 50/LSF PO35E broadcast 2 October 1984, transcript, pp. 29–30.
60. Quoted Roberts, p. 110, emphasis added.
61. Stephen Clissold (iii), *Djilas: The Progress of a Revolutionary* (London: Maurice Temple Smith, 1983), pp. 101–107.
62. Boughey interview, cited; Author's interview with Sir Douglas Howard, 1 December 1977.
63. On communist and pro-Soviet sympathisers in Britain at this time, see Boyle, pp. 211–14; on official policy in London and reactions to German intercepts and to SOE-Cairo's information, see Hinsley, pp. 138–56.
64. Davidson (i), pp. 129, 119.
65. Sweet-Escort, p. 164.
66. Hinsley, pp. 141, 147, 150–51, 159. Mihailovich in December, 1943 was with the British

Mission in Serbia. If negotiations took place in Zagreb then it would have been Chetnik commanders in peripheral areas who were involved. The negotiations reportedly concerned the organising of volunteer bands to fight in Northern Greece. Nothing ever came of such a scheme, which is hardly surprising, considering how hard-pressed Mihailovich's forces were at that time.

67. Quoted from Zivan L Knezevich, *Why the Allies Abandoned the Yugoslav Army of General Mihailovich*, Part II, Item 9, pp. 10–11. Mimeographed documentation, Library of Congress, D802.Y8K6: (Messages from General Headquarters invariably bore the Commander-in-Chief's signature, whatever their contents. Only rarely did senior commanders read what went out over their names).
68. *Ibid.*
69. Public Record Office, London: FO 371/37588.
70. Martin, p. 119, cites examples.
71. S W Bailey, 'British Policy Towards General Draza Mihailovitch', in Phyllis Auty and Richard Clogg, eds., *British Policy Towards Wartime Resistance in Yugoslavia and Greece* (London: Macmillan, 1975), p. 75.
72. Interview, Brigadier C D Armstrong, with Author, Camberley, Surrey, 8 November 1976; Sweet-Escort, pp. 170–71, throws light on the close working relationship between SOE-Cairo and its 'private air force'.
73. PRO, FO 371/37590.
74. Interview, Kenneth Pickthorne, with author, *circa*, October–November 1944.
75. See Martin, *op. cit.*, Chronological Compendium, pp. 70–82; sources are quoted therein.
76. PRO, WO 202/140.
77. Chapter 1.
78. Pincher, p. 114; de Mowbray, p. 21; Letter from Rendel to Howard, Head of the Southern department, Foreign Office, 12 April 1943 in PRO, FO 371/37583, quoted Elisabeth Barker, 'Some Factors in British Decision-Making over Yugoslavia 1941–4', in Auty and Clogg, p. 29; Letter from Rendel to Pierson Dixon of the Southern Department, Foreign Office, of 4 November, 1942, PRO, FO 371/33472, quoted Barker, p. 29.
79. Barker, *op.cit.*, p. 39; see also Winston S Churchill, *The Second World War, Vol. V: Closing the Ring* (London: Cassell, 1952), pp. 411–12.
80. See Clissold (i), pp. 145–6, Item 62. The Comintern on 5 March 1942 warned that Britain might have 'some [?justification] in suspecting the Partisan movement of acquiring a communist character, and aiming at the sovietisation of Yugoslavia . . .'
81. Clissold (i), p. 150, Item 72, Comintern instructions, November, 1942.
82. Churchill, *op. cit.*, pp. 411–20.
83. Howard, cited note 1.
84. PRO, FO 371/37616.
85. *Ibid.*
86. WO 202/140, Sheet No. 424, 19 December 1943.
87. Churchill, p. 413.
88. Paraphrased quotation by Lieutenant Colonel Robert H McDowell, in McDowell, *Report on Mission to Yugoslavia, Ranger Unit* (APO512, US Army, Headquarters 2677 Regiment OSS (PROV), 23 November 1944), declassified 19 January 1984, p. 14.
89. PRO, WO202/145, sheet No. 355, 13 December 1943.
90. PRO, WO202/140, Signal, Armstrong to SOE-Cairo, 20 December, 1943.
91. Interview, Jasper Rootham with Author, London, 3 November, 1976: see also Jasper Rootham, *Miss Fire* (London, Chatto and Windus, 1946).
92. Rootham interview.
93. Interview, George Musulin, commander of the Halyard Mission—the evacuation team—with Author, 6 June, 1977.
94. McDowell, *op cit.*, pp. 2, 3, 8, 9, 28, 29.
95. McDowell, pp. 9, 10–14, 16–21.
96. *Ibid.*, p. 38.
97. This is a verbatim rendering of a statement made to the author in early 1946 by Constantin Fotich, Yugoslav Ambassador to Washington. It was obvious, in turn, that Fotich was reporting on this sentence in the communication in a verbatim manner.

98. D T Hudson (iii), speaking in *SOE: The Sword and the Shield*, transcript, p. 37.
99. See also Nora Beloff, *Tito's Flawed Legacy: Yugoslavia and the West: 1939–1984* (London: Gollancz, 1985), and M. Deroc, *British Special Operations Explored: Yugoslavia in Turmoil 1941–43 and the British Response* (New York: Columbia University Press, 1988). Documents in the Public Record Office (UK) are cited with permission of the Controller of Her Majesty's Stationery Office.

CHAPTER 3

The Seamless Robe: The Russian Orthodox Church as a 'Transmission Belt'

PHILIP WALTERS

The world wants to be deceived.
Sebastian Brant, 1494

Since the end of the 1940s, the leaders of the Russian Orthodox Church in the Soviet Union have spoken out consistently in favour of world peace and have portrayed Soviet policies as essentially peace-loving. The Church has also been involved in misleading Western public opinion about the status and freedom of religious communities in the Soviet Union and about the role that the Church is playing in Soviet society. By presenting a picture of the Russian Orthodox Church as an independent source of influence within the Soviet social and political dispensation, and implying that the continuation of this delicate and moderating role is dependent upon the co-operation of Churches in the non-communist world, the Orthodox leaders have succeeded in limiting criticism of Soviet domestic and foreign policy by Western or Third World Church leaders and organisations. In this way the leaders of the Russian Orthodox Church have become 'transmission belts' of Soviet messages to the outside world.

This chapter examines the history of the Church under socialism, the means by which the voice of the Church is heard, the denials of religious persecution in the Soviet Union, the question of whether or not the Russian Orthodox Church does have an independent voice, the contrast between appearance and reality, the targets of deception, and the outlook. Although this review concentrates on the Russian Orthodox Church, an essentially similar analysis could be made concerning the other main religious organisations within the Soviet Union, notably the Baptists and the Muslims.

* * *

The Soviet Constitution guarantees to citizens all basic freedoms, including that of confessing any religious faith. The aim of the one permitted political party, the Communist Party, has however always been to combat religion by means of education, propaganda and outright persecution, while the basic Soviet Law on Religious Associations of 1929, which still remains in force essentially unchanged today, limits such religious activity as is explicitly permitted to the bare minimum: the holding of religious services in registered buildings.[1]

After an initial period of hostility towards the Bolsheviks, the Russian Orthodox Church under Patriarch Tikhon (1917–1925) adopted an attitude of political neutrality. This did not however stop the persecution of religion, and in 1927 the patriarchal *locum tenens* Metropolitan Sergi felt himself compelled to issue a Declaration of Loyalty to the (Soviet) Motherland, 'whose joys and successes are our joys and successes, and whose setbacks are our setbacks'.[2] Since that date, the Church has given positive support to Soviet domestic and foreign policies while claiming that religious believers suffer no discrimination for their faith.

During the 1930s, all religious denominations in the Soviet Union suffered savage and comprehensive persecution. On the eve of the Second World War, only a few hundred Orthodox churches remained open on Soviet territory and thousands of priests and believers and most of the bishops were in prison.

Not until the Second World War was the Russian Orthodox Church given a chance to give practical demonstration of its loyalty to the Soviet State. The Nazis invaded the Soviet Union in 1941 and Metropolitan Sergi immediately called on the Orthodox faithful to rally to the defence of the Motherland. Stalin called Metropolitan Sergi and two other bishops to a private meeting in 1943, and four days later the Church was allowed to elect Sergi Patriarch. All kinds of basic concessions were now allowed to the Church in return for its services in rallying the faithful to the defence of the Motherland. Churches and theological educational establishments were opened, and an official church journal founded.

After the war, the Soviet leadership found that new areas were opening up in which the Church could be of use to the State: in promoting the Soviet understanding of 'peace' in the world at large; and in preventing or limiting criticism of the Soviet Union by Christian organisations in the non-communist world.[3]

Obviously the general notion of 'peace' is one which churchmen can endorse without apparent hypocrisy. It would nevertheless be too much to say that Soviet churchmen chose to support this particular aspect of Soviet policy of their own free will.[4] Their appearance on the international stage serves Soviet aims in three interrelated ways. They consistently claim that the religious communities suffer no persecution in the Soviet Union. They provide an endorsement of Soviet policies by a body which is ostensibly independent both of Marxist ideology and of direct Party control.[5] And they set official Soviet policies in a theological context which disarms a good deal of potential criticism from a broad range of western audiences.[6]

In spite of the services rendered to the State and Party during and since the Second World War, the right of meeting for an act of worship in a registered building remains the sole right which religious communities in the Soviet Union are permitted to enjoy. All other activities which are considered natural for religious communities in the West, such as evangelical and charitable work are either illegal or systematically discouraged by the authorities. Moreover, control by the secular authorities over the internal life of the Church is almost complete.

Through published and secret laws and through arbitrary administrative action by local party representatives and anti-religious enthusiasts, ordinary believers are systematically treated as second-class citizens. Religious activists are liable to defamation in the press without right of reply.[7] They may find it difficult to obtain permission to live in a particular city, or to find adequate housing.[8] Their personal safety or lives may be threatened[9] and their property damaged without intervention by the authorities.[10] They may find it difficult to obtain more than menial jobs.[11] They may be unable to find anyone at all who will employ them, in which case they will be liable to a charge of 'parasitism' since it is theoretically illegal to be unemployed in the Soviet Union.[12] If they have responsible jobs or are in higher education, and are discovered to be believers, they will probably be dismissed.[13] For the most troublesome offenders, there is the psychiatric hospital, or labour camp or prison under various articles of the Criminal Code, notably Article 70 ('anti-Soviet agitation and propaganda') or Article 190–1 ('slandering the Soviet system').[14]

A secret report by V Furov, Deputy to the Chairman of the State's Council for Religious Affairs, received in the West in the 1970s, indicates the comprehensive nature of state control over religion.[15] Bishops are divided into three categories of acceptability to the State organs in proportion to their perceived willingness to support Soviet aims in the world at large and to deny that religious persecution takes place, and in inverse proportion to their pastoral zeal. The appointment of bishops proceeds only when the candidates have been investigated by the Council.

> The Council for Religious Affairs and its employees systematically study the servants of the cult, hold individual conversations with them, and train them in a spirit of respect for Soviet laws; all political work with the clergy is carried on in the interests of the State.'[16] The Council's work with the clergy is aimed 'not only at keeping them within the bounds of the law, but at diminishing their activity and limiting their influence on believers'.[17] 'If a priest gives sermons, they must be strictly Orthodox in content, containing expositions of the Gospel or Epistles in the Spirit of the Church Fathers and Teachers. Sermons must contain no political or social issues or examples.'[18] 'The Council . . . exercises its influence over many aspects of the activity of the Church's academies and seminaries': such influence includes the investigation of all teaching staff and material. 'Of course, this is done through the hands of the church functionaries themselves.'[19] '. . . The clergy of the Orthodox Church, although loyal to the Soviet State, still remains a body with an ideology which is incompatible with our world view.'[20]

* * *

The official organ of the Russian Orthodox Church, the monthly *Journal of the Moscow Patriarchate*, began publication in 1943. Apart from an annual theological journal and an annual church calendar, it is the only periodical publication the Church is allowed. It exists largely for propaganda purposes: like the theological educational establishments of the Church, the small number of functioning monasteries and convents, and the hierarchy and central administration of the Church itself, the Journal has no foundation in Soviet law and exists only as long as the authorities deem it expedient to allow it to do so. As an official Soviet report notes with satisfaction, it does not reach most Orthodox believers in the Soviet Union because of its extremely small print-run.[21] In its English language version, however, it is widely distributed abroad. From the end of the 1940s, its largest section has been 'In Defence of Peace', which in the last issue for 1952, for example, occupied 24 of the 60 pages of the journal.[22] The presence of this section is largely due to pressure from the secular authorities.[23] Archbishop Luka, author of an article which appeared in 1948, speaks in a letter to his son of the pressure put on him to write it.[24]

Hierarchs of the Russian Orthodox Church travel widely outside the Soviet Union.[25] It should be noted that for ordinary Russian Christians, as for any ordinary Soviet citizen, it is virtually impossible to travel abroad, and inasmuch as leading churchmen do so regularly, it is obviously to the direct advantage of the secular powers in the Soviet Union that they should continue to do so.

There is a special Department of External Church Relations which co-ordinates the activities of the Church in its contacts with foreigners. Father Gleb Yakunin, an Orthodox priest who has criticised the leadership of his Church on many occasions, writes that this Department has strong links with the KGB,[26] although its two most outstanding post-war chairmen, Metropolitans Nikolai (1946–1960) and Nikodim (1960–1972) were far from being mere agents of the government. The personalities of Metropolitans Nikolai and Nikodim have perplexed and intrigued observers of the Soviet religious scene. There is general agreement that both men were of great spiritual strength and sincere Christian faith, as well as having a realistic understanding of the price their Church had to pay to be allowed by the Soviet authorities to continue to function. Both were deeply concerned to promote genuine spiritual growth within the Church inside the Soviet Union; and both—Nikodim perhaps more readily than Nikolai—were prepared to co-operate with the State in the international arena.[27]

The Christian Peace Conference, founded in Czechoslovakia in 1958, at first represented a real attempt to bring together churchmen from East and West, and it was a forum for genuine debate.[28] Although the Russian Orthodox Church was represented at it, it was presided over by Czech churchmen.[29] After the views of the two participating camps became irreconcilable, as a result of the Soviet invasion of Czechoslovakia in 1968, however, the CPC has lost credibility and has become a forum for the expression of explicitly pro-Soviet views; and in 1971 Metropolitan Nikodim became its president.[30]

The Russian Orthodox Church became a member of the World Council of Churches (WCC) in 1961, after enjoying observer status for a number of years. The speed with which the eventual decision to apply for membership was taken must show that both Church and State wanted this to happen. The Church itself had a genuine desire to become involved in the world ecumenical movement, and probably also felt it might need the support of an international body as a defence against the renewed persecution of religion which was at that time gaining momentum within the Soviet Union. That the Soviet State also endorsed this application, despite its continuing distrust of ecumenism,[31] shows what a high priority it placed on having prominent churchmen abroad to represent Soviet interests.[32] Many Westerners were optimistic that full and frank discussion of world interests would now be possible and thought that the Russian Church would have little opportunity to 'play politics'.[33] We shall look a little later at developments in the WCC with regard to the Churches in the Soviet Union.

Many peace conferences have been organised within the Soviet Union, often ostensibly by the Russian Orthodox Church itself, gathering together believers of a wide variety of denominations both from within the Soviet Union and from the First and Third Worlds.[34] At one such conference, in the 1950s, Metropolitan Nikolai denounced alleged American atrocities in the Korean War in such wildly exaggerated terms that one observer has surmised that he might have hoped that nobody would take them seriously;[35] but there is no doubt that the tone and conclusions of most of these conferences have been notably pro-Soviet. One incidental benefit of conferences held within the Soviet Union is that delegates from abroad can be lavishly entertained by what seems to be a free and flourishing Church. It should be noted that there is no obvious shortage of open and functioning Orthodox churches in Moscow (over 40); but that in thousands of Soviet cities, which delegates from abroad never see, there are likely to be no functioning churches at all.

* * *

Soviet church leaders persistently claim that although there may have been difficulties in the past for the religious communities in the Soviet Union, relations between Church and State are steadily improving, and no religious believer suffers for his or her faith. When substantial evidence to the contrary is presented, church leaders try to prevent discussion of the issue. Religious believers who criticise Soviet policies on religion from within the Soviet Union have been denounced variously by Church leaders as politically or financially motivated, possessed of personal grudges or mentally unstable or, more generally, simply as troublemakers and lawbreakers quite unrepresentative of the majority of churchgoers.

For Vladimir Kuroedov, Chairman until 1984 of the Council for Religious Affairs, the secular body which supervises the activities of the Churches in the Soviet Union, the fact that a Peace Conference took place in Moscow in 1977 hosted by the Russian Orthodox Church and attended by foreign delegates 'demonstrated that the question of the infringements of human rights in socialist countries, fanned by imperialist propaganda, is a fictitious one'.[36] William Fletcher describes as one of Metropolitan Nikolai's foremost duties that of misrepresenting the true nature of Soviet policy towards religion, and quotes him as follows:

> 'Who of the impartial and honest people abroad is not aware of the fact that the Russian Orthodox Church, sheltered by the great Stalin Consti-

tution which guarantees freedom of conscience and religious worship, enjoys complete freedom in Her own internal affairs?'[37]

Shortly after Metropolitan Nikolai's dismissal as Chairman of the Department of External Church Relations in 1960, he was visited by a Western Orthodox churchman who said to him:

> 'Here you are telling me about persecution and the difficult situation for the Church in Russia, but there has just been a delegation from the Russian Church to the West . . . and in answer to . . . questions they replied that the Church in Russia is free and there is no pressure or persecution at all.' Metropolitan Nikolai smiled sadly and replied; 'if I had been in their place, I would very likely have said the same as them.'[38]

In a personal conversation with Metropolitan Nikodim, a Western observer of Russian church affairs asked why he had not been troubled by having to tell lies at a recent press conference to the effect that there is no religious persecution in Russia. Nikodim replied, 'That's just the way you react in the West. Our people have got used to this kind of thing and think it's quite normal.' The same observer goes on to recall an effort made by a Western Orthodox priest to persuade Metropolitan Nikodim not to deny publicly the persecution of religion in the Soviet Union, since there were by that time so many documents in the West proving the contrary. Nikodim replied: 'Let's agree to do this: you publish the documents, and we'll refute them.'[39]

Let us look in more detail at two important occasions when the Church concealed and diverted attention from religious persecution in the Soviet Union. When the Russian Orthodox Church joined the WCC in 1961, all religious organisations in the Soviet Union were suffering under the most comprehensive persecution campaign since the 1930s. Launched by Khrushchev in 1959, and continuing for about five years, the campaign resulted in the closure of two-thirds of the then legally functioning Orthodox churches, the arrest and trial of priests and believers, and the forcing of the Churches to accept new legislation whereby they restricted their own range of activities to a very narrow spectrum. When it entered the WCC, the Orthodox Church claimed it had 22,000 functioning churches. This was the total before the Khrushchev campaign began; but it was never challenged by the WCC, nor was it corrected by the Church itself as persecution intensified.[40] Similarly, no mention was made at the WCC Assembly of the curious fact that Metropolitan Nikolai, who had shaped the Orthodox Church's application for membership, was missing from the delegation. He had in fact just been removed from office in response to government pressure and was to die, some claim murdered, a week after the conclusion of the Assembly.[41] Western attention was successfully diverted away from the Khrushchev anti-religious campaign to the 'positive' fact of the arrival of the Russian Church within the WCC. Later in the 1960s, Soviet church delegations abroad were active in minimising the impact in the West of the news of the schism in the Baptist Church which had taken place in 1961 at the height of the Khrushchev campaign, as a result of the new restrictive legislation forced upon the Churches.

In 1975, the Soviet Union was one of the signatories of the Helsinki Final Act, which among other provisions committed the signatory states to respect human and religious rights within their borders. Encouraged by this move, in November of that year Fr Gleb Yakunin and the Orthodox layman Lev Regelson sent an Appeal, reporting on the infringement of believers' rights in the Soviet Union, to the General Assembly of the WCC in Nairobi.[42] The response of the Russian Orthodox delegation to the Appeal was swift. Metropolitan Yuvenali, Chairman of the Department of External Church Relations, attacked the character of the authors rather than attempting to answer the points they raised. Nevertheless, the Appeal was discussed by the Assembly—the first public debate on this topic ever held by the WCC. The subject was brought up by concerned delegates under the only remotely relevant item on the agenda, one dealing with disarmament and the Helsinki Agreements. In an atmosphere of some confusion, a resolution was passed deploring 'restrictions on religious liberty particularly in the USSR'. Then questions of procedure were raised and the resolution was sent back to a committee which included a member of the official Soviet delegation.[43]

The result was a much less specific resolution which did not mention the Soviet Union by name. In a statement issued by the Soviet news agency TASS, Metropolitan Nikodim construed the whole affair as a provocation by certain critics who wanted to create an anti-Soviet atmosphere, and claimed that 'the anti-Soviet clamour had been programmed and prepared beforehand'.[44] A comment also appeared in the *Journal of the Moscow Patriarchate*. This did not mention Fr Gleb or Regelson by name, but gave the impression that the affair had been concocted by some Western delegates. The representatives of the Russian Church had 'tried to clarify the true state of affairs and describe the real position of the Church in the Soviet Union where, as is well known, believers and non-believers participate without any restrictions or discriminations in constructing a new life . . .' The *Journal* explained the fact that the Russian Church delegation had abstained from voting even on the amended resolution by claiming that the debate was supposed to be about the Helsinki Agreements, and disarmament in particular, and that 'this kind of problem ought to be discussed without bringing in considerations of a non-churchly (*sic*) nature which destroyed the spirit of mutual Christian understanding and were devoid of the necessary objectivity . . .'[45]

Inasmuch however as the WCC, as a result of the debate, agreed to set up a mechanism for investigating infringements of religious liberty in all the signatory countries of the Helsinki Agreements, Fr Gleb and Regelson were encouraged to write again to the General Secretary of the WCC on 6 April 1976, giving a careful and detailed breakdown of the main areas in which discrimination against religious believers is built into the Soviet system. They also warned the WCC that the Soviet Council for Religious Affairs had already taken measures to undermine any attempt by the WCC to adopt a resolution of protest against the limitation of believer's rights in the Soviet Union. Fr Gleb went on in 1976 to found the 'Christian Committee for the Defence of Believers' Rights in the USSR' which sent over 3,000 pages of documentation to the West recording details of religious discrimination and persecution in the Soviet Union.[46] Fr Gleb was arrested in 1979 and sentenced to 10 years deprivation of liberty for 'anti-Soviet agitation and propaganda'.[47]

William Fletcher writes in 1973 that 'with some exceptions, representatives of the Russian Churches have supported their country's claims of religious toleration, and the result has been the continuation of a degree of ignorance among Western Christians concerning the actual religious situation in the USSR.'[48] This is substantially true even today, despite the efforts of such as Fr Gleb to convey the facts to the world at large. Why should this be so? Part of the answer must lie in the fact that the hierarchs of the Churches of the Soviet Union have succeeded in convincing Western audiences that they are in agreement with the direction of Soviet policy at a very fundamental level, and that to intervene in cases of alleged religious persecution is harmful to the basic interests of a Church, which has adjusted itself to Soviet reality. We shall be examining this phenomenon in more detail later in the chapter.

* * *

'I am addressing you here not only as a citizen of my peace-loving Soviet country, I am addressing you also as a bishop of the Russian Orthodox Church and in Her name.'[49] With these words at a World Peace Congress in Warsaw, Metropolitan Nikolai drew a distinction between his two roles. Metropolitan Antoni of Minsk went rather further when speaking in London in 1978: 'We represent part of the public opinion of the Soviet Union . . . We express our point of view, and the Soviet Government always listens attentively to what we have to say.'[50] The implication here is that the Church might in certain circumstances have its own view to put forward on matters of Soviet policy. It is evident that the aim of Church leaders from the Soviet Union, and the aim of the Soviet Government, is to give the impression that the Churches are an autonomous forum for public opinion within the USSR. The fact of the matter is, however, that the Church leaders, in the overwhelming majority of cases, promote policies so closely in line with those of the Soviet government as to be indistinguishable from them. The exceptions to this are so rare that they can be taken to prove the rule.[51] From the time of the earliest efforts of the Russian Orthodox Church to become involved in the ecumenical movement, prominent members of the latter expressed their fears that the Church might turn out to be simply a vehicle for the expression of a pro-Soviet political line;[52] and subsequent experience has shown their fears to be well-founded.

The leadership of both State and Church in the Soviet Union constantly indicate that the Churches are in fundamental agreement with the directions of Soviet policy.

> Russian Orthodoxy today is characterised not only by its functioning within the framework of Soviet law, but by active support by the Church for the internal and external policies of the Soviet Government which is acting in the interests of all workers in our country, including believers.[53]

The reason given is that socialist society is essentially peace-loving and, indeed, provides the only road to achieving world peace. In its survey of the peace work of the Church since the 1940s, the *Journal of the Moscow Patriarchate* writes that it was in Europe 'that socialism first triumphed . . . Hence it is not by chance that the all-embracing struggle for security and co-operation amongst nations began in Europe.'[54]

Members of all religious groups in the Soviet Union have therefore 'unanimously supported the peace initiatives of the Soviet government'.[55] Patriarch Pimen's utterances in praise of the Soviet system and Soviet leaders have in general been marked by an excessive sycophancy.[56] Curiously enough, precisely this intimate support by the Church for the Soviet Government is used by the former to refute allegations that it is merely a tool of the Kremlin: Soviet policies are essentially aimed at 'peace, friendship and fruitful co-operation with all peoples and nations', and hence 'it is high time to remove such notions as "an agent of the Kremlin" and such like . . . The negative stereotypes cannot cross out the truth of our life. It is self-evident in the rich and fruitful experience of the sixty years of the USSR which we . . . solemnly celebrate . . .'[57] Given this kind of conceptual framework, it is hard to imagine when the Church might find it necessary to represent its own independent view on any matter to the Soviet authorities or try to influence them in any way. Any influence must surely be directed abroad at Western organisations which have not yet appreciated the correctness of the Soviet line on international relations.[58]

In its pronouncements on world peace, then, the Russian Orthodox Church has consistently attacked warlike and disruptive policies followed by the imperialist powers and the essentially aggressive international policies which are a natural consequence of the capitalist structure of society. In his opening address to the World Conference of 'Religious workers for Peace, Disarmament and Just Relations Among Nations' in Moscow on 6 June 1977, Patriarch Pimen picked out the following 'positive' developments: the defeat of Fascism in Greece, Spain and Portugal; the reunification of Vietnam; the liberation of Mozambique and Angola; the successful struggles of the peoples of Zimbabwe and Namibia; the growing role of the United Nations in the struggle against militarism; and the declaration of the United Nations on a new international economic order. He held up for censure: the arms race; imperialist interference in the internal life of certain countries; exploitation of underdeveloped countries by transnational corporations; and support by the latter for reactionary racist regimes. In general, when speaking for example on the subject of observance of the Helsinki Agreements, church leaders from the Soviet Union have always given primacy to the discussion of means of achieving world peace through disarmament and have dismissed any efforts made by non-Soviet countries to raise the question of human rights as constituting an essential aspect of 'peace'—that is, peace and reconciliation through the establishment of justice inside particular countries.[59] However, when alleged violations of human rights occur in non-socialist countries, Soviet church representatives have not kept silent. In 1968 Patriarch Aleksi of the Russian Orthodox Church expressed his anxiety about those in Greece 'unjustly imprisoned for their views and their desire for freedom and democracy'. One commentator in the Western press gently reminded the Patriarch about living in glass houses.[60]

Perhaps the most obvious example of support by the Churches of the Soviet Union for Soviet foreign policy came at the time of the invasion of Czechoslovakia in 1968. The Czech founder of the Christian Peace Conference, Dr Josef Hromadka, expressed his 'deep bitterness' at the invasion, and questioned whether the intervention was not likely to discredit socialism in the eyes of the younger generation. A reply to Dr Hromadka, signed by representatives of the Russian Orthodox Church, the Baptist Church and the Lutheran Church in the Soviet Union condemned his comments as 'condescending and insulting' to the Soviet people.[61] The WCC too protested at the invasion, and in response Patriarch Aleksi maintained that there were no grounds for calling Soviet troops 'occupying forces' since the Soviet Union was bound to Czechoslovakia by bilateral treaties of co-operation and friendship which covered all aspects of their relationship, political as well as ideological.[62]

As its official journal states, 'the Russian Orthodox Church has for decades been involved in the global peace movement *in accordance with the foreign policy of the USSR*' (emphasis added).[63] An authority on the Orthodox Church cites 17 quotations from the *Journal* between 1953 and 1965 showing how the Church supports specific Soviet policies. 'The above quotations from primary sources', he writes, 'clearly show that the Moscow Patriarchate does not possess its own official point of view on political events. All public statements chronologically correspond to the aims and problems tackled at the same time by the Soviet regime.'[64]

When Soviet foreign policy changes, so does that of the Church. After the Second World War, the Russian Orthodox Church was entrusted with the task of extending its influence over the Orthodox Churches of Eastern Europe. The Serbian Orthodox Church accepted Moscow's leadership in 1945. When President Tito broke politically with Moscow in 1948, the Russian Orthodox Church severed its links with the Serbian Church. It restored them in 1957 when Khrushchev was reconciled with Tito. A similar phenomenon can be observed in relations between the Russian Orthodox Church and the Vatican. In the early 1950s, the Russian Orthodox Church attacked the Roman Catholic Church in the same breath as capitalist warmongers. After the death of Stalin and the accession of Pope John XXIII, there was a thaw in Kremlin-Vatican relations. 'This transformation of attitudes was soon reflected in the Russian Orthodox Church.'[65]

The Orthodox Church hosted a World Conference of Religious Workers for Saving the Sacred Gift of Life from Nuclear Catastrophe in May 1982. Despite the misgivings of some delegates in the early stages of the conference that it was going to be simply another 'political forum heavily tilted against the West', the final documents were subjected to some real discussion and emerged unexpectedly even-handed. The Conference encouraged many of the Western delegates who attended it.[66] One delegate considered that the short passage of political content in Patriarch Pimen's speech represented 'the minimum he could get away with in return for the approval of the State to organise such an inter-religious conference'.[67] Of course, the final documents could contain no specific criticism of the policies of the Soviet Union itself: it is what was omitted which was significant.

The same delegate says that he 'came away from the conference again convinced that there is a genuine abiding and deep concern for peace and disarmament on the part of the Soviet people and church leaders, and that it is not simply being fabricated to subvert us'.[68] No doubt this is true. The point is, however, that by means of using the Church to host such a conference and to act as a spokesman generally on the subject of peace, the Soviet Government is indeed succeeding in subverting world opinion in a different way: in persuading Western audiences that the Churches are speaking as autonomous and independent bodies within Soviet society, endorsing Soviet policies in this field as a free-will gesture of solidarity. As an example of this kind of misapprehension, we may quote a pre-conference report published in *The Times*.

The author of the report is of the opinion that the fact that the conference is taking place at all 'highlights the important unofficial role that the Russian Orthodox Church, the conference organiser, now plays in Soviet foreign policy and the extent to which the Church has managed in recent years to re-establish its influence in an atheistic society'. He goes on to argue that the 'Communist Party has a long and well-known record of opposition to religion and, for this reason, calls by religious leaders to halt the deployment of new weapons . . . carry greater credibility in the West than official pronouncements, and cannot easily be dismissed as Soviet propaganda'.[69] In short, once the Western audience, apparently like the journalist, has swallowed the fiction of Church autonomy and independence, the Church becomes a credible and effective 'transmission belt' of Soviet propaganda.

It is indeed difficult to comprehend exactly how the Soviet Government, with the help of the hierarchs of the leading Churches, has succeeded in keeping Western opinion so favourably disposed. Soviet leaders and church hierarchs claim that no religious persecution takes place, and Western churchmen and others in dialogue with the Soviets have not systematically pressed the point: yet thousands of documents of undisputed accuracy attest to the contrary. Soviet leaders and church hierarchs claim that the Churches have an independent voice in their enunciation of policies on peace: yet the record of the Churches has been to lend consistent support to specifically Soviet policies in the world at large; and when church leaders from the Soviet Union refrain from explicitly mentioning Soviet policies, as at the 1982 Conference, their reticence is greeted with as much warmth as if they had taken a decisive stand *against* certain aspects of the policies of their State, as is regularly done by Churches in the West and, where appropriate, even by the Churches in East Germany. The answer to the problem must be that the church leaders from the Soviet Union have indeed succeeded in weaving a seamless robe; that they have succeeded in persuading their Western debating partners that they are convinced at a very fundamental level about the correctness of the Soviet approach to domestic and foreign problems and to the problem of world peace; and that they have reached these convictions through the exercise of independent judgement. We will now look at various ways in which the Soviet authorities and church leaders collaborate in producing an impression of consensus and co-operation which is apparently impressive to many in the West.

* * *

As has already been noted, there is no need to doubt the fact that by and large the 'Soviet people' are in favour of 'peace', in the same way that most normal human beings are. The *official* Soviet understanding of 'peace', however, has little in common with a simple desire to put an end to war and violence in the world. It is important to realise that, like all Soviet spokesmen, Soviet churchmen are expected to give explicit endorsement to this particular Soviet understanding of 'peace'.[70] Inasmuch as they are religious leaders, however, they can lend their own theological glosses to the topic; and here lies an advantage from the point of view of the Soviet leadership: Western Christians may all too often think they are hearing the authentic, independent voice of Russian Orthodoxy on the subject. The result is a tendency to give churchmen from the Soviet Union the benefit of the doubt, and to listen to what they have to say with a greater receptivity than would be the case if the same views were being put across by identifiable representatives of the Soviet Government.

A former Soviet journalist now living in the West, commenting on a speech on peace by a professor from Leningrad Theological Academy, wonders how the combination of specifically Soviet political proposals and theological remarks in this kind of speech is achieved:

> I keep on wondering: how is this done, technically? Do they call in the professor and give him his nuclear rocket instructions, saying 'right, you'll

> be able to think up the theological formulae yourself, won't you, comrade priest?' Or do they simply hand over the speech all ready, put together by a working collective of theologians from the Central Committee of the Communist Party?[71]

Some Western experts have pointed out that Orthodox church leaders are indeed concerned to give their statements on 'peace' a specifically Christian content, quite apart from the political context in which they are couched;[72] but none has made special claims for the quality of the theological argument in such cases, and general opinion is that the specifically theological sections are in most cases subordinate to the political content of the message.[73]

Shortly after the Russian Orthodox Church joined the WCC, Metropolitan Nikodim sent a memorandum to the WCC General Secretary with critical observations on various topics raised in a recent WCC communiqué, including the topic of 'peace'.

> 'Every Christian is agreed', says Nikodim, 'that peace on earth must be achieved. But the authors of the report prefer to talk of an internal social peace, and it seems, moreover, that they do not even think of the possibility of a class*less* society already existing side by side with a class society . . . Any feeling of social solidarity with those classes whose very existence is bound up with the unjust distribution of social wealth is completely foreign to millions of Christians who live in the conditions of the new classless socialist society . . . The fact of the existence in society of antagonistic classes which resist one another . . . does not find the least justification in Christian teaching . . . One gains the impression that the authors of the report even assume the essential nature of the "ministry of reconciliation" in striving for class harmony . . . The authors' reasoning is so engulfed by the question of the peaceful establishment of "new and just relationships between nations, races and classes" that they are ready, apparently, to condemn in general all the revolutionary reforms of our age . . . one must remember that the Russian Orthodox Church, like many other Christian Churches, is living in conditions of social relations which correspond much more closely to Christian principles than those which had to give way to them as a result of the Great Revolution, which occurred in our country according to God's plan.'[74]

Metropolitan Nikodim's contention that the 1917 Revolution took place as part of God's plan indicates the extent to which the hierarchs of the Churches of the Soviet Union officially identify with the social and political evolution of their country. This policy has the effect of persuading Western opinion that it is inappropriate to criticise *any* aspect of Soviet policy in isolation, since to do so is to cast doubt on the validity and corrrectness of the whole Soviet experiment.

Whether all individual Christians within the Soviet Union agree with their leaders on the essentially Christian nature of the Soviet

State is irrelevant for the purposes of world diplomacy. What the Soviet leadership would like ideally would be a Church with no believers in it: simply the hierarchs to present the situation within the Soviet Union in terms of a gradual evolution towards goals which are intrinsically Christian. As Fr Gleb Yakunin says in his 1979 report on the present situation of the Russian Orthodox Church,

> In proportion as its spiritual and ideological energies are exhausted by the State, and its conversion to pragmatic positions increases, the status of the Moscow Patriarchate continues to rise in the eyes of the Government. It is becoming a more and more convenient instrument for the latter's goals, and its position in the State and social system is being consolidated.[75]

As we have seen, the Church supports the Soviet State in its peacemaking activities out of its alleged conviction that Soviet policies and Soviet reality are in their very essence peaceful. Church leaders have gone further, however. Some have even claimed to see in developing Soviet society an essentially Christian process at work. In 1967 Metropolitan Nikodim welcomed the achievements of the Revolution 'which Christianity had not realised with its own strength',[76] and in 1973 Patriarch Pimen spoke of 'the merits of the socialist way of life which, as we understand it correspond in great measure to Christian ideals'.[77]

But what of the fact that the Soviet State is essentially atheist and committed to promoting the eventual disappearance of religious superstitions? It is remarkable that some church leaders from the Soviet Union have been able to come to terms even with this phenomenon. In 1963 Metropolitan Nikodim, in the course of an address to a regional meeting of the Christian Peace Conference in Holland, spoke as follows:

> Many reject Communism because of its links with the 'mortal sin' of atheism. But they forget the atheism of any non-communist society. An objective study of atheism requires a strict analysis of motives leading to an atheistic outlook. We know that communist atheism represents a definite system of beliefs, which includes moral principals not contradicting Christian norms. The other atheism is blasphemous, amoral, emanating from the desire to live 'free' from the divine law of Truth . . .[78]

One important fact should incidentally be noted at this point. Even those church leaders who express to such an unequivocal degree the idea that Christian and communist aims are identical do not necessarily believe what they are saying. Since the time of Metropolitan Sergi's Declaration of Loyalty in 1927 the bishops have generally striven to keep inviolate the spiritual integrity of the Church and have maintained that while the duty of a Christian is to 'render to Caesar that which is Caesar's', the Kingdom of God is not of this world at all and that in comparison with the search for the Kingdom, the construction of Socialism is of negligible importance. Indeed, this insistence forms the basis for official attacks on the religious world-view as *essentially* anti-Soviet and reactionary, as we shall see shortly. However, the Soviet authorities have consistently made sure that those clergy who speak for the Church on the international scene are indeed those who are prepared to appear most enthusiastic in their role as loyal Soviet citizens who also happen to be religious believers; and it is evident that enough clergy are prepared to appear wholehearted enough to make the enterprise worthwhile from the point of view of the authorities.

Church leaders from the Soviet Union who appear on the international scene, then, do not see it as one of their tasks to point out differences between the Soviet authorities and Christians in the Soviet Union in their approach to social and political questions. Indeed, the usual implication is that such differences do not exist. What is more, when they speak to the world at large, these church leaders are not attempting to address an exclusively Christian or even religious constituency. Over the last decade, a great deal of stress has been laid on the fact that, in their peacemaking activities, they are calling not only on religious believers for their support but on all 'people of goodwill'. In giving credit to the Soviet Union, the other socialist countries and 'political and social activists in western countries' for initiating the Helsinki process and seeing it through to a successful conclusion, the *Journal of the Moscow Patriarchate* recognises the contribution of 'all peace-loving forces of our planet, including the religious community, and also including the efforts of all people of goodwill'.[79] In his opening address to the Moscow Peace Conference in 1977, Patriarch Pimen greeted 'representatives of major religions, people of goodwill', and went on to speak of the possibility of co-operation in pursuit of peace amongst people of different religions and people of goodwill in general.

> Fulfilling his religious duty or his non-religious ethical obligations reflecting the will of God, man is naturally seeking to serve the good of his neighbour, his society and all people. This aspiration cannot be suppressed by any sophistry or hostility, hatred or intimidation . . . Hence the significance and necessity of the harmonious assertion of life in its positive realities, in the triumph of creation over destruction, truth over falsehood, life over death.[80]

Mr Kosygin's message to the Prague Peace Conference in 1978 read in part: 'Problems connected with the preservation of peace on Earth . . . are important to all people, irrespective of their religious or other convictions . . .' Therefore 'all people of goodwill are united . . .'[81] The implications of the broad relevance of the Church's message on peace are summed up by a leading Soviet writer on religion, N S Gordienko.

> The position thus adopted by the Church, conditioned by our socialist reality, bears witness to the political unity of the Soviet people, including atheists as well as believers. This is an honest position corresponding to the interests of believers and of the Church a a whole. Hence it becomes obvious how completely groundless are the slanderous and provocative statements by Western anti-communist propaganda to the effect that support by the Russian Orthodox Church for the internal and external policies of the Soviet State is nothing more than an act of accommodation by the priests and believers, masking their real and fundamentally negative attitude to this question . . .[82]

It is clear that the Russian Orthodox Church is not prepared to put forward a specifically Christian peace plan, nor is it inclined to see any threat to peace in the atheist Soviet system. Its comprehensive loyalty to the Soviet political and social system permits a correspondingly comprehensive definition of what constitutes the main threat to peace. This threat is presented by a phenomenon even less specific than the international machinations of imperialists and capitalists: it has its origins in the very feeling of 'anti-Communism' itself. As Metropolitan Nikodim put it in 1966,' Imperialism . . . feeds the ideology of *anti-Communism,* which is dangerous for the cause of peace . . .[83] And as the *Journal* states in 1973, 'The great task for Christians, especially in Western European countries, is from now on the overcoming of anti-Communism and counter-revolutionary tendencies in their own ranks.[84] Moreover, 'The patriotic and peacemaking services of the Russian Orthodox Church are inseparable one from another.' It becomes impossible for any citizen of a Western country to criticise any aspect of Soviet reality without being branded by both Church and State as an enemy of peace. When asked in 1978 by an East German newspaper how he reacted to claims by Western sovietologists that the Church in the Soviet Union is persecuted, Patriarch Pimen replied in part that such claims represented efforts by circles in the West hostile to the Soviet Union to put about 'stereotyped concepts from the time of the "Cold War" about the "difficult" position of religion in socialist society and *thereby* put a brake on the process of détente in relations between states with different socio-political systems'[86]

* * *

An essential element in Soviet religious policy is what one recent author has called 'playing at appearances'. This consists in 'continuing to liquidate the Church and religion, while at the same time using propaganda to produce a lot of hot air about its *apparently* flourishing state'.[87] There is a total discontinuity between apparent Soviet policy towards religion for external consumption and what actually happens to the Church and believers inside the Soviet Union. In the view of another recent writer on the subject, ' "Peaceful coexistence" between Church and State under Communism is identical to "Peaceful Coexistence" on the international scene.'[88]

Although the Soviet authorities use church leaders as representatives in support of Soviet policies, they are only too aware that even the most articulate and loyal churchmen are likely to be inwardly loyal to the Kingdom of God rather than to the coming communist Utopia. An article in a leading Soviet atheist publication in 1969 attempted to demonstrate how it is that the Church can nevertheless serve the Soviet State.

> 'In order to understand such an unusual phenomenon as political support by the Church for a socialist State, it is necessary to keep in mind the difference between a religious organisation and the religious ideology.' Since the Church is part of the historical setting in which it operates, its political sympathies are moulded by its environment. But in those cases when the Church expresses the interests of bygone social strata and acts in accordance with the spirit of religious ideology, it is following a reactionary political course . . . Loyalty of a religious organisation to Soviet society is an absolute condition for it to function in our country. Religious associations which take up an attitude of opposition to the Soviet State will inevitably lose the trust and support of the believers and will cease to exist. This does not imply in any way that a politically positive attitude of the Church towards socialism is purchased by it at the cost of capitulation to the new State system.[89]

It is however difficult to see what else this state of affairs might 'imply', since the position of the Soviet secular authorities is that the Churches can be accepted in an ancillary role within Soviet society only inasmuch as they renounce that which makes them specifically Christian organisations—that is, an independence based on Christian theology.

However much freedom church leaders may appear to have to speak out on international issues when they travel abroad, we should not forget that this privilege does not extend to their activity within the Soviet Union. Article Six of the Soviet Constitution explicitly states that 'the leading and guiding force of Soviet society and the nucleus of its political system, of all state organisations and public organisations, is the Communist Party of the Soviet Union'.[90] Any unofficially organised groups independent of party control are thus unconstitutional from the very start, *whatever their aims may be*. A 'Group for Establishing Trust between the USSR and the USA' was set up in Moscow in 1982. Within weeks it had been effectively prevented from functioning. An interview in Summer 1982 with Sergei Batovrin, a member of the Group, is very revealing. He spoke of the need for ordinary people to speak out on the problem of nuclear war.

> Official groups 'for peace' cannot resolve [this problem], inasmuch as they are controlled by the state apparatus, and hence represent the political interests of the State . . . We tried to hold a demonstration. In accordance with the law, we told the authorities about it in detail, the number of people we expected, and the slogans we were going to use. The most important of these was 'Hiroshima—Never Again'. On the appointed day, at least 50 supporters of the Group were put under house arrest. During the summer, 'an international march of Scandinavian women for peace' passed through Moscow. Shatravka and Mischenko [two members of the Group] wanted to go there, but were arrested and tried under Article 190–1: our peaceful proposals were labelled 'slander against the Soviet system'.

Such accusations were, of course, a pretext: 'They can't say that they are persecuting the Group just because it is independent.'[91] If an unofficial group explicitly and exclusively concerned with 'peace' can expect this treatment, it is hardly surprising that the Church is unable to foster any independent activities of its own within Soviet society, be they of an educational, charitable, welfare or even 'peacemaking' nature.

In conversation with a Western expert in Russian Orthodox affairs, Metropolitan Nikodim justified the policy of loyalty to the Soviet State pursued by the Church since 1927 in terms of the Church's desire 'to persuade the State to recognise the Church legally as part of the national social organism and to make its peace with her, recognising in her a positive moral role and force in the State'.[92] Patriarch Pimen expressed a more realistic appreciation of the actual state of affairs in an interview in 1978: 'In putting into practice its religious life, the Russian Orthodox Church is guided by the teachings of the Gospel of our Lord Jesus Christ . . . But in their activities in society and at work the Orthodox Christians of our country, like all believers within the Soviet Union, follow the principles which were proclaimed 60 years ago when Soviet power was established in Russia.'[93] The atheist specialist Gordienko attributes two misapprehensions to the Russian Orthodox Church and attacks them, thereby demonstrating the unrealistic nature of the hopes of Metropolitan Nikodim quoted above. Gordienko denies the Church's claim that it has always played a positive role in the construction and development of Russian society: the reverse is true, according to him. 'Hence the present political orientation of the Russian Orthodox Church is no consequence of a centuries-old tradition, but the result of the evolution of this confession within socialist society.' Gordienko also denies the Church's claim that support by the hierarchs for the internal and external policies of the Soviet State makes them into positive participants in the social transformation of Soviet society. The truth is that Orthodox Christians participate in this process only insofar as they are good Soviet citizens and workers.

> But if we are going to talk about the Russian Orthodox Church as a whole, then we must assess the social aspects of its activity in a different way from the religious aspects. The social activity of the Church (for example, its participation in the peace movement, support for the Peace Fund and so on) is assessed positively by the Soviet people, including non-believers. But we have a different attitude to those activities of the Church which are directed towards the preservation and strengthening of religiosity in socialist society—here we see a hindrance to the spiritual development of Soviet citizens who are believers . . . So we can see that propagandists of scientific atheism . . . have a dual task. It is essential on the one hand to unmask the conjectures of Western clerics about the insincerity of the present political orientation of the Church, and on the other hand to refute those incorrect assumptions made by contemporary Orthodox theologians which we have cited above.[94]

Gordienko later refutes a third assumption made by these theologians—that the socialist revolution in Russia was really God's work. In reality it was achieved, and continues to be perfected, by those who 'are guided in their activities by the theories of scientific Communism'.[95]

As we have seen, any priest or believer in the Soviet Union who attempts to become socially or politically active will be heavily discouraged and, in extreme cases, punished by application of the law. The one exception is when the message is for consumption abroad: in such cases, as we have seen, the Church is paradoxically expected to fulfil an almost purely political role.[96] 'As is well known,' writes Furov in his *Report*, 'the modernism of Russian Orthodoxy has led its ideologies to the necessity of co-operating in all respects with the defence of peace and the strengthening of friendship amongst nations. The *Journal* constantly includes material emphasising the position of the Church on this.'[97] Even here, however, Furov warns of a danger: 'the desire to idealize

Christianity, to place it above all other theories, including the theory and practise of Marxism'.[98]

* * *

The Soviet deception operation regarding the Churches in the Soviet Union is aimed broadly at the Churches of the world, most obviously through Soviet representation at the World Council of Churches and through the Churches to the Western peace movements and more generally the peace-loving public in the West—'people of goodwill' as well as Christians.

As a result of its predominantly Third World membership, the WCC has come to be dominated by the voices of those Churches, and radical solutions to the social and political problems in Third World countries have been proposed with increasing regularity over the years. The Russian Orthodox Church naturally welcomes such proposals, which tend to coincide with the foreign policy aims of the Soviet State, particularly in respect of support for liberation movements.

As far as its own positive contribution to international debate in this forum goes, then, the Russian Orthodox Church is likely to find a receptive audience. As far as alleged infringements of religious rights *within* the Soviet Union is concerned, the WCC is hampered by its own convention whereby it will investigate problems of this nature within particular countries *only if the problems are raised by the official church delegates for those countries themselves*. That this policy may, in important ways, be inappropriate to Churches in the peculiar position of those in the Soviet Union, may be perceived by some within the WCC, but no special procedural case can be made as far as the Soviet Union is concerned. Thus, in November 1967, the General Secretary of the WCC made a statement on a document received from believers in the Soviet Union calling attention to 200 Baptists in prison: 'The WCC is studying the document closely . . . it is seeking direct contact with the competent authorities in the USSR, *particularly with leaders of the Baptist Church* (my italics), who have been asked to comment on the document and evaluate it.'[99] There was apparently no further mention of the document and one may presume that the officially approved Baptist leadership denied its content. At the end of January 1968, the Executive Committee of the WCC did however endorse protests on behalf of political prisoners in Greece and South Africa.[100] In their 1975 Appeal to the WCC, Fr Gleb Yakunin and Lev Regelson specifically deplored the fact that the WCC had consistently failed to speak out about infringements of believers' rights in the communist world.[101]

One substantial fear which seems to prevent the WCC making

statements which will be labelled 'anti-Soviet' is that the Russian Orthodox Church may withdraw its representation altogether. At the 1983 Assembly of the WCC in Vancouver, a resolution called for an end to aid for Afghan 'rebels' and for a withdrawal of Soviet troops only after a comprehensive settlement of the Afghan question guaranteed by East and West. An amendment which would have struck out the first clause and would have called for immediate Soviet withdrawal was narrowly defeated after Russian Orthodox delegates had hinted that its acceptance might mean their own withdrawal from the WCC.[102]

Another WCC worry was also evident: General Secretary Dr Philip Potter said, after the Afghan debate, that tough resolutions against Moscow might cause problems for Christians within the Soviet Union. It seems that Soviet delegates had hinted that they might suffer consequences from the Kremlin if the statement were too harsh. There is ample evidence, however, that in such circumstances church leaders from the Soviet Union are only too anxious for their Western counterparts to take the initiative in making strong statements on Soviet conduct. As Professor Pospielovsky notes, Metropolitan Nikodim 'once reprimanded me personally for keeping silent at a press conference where he himself had made a mendacious statement and I knew the truth behind it. He said: "Every time you in the West, particularly the Western Churches, criticise us, you're helping us. You force the Soviet authorities to be more cautious in putting pressure to bear on us. Such criticisms allow me to press the Soviet Government officials for concessions".'[103]

The Western Churches, and through them, 'people of goodwill' in the West provide a constant target for Soviet disinformation on the religious scene.

> As foreign tourism grows, and also a definite interest amongst foreign citizens in the situation of religion and the Church in the USSR, we make growing demands on the servants of cults who are entrusted with contact with foreigners. It must be stated that, in the main, they justify this trust.[104]

As an example of the many testimonies by visitors to the Soviet Union which appear every year in the Western press, we may quote from a 1976 report by a member of a delegation of churchmen from Australia hosted in Russia by the Russian Orthodox Church. Reference is made to hospitality, generous gifts, vodka and caviar, and the spiritually uplifting nature of Orthodox worship. The visitor incidentally records his 'impression that in various ways the

Government had assisted the Church to make our own visit possible', and goes on:

> Today, relations between the Church and the State would seem to have reached an often warm understanding of each other's role . . . The Church does not have schools or Sunday schools, but bases its work on the sound maxim that the worshipping community is the most powerful instrument of evangelism . . . As the State looks after social services, the Church is not involved in many of the activities to which we are accustomed in the West . . . The [seminary] course is very wide and includes Soviet philosophy and history . . . [The general impression is that] in spite of considerable difficulties over the past half-century, today religion is very much alive in Russia; and that a formula for coexistence has been achieved between an atheist Communist State whose constitution allows for freedom of religious belief, and the Churches who are prepared to accept the limitations of the law and develop their lives within these guidelines.[105]

* * *

When Billy Graham spoke at the Peace Conference in Moscow in May 1982, remarking that he had seen no religious persecution in the Soviet Union, he fulfilled a prediction. The 'Siberian Seven', a group of Pentecostal believers, then in their fourth year of refuge in the American Embassy in Moscow—a refuge aimed at persuading the Soviet authorities to permit their emigration on religious grounds—had predicted that Graham would make just such a statement.

> The appearance of Billy Graham at the Conference will be used by the Soviet Government for propaganda that there is complete religious freedom for Christians. Billy Graham will hear about our 'religious freedom' from people sent by the Kremlin. And they will not tell him about those Christians who decay in prison cells because they choose to follow God's commandments rather than to make compromises with the atheistic system of this country.[106]

The Soviet authorities, never slow to convey to their own citizens the reactions of Westerners to their presentation of Soviet religious reality, made much of Billy Graham's visit in the media and other publications.[107]

It is hard to see exactly how the extent of the Soviet deception operation regarding religious freedom in the Soviet Union can be demonstrated accurately to Western audiences, given the 'seamless robe' nature of the operation. Deception is achieved by dissimulation—the hiding of the reality of persecution by State control over information and foreign observation, and by simulation—the promotion of a false picture by stage-managed tours inside Russia and by the statements of hierarchs on their equally well stage-managed trips to the non-communist world. It is from this basic illusion of a free, independent Russian Orthodox Church that the secondary deceptions flow. The Church's statements on 'peace', on the superior status of the Soviet social system, on international affairs, and on the un-Christian nature of anti-Communism rely for their credibility on their apparently independent, Christian source. If the first deception fails, all others lose their power, for it is the notion of a free Church that makes its official spokesmen into an effective 'transmission belt'.

A member of the International Fellowship of Reconciliation who visited Moscow in March 1983 was critical of the one-sided nature of official Soviet peace proposals, seeing the small independent peace group's positions as 'much closer to our own'. He laments the harassment of members of the latter group, saying:

> It is very important for people in the West to challenge this double standard. If people on one side are allowed to demonstrate freely for disarmament and the establishment of trust, and people on the other side are harassed, punished and imprisoned for doing the same, then what kind of peace movement can that be? There must be a universal moral and ethical standard that applies equally everywhere.[108]

The application of a universal 'moral and ethical standard' to the problems of civil and religious rights was of course one of the aims of the Helsinki Final Act. The fact that the Soviet Union was one of the countries which ratified this Act gave impetus, as we have seen, to the dissent movement within the Soviet Union, and specifically to the movement for religious rights, which began in the 1960s as a result of the discontinuity which many Soviet Christians began to perceive in the post-war period between the image of a strong and flourishing Church for foreign consumption and the reality of a deeply deprived Church at home.

The image of what the Church could perhaps become, if its leaders stood up more vigorously for their rights, impelled dissenters to start urging their hierarchs to be much more determined in resisting state pressure. In 1965 Fr Gleb Yakunin and another priest, Fr Nikolai Eshliman, wrote to the Patriarch and the Soviet authorities protesting about excessive interference in church matters by the Council for Religious Affairs and urging the Patriarch to take action.[109] Aleksandr Solzhenitsyn, for the first time publicly identifying himself as an Orthodox Christian, intervened in the same vein with his *Lenten Letter* to the Patriarch in 1972.[110] In 1979 Fr Gleb Yakunin put forward proposals to the effect that Orthodox Christians ought to form an 'unofficial' wing within the Orthodox Church which would grow in size as centralised Soviet control of and pressure on the 'official' Church increased.[111] The Christian Committee for the Defence of Believers' Rights in the USSR, as we have seen, has in recent years provided ample evidence of discrimination against religious believers. There is, then, a growing body of material produced unofficially by Christians within the Soviet Union indicating that church-state relations in that country are far from being as officially presented. Although there are some signs of growing awareness in the West of the full complexity of the problem, this material has yet to receive systematic evaluation and assimilation by the Western audiences to whom it is directed.

The arrival of Mr Gorbachev and his policy of '*glasnost*' has improved the situation for religious believers in a welcome, but so far fairly limited, way—for example, a number of prominent religious prisoners have been released and from time to time complaints about specific cases of discrimination against individual believers or congregations are aired in the press. These changes are relatively superficial, however, '*glasnost*' in religious affairs has not yet led to fundamental '*perestroika*': there has been no fundamental change either in the hostile attitude of the authorities to religious faith as such, or in the basic strategy of manipulation analysed in this chapter.

Endnotes

1. The texts of the laws of 1918 and 1929, together with other legal material, are given in English in William B Stroyen, *Communist Russia and the Russian Orthodox Church 1943–62* (Washington, DC: Catholic University of America Press, Inc., 1967), pp. 117–45.
2. The text of Sergi's Declaration is given in English in William C Fletcher (i), *A Study in Survival: The Church in Russia 1927–43* (London: SPCK, 1965), pp. 28–32. The Russian text makes it clear that Metropolitan Sergi intended to indicate that Christians owe loyalty to the Motherland rather than to the Soviet state as such, and he later expressed his unhappy surprise that this distinction had not been appreciated by Christians in the West. Whatever the text says, however, it has had no influence on how it has subsequently been interpreted by the Soviet authorities.
3. See William C Fletcher (ii), *Religion and Soviet Foreign Policy 1945–70* (London: Oxford Univ. Press, 1973).
4. William C Fletcher (iii), *Nikolai: Portrait of a Dilemma* (New York and London: Macmillan, 1968), p 110. For Soviet interpretations of the word 'peace' and its exploitation in deceptive propaganda activities, see Chapter 10.
5. Fletcher (ii), pp. 30–3.
6. Elena Pozdeeva, *Khristianskaya Mirnaya Konferentsiya i ee rol v svobodnom mire*, Radio Liberty Research Paper 432/76, 3 November 1976, p. 2.
7. In 1977 the Soviet journal *Literaturnaya gazeta* devoted a long article to a personal attack on certain prominent religious activists, including Aleksandr Ogorodnikov, Anatoli Levitin, Lev Regelson and Fr Gleb Yakunin: 'Svoboda religii i klevetniki', *Literaturnaya gazeta*, 13 April and 20 April 1977. At least four letters were written by Christians to the journal to protest and set the record straight, but none of these was published.
8. The Russian Orthodox dissenter Vladimir Osipov was punished for taking part in

unofficial cultural gatherings by seven years in a labour camp. After he was released, it was extremely difficult for him to find anywhere to live. See Vladimir Osipov, 'V poiskakh kryshi', July 1970, Arkhiv Samizdata No. 526.

9. Between October 1980 and November 1981, three prominent activist Roman Catholic priests were killed in Lithuania, and three others were assaulted over the same period.
10. While five members of the Pentecostal Vashchenko family were taking refuge in the American Embassy in Moscow, in pursuance of their desire to emigrate from the USSR, the children back at home in Chernogorsk suffered from various forms of intimidation. On one occasion their gates were forced by militiamen, who trampled their vegetables.
11. This explains why some of the most gifted and creative people in the USSR are to be found working as road-sweepers, nightwatchmen and so on. Aleksandr Ogorodnikov, founder of the Christian Seminar on Problems of the Religious Renaissance, 'was employed as a janitor in a tuberculosis clinic . . .' Vladimir Poresh, 'Progulka po Moskve,' translated in *Religion in Communist Lands*, Vol. 8, No. 2, 1980, pp. 103–6.
12. In November 1978, Ogorodnikov was arrested while on his way to register for a job. In January 1979 he was sentenced to one year in a labour camp on a charge of parasitism.
13. Tatyana Shchipkova had taught French at Smolensk Pedagogical Institute for 17 years before it became widely known that she had become a Christian. She was criticised in her department for 'religious propaganda among students,' and one of her colleagues expressed the opinion that 'teachers simply do not have rights to a large number of things, for example, certain sorts of clothes or hairstyles, and even more so, certain beliefs.' She was finally dismissed, however, for being 'underqualified'. Tatyana Shchipkova, 'Imeet li pravo sovetskii prepodavatel na svobodu sovesti?' translated in *Religion in Communist Lands*, Vol. 8, No. 2, 1980, pp. 106–9. She was subsequently charged with 'malicious hooliganism' (resisting a militiaman who was searching her flat) and sentenced to three years in a labour camp.
14. The Baptist rock musician Valeri Barinov was arrested in October 1983 and held in a psychiatric hospital in Leningrad. The doctor in charge of the case told Barinov's wife that while he was 'not really ill', his views and beliefs differed so much from 'normal Soviet views' that he was 'in need of treatment'. 'Christian Musician in Psychiatric Hospital', *Keston News Service*, No. 185, 20 October 1983, p. 2.
15. A photocopy of the *Report* is in the archives of Keston College. Parts of it have been published in Russian under the title 'Iz otcheta Soveta po Delam Religii—chlenam TsK KPSS' in *Vestnik RKhD*, Paris (No. 130, 1979, pp. 275–344; No. 131, 1980, pp. 362–72; No. 132, 1980, pp. 197–205). Translations of extracts in English appear in *The Orthodox Monitor*, Nos. 9–10, July–December 1980, pp. 58–80. Extracts from the 1974 report have appeared as *Rapport Secret au Comité Central sur l'Etat de l'Eglise en URSS*, Seuil, Paris, 1980.
16. 'Iz otcheta . . .', *Vestnik RKhD*, No. 130, 1979, p. 297.
17. *Ibid.*, p. 306.
18. *Ibid.*, p. 311.
19. *Ibid.*, p. 319.
20. *Ibid.*, p. 315.
21. *Ibid.*, p. 337.
22. Gleb Rar, *Plenennaya tserkov*, (Frankfurt-am-Main: Posev, 1954), p. 69.
23. Dimitry Pospielovsky (i), *The Russian Church under the Soviet Regime 1917–82*, 2 vols, (New York: St Vladimir's Seminary Press, 1984), p. 317.
24. Mark Popovsky, *Zhizn i zhitie Voino-Yasenetskogo arkhiepiskopa i khirurga*, (Paris: YMCA Press, 1979), pp. 404–7, esp. footnote 103.
25. See Stroyen, pp. 54–68 for survey of such travel between those dates.
26. Fr Gleb Yakunin, *Doklad . . . o sovremennom polozhenii Russkoi Pravoslavnoi Tserkvi i o perspektivakh religioznogo vozrozhdeniya Rossii*, 15 August 1979, in *Documents of the Christian Committee for the Defense of Believers' Rights in the USSR*, (San Francisco: Washington Street Research Center) Vol. 11, pp. 1128–68, here p. 1159. Partial English translation in D Konstantinow, *Stations of the Cross: the Russian Orthodox Church 1970–80*, (London, Ontario: Zaria, 1984).

27. On Nikolai see Fletcher (iii) *op. cit.*; Arkhiep. Vasili (Krivoshein), 'Poslednie vstrechi s Mitropolitom Nikolaem (Yarushevichem),' *Vestnik RKhD*, No. 117, 1976, pp. 209–19. On Nikodim see Fletcher (ii), pp. 123–4; Pospielovsky (i), pp. 359–63; D Pospielovsky, J Lawrence, P Oestreicher, 'Metropolitan Nikodim Remembered,' *Religion in Communist Lands* Vol. 6, No. 4, 1978, pp. 227–34; D Pospielovsky (ii), 'Mitropolit Nikodim i ego vremya,' *Posev*, No. 2, February, 1979, pp. 21–6.
28. Stella Alexander, Review of Laszlo Revesz, *The Christian Peace Conference: Human Rights and Religion in the USSR*, (London: Institute for the Study of Conflict, Conflict Studies No. 91, January 1978), *Religion in Communist Lands*, Vol. 6, No. 4, 1978, pp. 250–2.
29. Fletcher (ii), pp. 35–9.
30. Elena Pozdeeva, *op. cit.*
31. See for example 'Russian Journal's Attack on WCC,' *Church Times*, London, 21 November 1969, p. 1.
32. J A Hebley, *The Russians and the World Council of Churches*, (Belfast–Dublin–Ottawa: Christian Journals Ltd., 1978), pp. 93–4; John Lawrence, 'East and West–the New Opportunity', *The Ecumenical Review*, April 1962, pp. 329–35, here pp. 329–30.
33. John Lawrence, *op. cit.*, p. 331.
34. For a chronology of these, see Hebly, *op. cit.*, p. 67.
35. Pospielovsky (i), pp. 313–5.
36. V A Kuroedov, *Religiya i tserkov v sovetskom obshchestve*, (Moscow: Politicheskaya Literatura, 1984), p. 207.
37. *Zhurnal Moskovskoi Patriarkhii*, September 1949, quoted in Fletcher (iii), p. 111.
38. Arkhiep. Vasili, p. 216.
39. Pospielovsky (ii), p. 23.
40. Fletcher (ii), p. 127.
41. Fletcher (iii), pp. 200–2.
42. See David Kelly, 'Nairobi: A Door Opened', *Religion in Communist Lands*, Vol. 4, No. 1, 1976, pp. 4–17 (including a translation of the Yakunin and Regelson Appeal); Helene Posdeeff, 'Geneva: the Defence of Believers' Rights', *Religion in Communist Lands*, Vol. 4, No. 4, 1976, pp. 4–15; Michael Bourdeaux, Hans Hebly, Eugene Voss, eds., *Religious Liberty in the Soviet Union. WCC and USSR: a post-Nairobi Documentation*, (Keston College, 1976). For a full discussion of the Yakunin-Regelson affair in the context of WCC policy on religion in Eastern Europe to date, see Michael Bourdeaux, 'The Russian Church, Human Rights and the World Council of Churches', *Religion in Communist Lands*, Vol. 13, No. 1, 1985, pp. 4–27.
43. Bourdeaux, pp. 10–13.
44. TASS, 29 January 1976.
45. 'V Generalnaya Assembleya Vsemirnogo Soveta Tserkvei v Nairobi,' *Zhurnal Moskovskoi Patriarkhii*, No. 6, 1976, pp. 58–61, here p. 61.
46. Bourdeaux, *op. cit.*
47. Jane Ellis, 'The Christian Committee for the Defence of Believers' Rights,' *Religion in Comunist Lands*, Vol. 8, No. 4, 1980, pp. 279–98.
48. Fletcher (ii), p. 95.
49. *Zhurnal Moskovskoi Patriarkhii*, December 1950, quoted in Fletcher (iii), p. 115.
50. 'Russians Say "No" to Unilateral Disarmament', *Catholic Herald*, London, 12 May 1978, p. 3.
51. As one example of a critical statement of this kind, Metropolitan Nikodim cited a statement issued by the Christian Peace Conference in Prague regretting the resumption of nuclear tests by the Soviet Union in 1962. Hebly, p. 74. Hebly comments that silence by church representatives on certain issues can often be taken as amounting to an expression of disagreement. *op. cit.* p. 69.
52. Hebly, pp. 49–66.
53. N S Gordienko, Evolyutsiya russkogo pravoslaviya (20–80^{e} gody XX stoletiya), Znanie, Moscow, series 'Nauchnyi ateizm', No. 1, 1984, p. 21.
54. 'Retrospektiva i perspektivy mirotvorchestva,' *Zhurnal Moskovskoi Patriarkhii*, No. 5, 1976, p. 35.
55. Patriarch Pimen to a meeting in Zagorsk of representatives of all religious groups,

1981, quoted in Mitrop. Yuvenali, 'Russakaya pravoslavnaya tserkov v borbe za mir i razoruzhenie', *Mir i razoruzhenie*, (Moscow: Nauka, 1982), pp. 147–57, here p. 156.

56. See the reactions to his 1973 remarks that 'in the USSR there are neither rich nor poor, privileged nor persecuted', in *Glaube in der 2 Welt: Materialdienst*, October, 1973, p. 5, and in Prot A Shmeman, 'Mera nepravdy,' *Vestnik RSKhD*, No. 108/9/10, 1973, pp. 142–3; and to his 1984 remarks on Andropov at the latter's funeral in D Pospielovsky (iii), 'The Russian Church: How to Help,' *The Orthodox Church*, October, 1984, p. 4.
57. 'The Moscow Patriarch on Church Concern for Peace,' *Ecumenical Press Service* 83.01.88.
58. In the view of the Russian Orthodox Church, the WCC before 1961 had a policy of 'either tacitly or openly condemning the socio-economic and political innovations in countries on the path of socialist construction', and produced declarations 'couched in the spirit of the "Cold War" ', but these have ceased since the Russian Orthodox Church became a member. *Zhurnal Moskovskoi Patriarkhii* No. 2, 1984, pp. 59–67, quoted in *Annotations*, Radio Liberty Research Paper 174/84, 30 April 1984.
59. See the comments, cited earlier, by Soviet Church delegates to the WCC explaining that they had not voted even for the amended resolution brought up at Nairobi under an agenda item dealing with disarmament and the Helsinki Agreements on the grounds that the question of religious rights in the countries which had signed the Helsinki Agreements was not relevant to the topic.
60. 'Pot and Kettle', *Catholic Herald*, London, 19 January 1968.
61. 'Leftist Church Group Divided over Czech Invasion', *New York Times*, 16 February 1969.
62. 'Russia's Christian Leader Backs Czech Invasion', *Catholic Herald*, London, 4 October, 1968.
63. *Zhurnal Moskovskoi Patriarkhii*, No. 2, 1978, p. 4.
64. The Very Rev Dimitry Konstantinow, *The Crown of Thorns: Russian Orthodox Church in the USSR 1917–67*, (London, Canada: Zaria, 1978), pp. 83–9.
65. Fletcher (ii), p. 107. See also Fr Gleb Yakunin and Lev Regelson, 'Appeal to the Delegates of the 5th Assembly of the World Council of Churches', 16 October 1975, translated in *Religion in Communist Lands*, Vol. 4, No. 1, 1976, pp. 9–14, here p. 11.
66. See *Occasional Papers on Religion in Eastern Europe*, Lafayette College, Easton, PA, Vol. II, No. 4, July 1982; Richard Chartres, 'The Moscow Peace Conference, May 1982', *Religion in Communist Lands*, Vol. 10, no. 3, 1982, pp. 337–9.
67. H Lamar Gibble, 'Report', *Occasional Papers* . . ., Vol. II, No. 4, July 1982, p. 2.
68. *Ibid.*, p. 4.
69. Michael Binyon, 'Peace Issue Marks Church Revival', *The Times*, 10 May 1982, p. 6.
70. See this volume, pp. 241–262.
71. L Donatov, 'Gosti i khozyaeva,' *Russkaya mysl*, 17 February 1983, p. 5.
72. Walter Kolarz, *Religion in the Soviet Union* (London: Macmillan, 1961), pp. 63–4; Konstantinow, pp. 79 ff.
73. Fletcher (iii), pp. 101–6; Hebly, p. 67; Pospielovsky (i), pp. 469–71.
74. Metropolitan Nikodim, *To the General Secretary of the WCC, Dr W A Visser't Hooft* (Moscow, June 1963), pp. 16–19. Copy in archive of Keston College, Keston, Kent, England.
75. Fr Gleb Yakunin, *op. cit.*, Vol. 11, p. 1161.
76. Quoted in Gerhard Simon, *Church, State and Opposition in the USSR*, (London: C Hurst & Co., 1974), p. 124.
77. *Patriarch Pimen of Moscow and All Russia in Geneva*, Radio Liberty Research Paper 304/73, 26 September 1973, p. 2.
78. Archbishop Nikodim, 'Mir i svododa,' *Zhurnal Moskovskoi Patriarkhii* No. 1, 1963, pp. 39–44, here p. 42. See also Konstantinow, pp. 99–104.
79. 'Retrospektiva . . .', *Zhurnal Moskovskoi Patriarkhii*, No. 5, 1976, p. 36.
80. Patriarch Pimen, *Opening Address at the World Conference: Religious Workers for Peace, Disarmament and Just Relations among Nations* (Moscow, 6 June, 1977). Copy in Archive of Keston College, Keston, Kent, England.
81. A N Kosygin, 'Privetstvie . . . uchastnikam Kongressa', *Zhurnal Moskovskoi Patriarkhii*, No. 9, 1978, p. 27. See also Elena Pozdeeva, p. 7.

82. N S Gordienko, pp. 23–4.
83. 'Rech mitropolita . . . Nikodima na plenarnom zasedanii Soveshchatelnogo Komiteta KhMK v Soffii . . .', *Zhurnal Moskovskoi Patriarkhii*, No. 12, 1966, p. 50; See also Elena Pozdeeva, p. 5; Brian Norris, 'The Christian Peace Conference Criticised,' *Religion in Communist Lands*, Vol 7, No. 3, 1979, pp. 178–9.
84. 'Zayavlenie po voprosu o evropeiskoi bezopasnosti', *Zhurnal Moskovskoi Patriarkhii*, No. 7, 1973, p. 44, quoted in 'Retrospektiva . . .', p. 39.
85. 'Zayavlenie Svyashchennogo Sinoda Russkoi Pravoslavnoi Tserkvi ot 20 marta 1980 goda,' *Zhurnal Moskovskoi Patriarkhii*, No. 5, 1980, p. 5.
86. 'Intervyu Svateishego patriarkha Moskovskogo i vseya Rusi Pimena: V sluzhenii miru na zemle vidim velikuyu tsel svoyu,' *Golos rodiny*, No. 5, February 1978.
87. G K Nikitin, 'Sobrannost i trezvenie,' *Vestnik RKhD*, No. 126, 1978, pp. 275–83, here pp. 276–7.
88. Yu Kublanovsky, 'O "Zhurnale Moskovskoi Patriarkhii" ', *Posev*, No. 6, 1984, pp. 53–5, here p. 53.
89. P Kurochkin, 'Evolyutsiya sovremennogo russkogo pravoslaviya', Part III: 'Politicheskaya orientatsiya', *Nauka i religiya*, No. 4, 1969, pp. 48–52.
90. *Konstitutsiya Soyuza Sovetskikh Sotsialisticheskikh Respublik* (1977), Article 6, in *Konstitutsiya i zakony Soyuza SSR*, Moscow, 1983.
91. 'O "gruppe doveriya" ', *Posev*, No. 11, 1983, pp. 18–22.
92. Pospielovsky (ii), p. 21.
93. 'Intervyu Svyateishego patriarkha . . .'
94. N S Gordienko, pp. 24–6.
95. *Ibid.*, p. 31. A spirited attempt in 1960 by Metropolitan Nikolai and Patriarch Aleksi, in the face of growing persecution of the Church under Khrushchev, to glorify the historical role of Orthodoxy in the development of the Russian State and nation caused a '*skandal*' amongst those who were present and was vigorously challenged from the side of the Party. See Fletcher (iii), pp. 187–93. Criticism of the Church along the same general lines as those followed by Gordienko increased in the Soviet press and media recently as the Millennium of Orthodoxy in Russia (1988) drew nearer. The aim was to dissipate in advance any capital the Church might make out of the occasion. See Sergei Voronitsyn, *Propaganda Mounting Against the Thousandth Anniversary of the Christianisation of Rus'*, Radio Liberty Research paper 302/84, 9 August 1984.
96. For attempts by Metropolitan Nikolai to justify his Church's involvement in politics in the international arena, see Fletcher (iii), pp. 113–14.
97. 'Iz otcheta . . .', *Vestnik RKhD* No. 130, 1979, p. 331.
98. *Ibid.*, p. 332.
99. *Ecumenical Press Service*, 9 November 1967, p. 7.
100. *Ecumenical Press Service*, 22 February 1968, p. 2. See Fletcher (ii), p. 127.
101. Fr Gleb Yakunin and Lev Regelson, 'Appeal . . .,' *Religion in Communist Lands*, Vol. 4, No. 1, 1976, pp. 11–12.
102. See J A Emerson Vermaat, 'The World Council of Churches and the Afghanistan Crisis,' *Conflict Quarterly*, Vol. V, No. 3, Summer 1985.
103. D. Pospielovsky (iii), p. 4.
104. 'Iz otcheta', *Vestnik RKhD*, No. 130, 1979, p. 304.
105. Ian Shevill, 'Russian Church is Misunderstood', *Church Times*, 13 August 1976, pp. 8, 9.
106. Liubov Vashchenko, *Letter*, 7 April 1982, in *Keston News Service*, No. 147, 22 April 1982, p. 7.
107. See for example V A Kuroedov, pp. 213–6.
108. Jim Forest, 'USSR', *Churches Register*, Vol. II, No. 2, 1983, pp. 18–19.
109. N Eshliman and G Yakunin, *Letter* to Patriarch Aleksi, 21 November 1965; *Letter* to N V Podgorny, 15 December 1965. To be found in abridged translation in Michael Bourdeaux, *Patriarch and Prophets: Persecution of the Russian Orthodox Church Today*, (Oxford and London: Mowbrays, 1970), pp. 189 ff.

110. A I Solzhenitsyn, *Letter* to Patriarch Pimen, Lent 1972. English translation in Simon, *op. cit.*, pp. 202–5.
111. Fr Gleb Yakunin, pp. 1160–8.

CHAPTER 4

Bluff and Deception in the Khrushchev Era

HANNES ADOMEIT

There is no need for the Soviet Union to shift its weapons for the repulsion of aggression, for a retaliatory blow, to any other country, for instance, Cuba.
TASS, 12 September 1962.

At the United Nations General Assembly, in September 1960, Khrushchev enlivened the ordinarily rather dull proceedings by banging his shoe on the table to make a point. He did not, apparently, manage to put the speaker at the rostrum, British Prime Minister Harold Macmillan, off his stride. He also did not gain wider acceptance for his *troika* proposal—the conversion of the office of Secretary General into a commission of three representatives. But as the extensive coverage of this incident in the Western press and television at that time and the replay of this event on newsreel, time and again demonstrated, he may have been more successful in achieving a larger objective: to convey to the Western (United States and Western European) public and to Western policy makers the image of an impatient and impetuous political leader unconstrained by conventional diplomatic *politiesse*, determined to get his way against all odds and perhaps even capable of wild irrationality. In short, he may have succeeded at least to some extent in giving rise to the idea that, in a game of 'chicken', he would be the most reckless competitor.

Appearances, of course, do not always conform to reality. In this instance, they may not have conformed to it at least in one respect. According to a well-placed eye-witness, Khrushchev had both his shoes on at the time.[1] Perhaps this observation is untrue. But as with many other anecdotes which may be incorrect on the specific point, the observation can be considered as quite 'true' in a more general sense: conduct designed for maximum impact on international opinion must be regarded as one of the foremost characteristics of Khrushchev's style in foreign policy. Part and parcel of this—in the broad sense—'diplomacy of shoe-banging' are the threats and ultimatums, the brandishing of nuclear weapons, the manipulation of 'bomber gaps' and the 'missile gaps', and the innumerable public appearances at which Khrushchev demonstratively wore the uniform of a Marshal of the Soviet Union. All of this was carefully calculated and calibrated so as to coax, trick and frighten the Western public and policy makers into making concessions.

But it was more than that. It was part of a larger design of deception which consisted of three major facets: first, deliberate exaggeration of the international *political* trends favouring the Soviet Union and the socialist community; second, over-emphasis on the trends in modern weapons technology and, more broadly, the *military balance* in favour of the Soviet Union; and third, overstatement as to the degree to which the Soviet leaders were determined to intervene militarily in local crises and to *accept risks* of confrontation with the adversary superpower.

These three facets will be analysed in more detail by focusing on the evolution of this strategy during the Suez crisis, its culmination during the Berlin crisis and its collapse during the Cuban missile crisis. But bearing in mind the axiom that 'foreign policy begins at home', as well as Lenin's observation that 'There can be no more harmful idea than the separation of domestic from foreign policy',[2] the three facets of Khrushchev's strategy will first be related to their proper domestic context.

* * *

The domestic context of Khrushchev's foreign policy was to a considerable extent shaped by a protracted succession struggle after Stalin's death. In retrospect, it is fair to summarise that the extent of Khrushchev's power lay somewhere between the autocracy of Stalin and the oligarchic rule under Brezhnev, Andropov and Chernenko; between one-man dictatorship and 'collective leadership'; and between an almost complete lack of challenge to supreme authority and a cumbersome, gradualist committee-type of decision-making. The scope of Khrushchev's power, therefore, changed according to changing domestic and international circumstances. He was intent on attaining a power position closer to dictatorship then collectivism.[3]

Nevertheless, in June 1957 he was barely able—with the help of Marshal Zhukov—to reverse a 'mathematical majority' against him in the Presidium and defeat an alleged 'anti-Party', that is, anti-Khrushchev, group consisting, among others, of Molotov, Malenkov, Bulganin, Kaganovich, Shepilov, and Pervukhin. Thereafter, the pressure on him seemed to ease. This may, at least in part, have been due to the higher degree of tension in East–West relations, the deterioration of the relationship with China, the crises in Lebanon, Laos, the Congo and the Taiwan Straits, and the confrontation with the United States over Berlin and Cuba, which may have produced a 'rallying around the flag'.

Yet the victory he achieved was incomplete and the foundations of his rule remained unstable. Khrushchev himself contributed to the weakening of the Party's authority at the 20th CPSU Congress, in his 'Secret Speech', with its damaging revelations about the 'gross violations of socialist legality' committed by Stalin in the period of the 'cult of personality'. In these circumstances, it was obvious that success and failure, both in the socio-economic development and the international power position of the country, would be the main criteria for separating true statesmanship and 'scientific' policies from 'subjectivism' and 'hare-brained scheming'.

The conditions in domestic politics, however, were not at all conducive to rapid progress. They did not favour a strategy of dramatic breakthrough. Given the previous stifling of initiative under Stalin and the persistence of an all-pervasive bureaucracy and built-in disincentives to innovation and reform, change could only be achieved gradually. The goals set in the Seven-Year Plan (1959–65) and the CPSU programme of 1961 did *not* conform to this requirement.[4] To the embarrassment of Khrushchev's successors, the Party programme bluntly asserted:

> *In the current decade* (1961–70), the Soviet Union, creating the material and technical base of Communism, will surpass the strongest and richest capitalist country, the USA, in per capita production.[5]

All this was wildly over-optimistic, and results in all sectors fell far short of forecasts. At the beginning of the 1960s, therefore, a wide gap was beginning to develop between grandiose plans and socio-economic reality. 'Life itself' was conspiring against Khrushchev and putting pressure on him to be more successful in foreign policy.

* * *

The preconditions for the 'progress of socialism' were, or at least could be interpreted to be, much more favourable in international politics than in domestic affairs. Two major trends seemed to be advantageous from the Soviet point of view: first, the Soviet advances in military technology, notably the development of capabilities for delivering nuclear weapons over long ranges, and second, the rapidly accelerating process of de-colonisation in the

Third World. It was these trends, above all, to which Khrushchev pointed in order to convey the impression that the global 'correlation of forces' between the two opposed world systems had decisively shifted in favour of socialism.

The potential benefits of the first trend were indeed quite significant. The improvements in Soviet nuclear delivery capabilities, including the deployment of long-range bombers and intercontinental missiles, had the effect, for the first time in the history of the Soviet-American strategic relationship, of making United States territory vulnerable to Soviet strikes. The European allies of the United States, of course, had—since the beginning of the 1950s—been exposed to the threat of nuclear strikes because of the Soviet Union's medium-range bomber capabilities. For all practical purposes, they had been held hostage to friendly American behaviour toward the Soviet Union. But the extension of the nuclear threat directly to the United States, potentially offered a twofold political advantage: it could serve to restrain the United States in crises and undermine West European confidence in American security guarantees. It is probable that it was for these reasons that Khrushchev, very early on, embarked on a deception of the West over the Soviet Union's true military capabilities.

The beginning was made in 1955. In the Spring of that year, several articles written by Soviet military leaders appeared in the Soviet press, claiming in effect that the Soviet Union now had at its disposal the weapons necessary 'to anticipate a surprise attack and to strike before the aggressor can take advantage of his own preparations for a military strike'.[6] On 3 July, just two weeks before the opening of the Summit Conference in Geneva, the Soviet military staged an elaborate air show at Tushino Airfield outside Moscow. For several weeks in advance of the show, flights of new military aircraft appeared daily over Moscow, apparently to convince foreign observers that a Soviet strategic nuclear capability was a reality. It seems likely, in retrospect, that every available aircraft of the three main new types, the MYA-4, TU-16 and TU-95, was put into the air for these displays. The appearance of even relatively small numbers of these aircraft at Tushino did, in fact, serve to convince many Western observers that a 'bomber gap' was developing and that the United States was in danger of losing its air supremacy.[7]

The 'bomber gap' never did become reality. But the interaction of American apprehensions and Soviet deception may have served to give a boost to the US bomber programmes.[8] Soviet efforts, in contrast, were not directed at attempting to counter the American preponderence of long range aircraft but at significantly expanding the number of their own *medium range* bombers (thereby enhancing even further the threat posed to Europe) and at pushing forward with the development of intercontinental ballistic *missiles*.

The transition from the 'bomber gap' to the 'missile gap' was rapid, and in all likelihood neither 'myths'would have arisen in the absence of stringent Soviet efforts to create them. Thus, as early as at the 20th Party congress, in February 1956, Marshal Zhukov stated that the

> Soviet Armed Forces, due to the constant attention of the Party and Government in securing the defence capability of the nation, have been completely reorganised. They now have diverse atomic and nuclear weapons, mighty guided missiles, among them *long range missiles*.[9]

Two months later, during a visit to Britain, Khrushchev predicted: 'I am quite sure that we will have a guided missile with a hydrogen bomb that can fall anywhere in the world.'[10] For Khrushchev, it may have been quite clear that the primary means of support which the Soviet Union would extend to the 'national-liberation movement' would be political, moral and economic, not military. This, however (probably quite deliberately), was not spelled out to the adversary. In contrast, when the Suez crisis began, he warned Britain and France—informally at a reception in August 1956—that if war broke out in the Middle East, 'the Arabs will not stand alone'; 'volunteers' would come to their aid.[11]

After Britain and France had intervened nevertheless, the Soviet Union refrained from any form of military intervention or weapons deliveries. Only *after* the height of the crisis had passed, and only after it had become obvious that the United States was putting severe pressure on its two Western European allies and Israel to stop the fighting, did Bulganin, on 5 November, issue his famous 'nuclear ultimatum' to Britain and France.[12] But in essence, there was no 'ultimatum'. There was no threat unilaterally to use force if France and Britain did not comply with Soviet demands. There was only the call, in Bulganin's letter to President Eisenhower, for *joint* intervention and the rather general warning in the same letter that 'If this war is not stopped, it contains the danger of turning into a third world war.'

The lessons which Khrushchev drew from the Suez crisis were in all likelihood instrumental in shaping his tactics of bluff, boasts, threat, and deception, and determining his approach in future crises, notably in Berlin and Cuba. It was claimed that the 'clear and firm position of the Soviet Union in defence of Egypt' and

Moscow's 'determination to take an active part in restraining the aggressors, in restoring peace in the Near East and in averting a new world war' had proved to have had a 'sobering influence on the ruling circles of England and France' and to have played a 'decisive role in the cessation of hostilities'.[13] Khrushchev may have believed that he had saved the Arabs from France and Britain, and could save the world by a similar mixture of threat and deception.

* * *

The contours of Khrushchev's strategy came sharply into focus on 26 August, 1957. On that day, the official Soviet news agency, TASS, announced that a 'super-long-range, multi-stage intercontinental ballistic rocket had recently been successfully tested and that the results indicated that 'it is now possible to send missiles to any part of the world'.[14] In the weeks that followed, the Soviet press published several articles on the technical characteristics of ballistic missiles, emphasizing their speed, high altitude, long range, relative accuracy, and ability to hit distant targets with little or no warning. On 4 October, 1957, TASS announced the successful launching into space of the first man-made earth satellite, Sputnik I.

This feat, in turn, was used by Khrushchev to link space and military technology. In a series of interviews with prominent Western correspondents, he tried to convey the idea that the advances which the Soviet Union had made in the space race had a direct impact on the US–Soviet balance in strategic weapons. Claiming that the boosters for Soviet satellites and warheads were interchangeable, he boasted in November 1957 that the USSR already had a stockpile of up to twenty ICBMs:

> The fact that the Soviet Union was the first to launch an artificial earth satellite, which within a month was followed by another, says a lot. If necessary, tomorrow we can launch 10, 20 satellites. All that is required for this is to replace the warheads of an inter-continental ballistic rocket with the necessary instruments.[15]

Until November 1958, Khrushchev merely asserted that the Soviet Union had succesfully tested ICBMs and possessed a stockpile of vehicles, but nothing was said directly about their production. In that month, however, in a speech on the seven-year plan, he announced: 'The production of the inter-continental ballistic rocket has been successfully set up.'[16] This announcement was followed up in the subsequent weeks and months with similar statements such as: 'When we say that we have organised the serial production of intercontinental ballistic rockets, it is not just to hear ourselves talk.'[17]

Khrushchev's claims fulfilled their purpose. They gave rise to anxieties in the United States that the Soviet Union was engaged in a major effort to deploy first-generation ICBMs; that a 'missile gap' was opening up in favour of the Soviets; and that the changed and changing strategic relationship was bound to have profound political consequences. According to an article in the *New York Times* at the beginning of 1959, reportedly based on interviews with 'numerous persons having intimate knowledge of the defence effort', Soviet ICBM inventories in the early 1960 were estimated to rise from about 100, to 500 in 1961 and to about 1,000 in 1962. The ICBM force of the United States, in contrast, was calculated only to reach a maximum of 130 missiles in 1962. The gap was not expected to be closed until after 1964, when sizeable numbers of Polaris and Minuteman missiles were scheduled to enter the United States force.[18]

The military and political significance of the assumed gap rested on the fact that at the end of the 1950s and the beginning of the 1960s, no anti-missile warning system existed in the United States. Furthermore, as that country's strategic forces had a relatively narrow base, presenting no more than about 50 targets, it was assumed that by 1961 the Soviet Union would be in a position to destroy most of the United States strategic power in a first strike.

However, the Soviet Union never did embark upon mass production of first-generation ICBMs. Whatever the reason—constraints of cost, serious technical deficiencies, or the 'ganging up' of the traditional branches of the Soviet services against new missile projects—the fact remains that they never came close to opening the much-talked-about and much-anticipated 'missile gap'. Although United States intelligence estimates were scaled down several times during 1960, it was not until the end of November, when the first reconnaissance satellites became operational and were programmed to look at the areas missed by the U-2, that the limited scope of the Soviet ICBM effort became apparent. In 1961, Western press reports still estimated the number of Soviet ICBM force to be between 35 and 75 missiles. But even these estimates exaggerated its size. Later, in 1964, the Department of Defense stated that by 1961 the Soviet Union had deployed only 'a handful' of ICBMs.[19]

To use a standard Soviet phrase, it was in all likelihood 'not accidental' that Khrushchev's false claims about the beginning of serial production of ICBMs coincided with his announcement that the Soviet Government was determined to end the Allied occupation of Berlin. Speaking at a meeting at the Polish embassy in Moscow, on 10 November, 1958, he charged that the Western powers had repeatedly violated the Potsdam agreement. The Soviet Union, he continued, thus had every reason to set itself free from its obligations. Two weeks later, in separate notes to the United States, Britain and France, the Soviet Government demanded that the occupation of Berlin be terminated and that West Berlin be converted into a demilitarised free city. Six months, the notes read, should be sufficient to achieve this conversion. And if the West were not to accept the Soviet proposals within that time, then the Soviet Union would conclude a separate agreement with the East German Republic to end the occupation regime.[20] The threat was withdrawn following the Eisenhower–Khrushchev Summit in September 1959, although the issue of Berlin itself remained unresolved.

It is easy to get confused about the welter of events in the 1959–60 Berlin crisis. But closer examination reveals an important pattern: a reversal in the character and fortunes of the Soviet political offensive in Central Europe. Khrushchev based his November 1958 ultimatum on deliberate deception over the true state of the Soviet Union's military power, a number of actual international conditions which objectively favoured the Soviet Union and developments, notably in the sphere of socio-economic competition, which were expected to be advantageous to World Socialism. By 1961, however, the situation had changed significantly. The strategy of deception was being undermined; international conditions were no longer so favourable; and expectations turned out to be unwarranted.[21]

* * *

The demise of deception began to take shape in January 1961, when President Eisenhower in his farewell address asserted that 'the "missile gap" shows every sign of being a fiction'.[22] At that time, the Atlas ICBM was operational. The first American nuclear submarine armed with Polaris missiles had put to sea. Long range bombers of the Strategic Air Command had begun to be armed with air-to-surface cruise missiles. Fiscal years 1961 (beginning 1 July, 1960) and 1962 showed dramatic increases in funds allocated to offensive strategic forces.

International conditions from the Soviet point of view had worsened, too. Relations between China and the Soviet Union had slowly but inexorably been moving towards an open break. Given his foreign policy ambitions, Khrushchev could not, and did not, remain indifferent to the shift in the balance of forces and the changes in Western perception. As a result, the Soviet Government adopted a number of measures and postures so as to contain the negative effects of the changes.

First, as the strategic posture of the Soviet Union *vis à vis* the United States was no longer as credible as before, the Soviet Union now attempted to enhance as much as possible the sense of vulnerability of the United States' European allies. Khrushchev issued warnings that, in the event of war, NATO military bases in Italy would be destroyed ('even if they are in orange groves'), and in Greece as well ('reportedly located among olive groves');[23] that six H-bombs would be quite enough to annihilate the British Isles, and that nine would take care of France;[24] that Germany would be 'reduced to dust',[25] and that the very existence of the population of West Germany would be placed in question.[26] Khrushchev apparently thought this deterrent posture to have been successful. In his memoirs he wrote:

> 'If a third world war is unleashed,' Adenauer often said, 'West Germany will be the first country to perish.' I was pleased to hear this, and Adenauer was absolutely right in what he said . For him to be making public statements like that was *a great achievement on our part. Not only were we keeping our number one enemy in line, but Adenauer was helping us to keep our other enemies in line, too.*[27]

Second, Soviet political and military leaders were toning down claims of military and military-technological superiority over the United States. Thus, on 21 June, Khrushchev spoke of *parity* in military power, not Soviet superiority, when he said that 'Even the representatives of the imperialist powers themselves now say that an equilibrium of forces has formed in the world between the Western states and the socialist countries.'[28] These assertions were a far cry from the bold superiority claims made earlier.

Third, rather than speaking of quantitative superiority in ICBMs, Khruschchev began stressing qualitative improvements in missile technology (which were not as easily subject to verification by satellite reconnaissance as the numbers of missiles) and increases in the destructiveness of Soviet nuclear weapons. Thus, on 27 July, Khrushchev told United States arms control negotiator John McCloy in Sochi that if war broke out, it would be decided by the *biggest rockets* and that the Soviet Union had them. He failed, however, to mention any numerical advantages. On the same occasion, he boasted that the Soviet Union had the ability to build and deliver by missile on United States territory a *100 megaton nuclear bomb.*[29] He repeated these and similar boasts on other occasions[30]

A fourth response was enhancement of the Soviet Union's conventional posture. In early 1960, Khrushchev had committed himself to what appeared at the time to be an entirely new doctrine of minimum nuclear deterrence.[31] He had stressed the importance of missiles and of the newly created branch of Soviet Armed Forces, the Strategic Rocket Troops. Conventional weapons such as aircraft and surface ships were defamed as 'outmoded' and 'obsolete'. In a third stage of reduction, announced in January 1960, and apparently quite in line with the new doctrine, the Soviet Armed Forces were to be cut by 1.2 million men.[32]

However, only part of this programme was implemented. In the summer of 1961, the troop reduction programme was halted, and force levels—due to the Defence Ministry retaining in service officers and men scheduled to be released from active duty—actually *increased.* The production of tanks was accelerated. Stress was put on the advantages that the Soviet Union enjoyed on the ground, in Central Europe. Thus, Khrushchev told McCloy that 'if the Western powers wanted to shoot their way through to West Berlin, they should remember that the Soviet Union had superiority in arms and divisions and was nearer to the field of battle.'[33]

A fifth and, in the present context, most important response was to look for ways of redressing the strategic balance. Given the tremendous expense and the long lead times involved in the acquisition of new strategic weapons, and given the fact that the United States strategic programmes were well under way and lavishly funded, the temptation must have been great to embark on a 'short cut' to the improvement of the Soviet strategic posture. Apparently, Khrushchev was confident that he had found such a shortcut.

* * *

It is not entirely clear when exactly the thought ripened in his mind to deploy medium and intermediate range missiles in Cuba, as well as IL-28 bombers capable of carrying nuclear weapons. According to the British specialist on Soviet military affairs, John Erickson, the first logistical preparations for the Caribbean venture were made as early as the Autumn of 1961. At that time, he asserts, there were large shipments of military equipment to Black Sea ports and these were unrelated to the current exercises of the Soviet Armed Forces in that area.[34] Planning was most likely in a more advanced stage by the Spring of 1962. It may very well be that because of his expressed view that the risks of the planned operation were too high, the head of the Strategic Rocket Troops and Deputy Defence Minister, Marshal K S Moskalenko—the military leader most directly involved in the operation—was relieved of both appointments in April 1962. Marshal Biryuzov was appointed in his place. (After the missile crisis, Moskalenko regained the post of Deputy Minister of defence.)[35]

Khrushchev himself recalled in his memoirs that the idea of sending missiles to Cuba occurred to him when he was on a state visit to Bulgaria from 14–19 May, 1962. In the wake of the Bay of Pigs invasion, he was concerned that 'We might very well lose Cuba if we didn't take some decisive steps in her defence.' After 'brooding over what to do', he conferred with others in the Soviet leadership upon returning to Moscow and after discussions, as he wrote, 'we decided to install intermediate-range missiles, launching equipment, and IL-28 bombers'.[36] Khrushchev's concern for Cuba notwithstanding, most Western analysts believe that he had wider Soviet strategic interests in mind when he contemplated installation of the missiles. These will be discussed shortly.

During the Spring of 1962, the Soviet Union had supplied Cuba with large numbers of weapons, including heavy artillery, tanks, aircraft and helicopters. But in late July and throughout August, the pace of the deliveries increased substantially. In addition to weapons, large numbers of military technicians began to arrive. In August, a high altitude American reconnaisance mission positively identified two surface-to-air (SA-2) missile sites on the island; the construction of six others was held to be possible. In September, additional Soviet–Cuban agreements on the supply of weapons were announced.[37]

It was in these conditions that concern in the United States heightened over the Soviet military build-up. President Kennedy, in statements on 4 and 13 September, declared that so far, no evidence of the presence of offensive weapons had been discerned. However, he continued, 'Were it to be otherwise, the gravest issues would arise.' If the Caribbean island were to 'become an offensive military base of significant capacity for the Soviet Union', he warned, 'then this country will do whatever must be done to protect its own security and that of its allies'.[38] The stage was thus set for a confrontation with the Soviet Union if Khrushchev were to proceed with his MRBM and IRBM deployment plans.

There was still time for him to reconsider. But this he failed to do. Instead, he went head with the venture, apparently assuming that the possible gains were worth the risks and that the chances of success were high. What was it that inspired such confidence? What was he trying to achieve? And how was he going to make sure that the operation would be a success?

The main reason for his confidence, the methods employed and the objectives he pursued, were in all likelihood intimately connected with the lessons of the Berlin crisis. In contrast to many

Western interpretations, it is very doubtful whether Khrushchev regarded the outcome of the crisis as a failure. The construction of the Wall, he may have reasoned, had come as an almost complete surprise to the West. There were no serious complications: the East German population did not rise in revolt and the Western Allies failed to challenge the unilateral Soviet and East German measures by force. The West had, for all practical purposes, 'swallowed the bitter pill' in August 1961.[39] Given the right conditions, the West would swallow another. Certainly, more pills of different potency could still be prescribed: the conclusion of a separate peace treaty, the curtailment of the Western Allied military presence in West Berlin, the limitation of political, economic, social and cultural links between West Germany and West Berlin, restrictions on the access roads, and changes in Western Allied and West Berlin broadcasting content to allow for a more friendly stance towards the Soviet Union and the GDR. All this, to conclude Khrushchev's probable reasoning, would be facilitated by significant changes in the East–West correlation of forces in favour of the Soviet Union or, failing that, changes in *perception* to that effect.

If Khrushchev did indeed consider the construction of the Wall to have been at least a partial success, and that further advances could be made, it would serve to explain the origins of his Cuban venture not only as to the substance (that is, the attempt to improve the Soviet strategic posture, both militarily and politically) but also as to the method and techniques employed, that is, the attempt to take the United States by surprise and the adoption of a strong deterrent posture in the hope of creating a successful *fait accompli*. He apparently realised quite well that it is easier to stop an opponent from doing what he is doing than to compel him to undo what has already been done. To quote from his memoirs:

> My thinking went like this: If we installed the missiles secretly and then if the United States discovered the missiles were there after they were already poised and ready to strike, the Americans would think twice before trying to liquidate our installation by military means.[40]

To be sure, completion of the planned six MRBM sites (each with four launch positions, which in turn were each equipped with two missiles) and three IRBM sites (each consisting of four launch positions with three missiles) as well as the stationing of 42 IL-28 bombers would most likely not have altered the strategic balance between East and West. Specifically, the addition of missiles in Cuba—although they would have outflanked the United States radar system—would not significantly have increased the probability of a Soviet first strike, reduced the impact of an American first strike or added much to the Soviet retaliatory capacity. However, the Soviet Union would have gained at least some strategic advantage. An 'element of flexibility' would have been added to its strategic posture.[41] And over a period of time, the Soviet strategic capabilities on the island could have been expanded and hardened. Above all, the psychological shock effects produced by American inaction among allies in Europe would most likely have been severe. Perceptions of United States weakness and unreliability and images of a Soviet Union in ascendancy, would almost inevitably have spread. As a result, the conditions for a renewal of the political offensive in Central Europe would have been, from the Soviet perspective, 'more mature'.[42]

In 1962, however, there were two crucial differences as compared to the preceding period of Soviet deception. The manipulation of the missile gap had involved the mere semblance of reality. As already mentioned, it was a tangible effort at redressing at least part of the strategic balance. Second, during the Berlin crisis in 1961 the Soviet Union had considerable room for manoeuvre thanks to conventional military superiority at the local level. In Cuba, however, the state of affairs was reversed. In this crisis it was the United States which possessed a preponderance of air and naval power and capacity to transport significant numbers of ground troops.

Hence, one of the foremost requirements of the operation was for Khrushchev to try to dispel any suspicion in the United States about the true nature of Soviet intentions. In line with this requirement, and in all likelihood fully aware of American concern over the accelerated flow of Soviet weapons and technicians to Cuba, as well as the publicly stated sensitivity of the Kennedy Administration over the possible presence of offensive weapons (such weapons certainly including long range bombers and medium and intermediate range missiles capable of carrying nuclear weapons), Soviet officials embarked upon an elaborate campaign of cover-up and deception. Repeatedly, they asserted that the Soviet involvement in Cuba had defensive purposes only.

Thus, early in September, Soviet Ambassador Dobrynin conveyed, on separate occasions, Khrushchev's assurances that the equipment being shipped to Cuba was 'defensive in nature and did not represent any threat to the security of the United States' to Robert Kennedy, one of the main protagonists in the management of the crisis, and Sorensen, one of President Kennedy's closest advisors.[43] Then, in an official statement published on 11 September, the Soviet Government stated that in light of the power of Soviet nuclear rockets there was 'no need for the Soviet Union to shift its weapons for the repulsion of aggression, for a retaliatory blow, to any other country, for instance Cuba'.[44]

In early October, Khrushchev and Mikoyan told Georgi Bolshakov, a public information official at the Soviet Embassy in Washington, that the Soviet weapons being sent to Cuba were 'intended' only for defensive purposes and were only anti-aircraft missiles that could not reach American targets. On his return, Bolshakov diligently passed the word around Washington, and apparently to the President and the Attorney General via a personal message from Khrushchev.[45]

On 13 October, Dobrynin assured Chester Bowles, a high-ranking official in the Administration, 'convincingly and repeatedly', according to Hilsman, that there were no 'offensive weapons' in Cuba. Bowles had pressed the Ambassador 'very hard' on this matter.[46] Three days later, on 16 October, Khrushchev gave the same assurances to Ambassador Foy Kohler in Moscow in the wake of a 'storm of public suspicion', as Hilsman put it, over the agreement to build a fishing port in Cuba. Khrushchev explained that Soviet 'purposes' there were wholly defensive. And finally, as late as 18 October, two days after the President had concrete evidence of the missiles' presence, Soviet Foreign Minister Gromyko personally assured Kennedy in a White House meeting that Soviet aid to Cuba was being 'pursued solely for the purpose of contributing to the defensive capabilities of Cuba', that 'training by Soviet specialists of Cuban nationals in handling of defensive armaments was by no means offensive', and that 'if it were otherwise, the Soviet Government would never become involved in such assistance'.[47]

The evolution and resolution of the crisis can quickly be summarised. On 22 October, President Kennedy addressed the nation on television. In order 'to halt this offensive build-up' he ordered a 'strict quarantine on all offensive military equipment under shipment to Cuba'.[48] On the diplomatic level, Khrushchev adopted delaying tactics for several days while, on the ground, work on the sites continued at full speed. Khrushchev apparently still hoped to create a *fait accompli*. By 24 October, two Soviet surface ships and one submarine were within a few miles of the quarantine line. Other surface ships were following closely behind. President Kennedy ordered depth charges to be used to force the submarine to the surface. A showdown at sea, however, was averted when the ships closest to the quarantine line stopped or turned around. The missiles, however, remained in Cuba, and it was still quite unclear whether Khrushchev would comply with the American demand for their withdrawal. A new peak of the confrontation was reached, therefore, when on 27 October a U-2 reconnaissance plane was shot down over Cuba.

This action demonstrated that the network of anti-aircraft missiles on Cuba had become operational. The seriousness of the crisis was underlined further by Robert Kennedy who told Dobrynin, later on the same day, that the United States 'had to have a commitment by tomorrow that those bases would be removed'; that this was not 'an ultimatum' but a 'statement of fact'; and that if the Soviet Union would 'not remove those bases, we would remove them'.[49] On 28 October, the crisis was suddenly resolved. In his reply to a (second) letter from President Kennedy (broadcast over Radio Moscow on that very day), Khrushchev gave assurances that construction of the missile sites would stop, and that the weapons 'which you described as offensive' would be crated and returned to the Soviet Union.[50]

* * *

What remains to be examined are the reasons why Khrushchev agreed to withdraw the missiles and how deception influenced the behaviour of the two super-powers in the crisis.

The most widely held explanation of Khrushchev's retreat, and one that has been expressed most strongly by Western defence analysts, is that it was United States strategic superiority, or the combination of strategic superiority and a preponderance in conventional power in the Caribbean, which left the Soviet leader with no choice but to shrink from a 'competition in risk taking'. The assumptions underlying this explanation are straightforward. They consist of the idea that decision-makers are rational actors; that superior military power creates political and psychological leverage over the adversary; and that military inferiority weakens the determination of one actor locked into a crisis to the same extent as military preponderance strengthens the resolve of the other.

Determination and resolve, however, are only in part a function of military power. After all, there are numerous cases in history which show that superior power does not always erode the will to resist. Thus, for instance, Allied bombing of German cities in the Second World War and United States bombing of North Vietnamese towns did much to stiffen the enemy's resistance and bridge the gaps between the leadership and the people. As these examples show, the will to resist and to take risks often depend on *irrational* factors, such as ideological and religious conviction, nationalist fervour, and considerations of honour, pride and prestige. Moreover, they depend on what one could call the *balance of interests*. They are related to the *stakes* involved for the two sides in winning, losing, or merely achieving a draw. And they are closely connected with the perceived *legitimacy* of the actions and counteractions taken by the two sides. It is on this last issue that the problem of deception during the Cuban missile crisis is of considerable importance.

This is the case because by explicitly and repeatedly denying that the Soviet Union was in any way in need of, planning for, or in the process of stationing offensive weapons in Cuba, the Soviet leaders had manoeuvred themselves into a difficult position. In essence, by their explicit denials, they were implicitly acknowledging the violation of certain tacitly understood rules of the game in the relations between the two super-powers and conceding *a priori* the legitimacy of American counter-measures. The very secret, hasty and—inadequately—camouflaged nature of the Soviet venture underlined this very fact. (It is instructive to consider, in contrast, the position adopted by Truman and Kennedy in the Berlin crises of 1948 and 1961, i.e. that the Western powers had every right to be in Berlin and that they would stay there, period.) Only after the deception in Cuba was uncovered and justification of the Soviet action became sorely needed, did Soviet officials assert that in principle there was no difference between United States missiles in Italy and Turkey, and Soviet missiles in the Caribbean. This analogy, however, came too late to convince the Kennedy Administration or American public opinion of the legitimacy of the Soviet operation.[51]

Equally unconvincing is the thesis put forward *ex post facto* by Soviet writers that it was quite unclear what was meant by offensive weapons, and that if American officials had put a clear question to Soviet diplomats about the presence of missiles in Cuba, they would have received a clear answer (i.e. presumably that the missiles were there but that they were entirely defensive in purpose). This is the line of reasoning adopted, for instance, by the son of the former Soviet Foreign Minister, Anatolii Gromyko, to explain why his father, in the meeting with the President on 18 October, had said nothing about the missiles.[52] This portrayal of Soviet motives, however, leaves out an important fact: at the end of the meeting, the President read out to the Soviet Foreign Minister the text of his declaration of 4 September. This declaration had specifically referred to 'ground-to-ground missiles' as weapons systems that would be considered part of an 'offensive capability'. Furthermore, the claim made by the Soviet Government only one week later to the effect, as noted, that the Soviet Union possessed missiles so powerful that there was no need for Moscow to seek sites elsewhere for launching nuclear weapons, 'for instance [in] Cuba', showed that the Soviet Government was acutely aware of American sensitivities. Soviet actions and verbal communications thus did not reflect ignorance or lack of concern but formed part of an elaborate plot of deception.

The exposure of the deception was undoubtedly an embarrassment for the Soviet leadership, and the retreat under American pressure a humiliation. The outcome of Khrushchev's risky initiative validated previous patterns of Soviet foreign policy behaviour. It confirmed standard Leninist warnings against any kind of adventurism, i.e. 'risky actions, foolhardy and dangerous endeavours which are embarked upon without regard for existing forces and the possibilities for the implementation of objectives, and which, therefore, are ultimately doomed to failure'.[53]

The failure which Khrushchev experienced in Cuba was in all likelihood one of the reasons for his downfall. His successors, not surprisingly, were to draw the conclusion that a more business-like approach in domestic and foreign affairs was called for. They were to act on a principle which Brezhnev explained, shortly before his death, to high-ranking officers of the Soviet Army and Navy as follows: in the conduct of policy, 'words are not enough. Policy is effective [only] when it is based on real economic and military power . . .'[54]

In the 1950s and early 1960s, the economic and military power available to the Soviet Union was clearly insufficient to achieve the far-reaching objectives which Khrushchev had outlined. Bluff and deception were employed to make up for this deficiency. This approach worked up to a point. But as Khrushchev's colleagues and successors realised, it was clearly not a sound basis for long-term policies.

Endnotes

1. As reported by Robin Edmonds, *Soviet Foreign Policy, 1962–1973: The Paradox of Superpower* (London: Oxford University Press, 1975), p. 22. The author is a former member of the British Foreign Office.
2. V I Lenin, *Sochineniya*, Vol. 15, in Russian (4th Edn., Moscow: Gospolitizdat, 1935), p. 67.
3. Khrushchev's claim to pre-eminence in the Soviet Political leadership has been described by, among others, Edward Crankshaw, *Khrushchev: A Career* (New York: Viking Press, 1967); Carl A Linden, *Khrushchev and the Soviet Leadership* (Baltimore, MD: Johns Hopkins UP, 1966); Boris Nicolaevsky, *Power and the Soviet Elite*, ed., Janet D Zagoria (London: Pall Mall Press, 1965); R W Pethybridge, *A History of Postwar Russia* (London: Allen & Unwin, 1966); and Michael Tatu, *Power in the Kremlin: From Khrushchev to Kosygin* (New York: Viking Press, 1971).
4. N S Khrushchev, Speech delivered on 6 January 1961, at a joint meeting of the party organisation of the Higher Party School, the Academy of Social Sciences and the CPSU Central Committee's Institute on Marxism-Leninism, *Pravda*, 25 January 1961.
5. *Pravda*, 2 November 1961 (emphasis in the original). The Draft Programme of the Communist Party of the Soviet Union was adopted in the June 1961 plenary meeting

of the CPSU Central Committee (*Ibid.*, 30 July 1961) and the final version at the 22nd Party Congress in October of the same year.

6. Marshal A M Vasilevskii in *Izvestia*, 8 May 1955.
7. This account of the Tushino deception draws on Lincoln P Bloomfield, Walter C Clemens Jr and Franklyn Griffiths, *Khrushchev and the Arms Race: Soviet Interests in Arms Control and Disarmament 1954–1964* (Cambridge, Mass: MIT Press, 1966), p. 39; see also, John Prados, *The Soviet Estimate: US Intelligence Analysis and Soviet Military Strength* (New York: Dial Press, 1962), pp. 41–5.
8. Prados, pp. 43–50. From 1957 the US National Intelligence Estimate revised downwards the projected size of the Soviet bomber force. On the role of the U-2 in this revision process see this volume, Chapter 14, note 7.
9. *Pravda*, 20 February 1956 (emphasis added).
10. Quoted *New York Times*, 24 April 1956.
11. At a reception celebrating the twelfth anniversary of Romania's liberation from Nazi-German occupation *New York Times*, 23 August 1967. Khrushchev's remarks were not reported in the Soviet Press.
12. Five communications concerning the Suez crisis went out from Moscow on 5 November 1956: letters by Bulganin (the Soviet Premier) to Prime Ministers Eden, Mollet and Ben Gurion and to President Eisenhower, and a message by the Soviet Foreign Minister, Shepilov, to the United Nations Security Council. The texts are published in *New York Times*, 5 November 1956. For the texts in Russian see: Dmitri T Shepilov, *Suetskii vopros* (Moscow, Politizdat, 1956) and *Pravda*, 6 November 1956.
13. Institut Mirovaia Ekonomika i Mezhdunarodnye Otnosheniia, ed., *Suetskii vopros i imperialisticheskaia aggressiia protiv Egipta* (Moscow: Izdatel'stvo 'Mezhdunarodnye Otnosheniia,' 1957), pp. 101–2.
14. The reconstruction of Khrushchev's claims concerning ICBM development and production draws on Arnold L Horelick and Myron Rush, *Strategic Power and Soviet Foreign Policy* (Chicago: University of Chicago Press, 1965).
15. In an interview with William R Hearst Jr, *Pravda*, 29 November 1957.
16. Contained in the discussion of the control figures for the development of the national economy for 1959 to 1965, *Pravda*, 14 November 1958.
17. At the 21st Party Congress, *Pravda*, 5 February 1959.
18. *New York Times*, 12 January 1959, as quoted by Horelick and Rush, p. 51; see also Prados, pp. 59–66, 76–8, 80–86. Evidently the Soviets enjoyed 'feedback' about this deception from their agent, Colonel William H Whalen, who served both United States and Soviet intelligence and provided the latter with United States National Intelligence Estimates between 1959 and 1961.
19. 'Department of Defense Statement on US Military Strength', 14 April 1964; *New York Times*, 15 April 1964. See also the discussion of the rise and fall of the missile gap myth by Joseph Alsop in *Foreign Policy* (Fall, 1970), pp. 84–7, and Prados, pp. 86–95, 112–19.
20. The full text of the notes and Khrushchev's comments on them were published in *Pravda*, 28 November 1958.
21. For an extensive discussion see the case study on the Berlin crisis of 1961 in Hannes Adomeit, *Soviet Risk-Taking and Crisis Behavior* (London: Allen and Unwin, 1982).
22. *New York Times*, 12 January 1961. For discussion of the dissipation of the missile gap myth in the United States see Horelick and Rush, pp. 80–86; on the Soviet responses see Adomeit, pp. 251–3.
23. *Pravda*, 12 August 1961.
24. Khrushchev in talks with the British Ambassador Sir Frank Roberts, as reported in *Washington Post*, 12 July 1961.
25. *Pravda*, 8 August 1961.
26. *Pravda*, 12 August 1961.
27. *Khrushchev Remembers: The Last Testament*, transl. and ed., Strobe Talbott, with a foreword by Edward Crankshaw and an introduction by Jerrold L Schecter (London: Deutsch, 1974), p. 53 (emphasis added).
28. In a speech in the Kremlin at the 20th anniversary of Hitler's attack on the Soviet Union, *Pravda*, 22 June and *Izvestiya*, 23 June 1961. Similarly, in a speech delivered a

week later, Khrushchev stated that 'Western statesmen maintained that the military power of the capitalist and socialist camps is now balanced'. *Pravda*, 29 June 1961.

29. As reported by the *New York Times*, 1 and 11 August 1961.
30. On 9 August, for instance, at a reception for cosmonaut Titov, *New York Times*, 11 August 1961, Khrushchev again appeared to stress qualitative advantages when he said on the same occasion, 'The Americans do not launch any sputniks. They hop and fall down into the ocean.'
31. In his authoritative speech on defence policy, *Pravda*, 15 January 1960.
32. According to the Law on a New Large Reduction of the Soviet Armed Forces, *Pravda*, 16 January 1960; *Krasnaya zvezda*, 20 January 1960.
33. *New York Times*, 1 August 1961.
34. In an interview with the author in Edinburgh on 24 June 1976 (personal notes).
35. Michael Tatu, *Power in the Kremlin* (London: Collins, 1969), pp. 236–7.
36. *Khrushchev Remembers*, p. 495.
37. Raymond L Garthoff, *Reflections on the Cuban Missile Crisis* (Washington, DC: Brookings, 1987), p. 8, n. 9, notes that Soviet–Cuban relations were strained during the first months of 1962, and that, as a result, military shipments were reduced at that time. Relations improved in June and military aid increased considerably thereafter.
38. *New York Times*, 5 September 1962 (first quote); 14 September 1962 (first quote); 14 September, 1962 (second). For discussions of the Cuban Missile Crisis see, *inter alia*, Garthoff, *op. cit*; Robert A Pollard, 'The Cuban Missile Crisis: Legacies and Lessons,' *Wilson Quarterly* (Autumn, 1982), pp. 148–58; Richard Ned Lebow, 'The Cuban Missile Crisis: Reading the Lessons Correctly,' *Political Science Quarterly* (Autumn, 1983), pp. 431–58; James G Blight *et al*, 'The Cuban Missile Crisis Revisited', *Foreign Affairs*, Vol. 66, No. 1 (Autumn, 1987), pp. 170–88; David Welch and James Blight, 'The Eleventh Hour of the Cuban Missile Crisis: an Introduction to the ExComm Transcripts', *International Security*, Vol. 12, No. 3 (Winter, 1987/88), pp. 5–29 (the edited transcripts are on pp. 34–92). The other standard works on the crisis include: Robert F Kennedy, *Thirteen Days: A Memoir of the Cuban Missile Crisis* (New York: W W Norton, 1969); Malcolm Mackintosh, 'Clues to Soviet Policy', *US News and World Report*, 2 November 1970; Herbert Dinerstein, *The Making of a Missile Crisis: October 1962* (Baltimore, Maryland: Johns Hopkins, 1976); Elie Abel, *The Missile Crisis* (Philadelphia: Lippincott, 1966); Graham T Allison, *Essence of Decision: Explaining the Cuban Missile Crisis* (Boston: Little, Brown, 1971); Harold Horelick and Myron Rush, *Strategic Power and Soviet Foreign Policy* (Chicago: University of Chicago Press, 1966).
39. This is a phrase used by Khrushchev in *Khrushchev Remembers*, p. 509. The same metaphor occurs frequently in Oleg Penkovsky, *The Penkovsky Papers*, Transl. by P Deriabin, with an Introduction by Frank Gibney and a Foreword by Edward Crankshaw (London: Collins, 1965). Khrushchev, for instance, is reported by the author as having been content that the Western countries had 'swallowed their first pill on 13 August 1961, when Berlin was closed off' and that he hoped that they would swallow a 'second pill': the conclusion of a separate peace treaty with the GDR (*Ibid.*, p. 161). Colonel Penkovsky was an officer of the chief intelligence directorate (GRU) of the Soviet General Staff who provided the West with secret information until he was arrested in October 1962 and shot. The authenticity of the material is no longer much in doubt. On this point see Adomeit, p. 280, fn. 178.
40. *Khrushchev Remembers*, p. 494.
41. This, according to recently declassified information was the view held by most members of the ExComm (Executive Committee of the National Security Council). See the analysis by Barton J Bernstein on Kennedy and the Cuban missile crisis, *International Herald Tribune*, 3 November 1975. See also, Garthoff, pp. 9–10, and Appendix D, pp. 138–46.
42. This is a phrase which Khrushchev had used earlier. Speaking in East Berlin, on 20 May 1960, on his return journey to Moscow, he stated: 'We are realists and we will never pursue a gambling policy. Under present conditions, it is worthwhile to wait a little longer and try to find a solution for the long-since ripe question of signing a peace treaty with the two German states. This will not escape our hands. We had better wait, and the matter will get more mature.' *Pravda*, 21 May 1961.
43. This enumeration draws on the account of one of the former officials in the Kennedy

administration, Roger Hilsman, *To Move a Nation: The Politics of Foreign Policy in the Administration of John F Kennedy* (Carden City, NJ: Doubleday, 1967), pp. 165–6, and on US House of Representatives, Committee on Foreign Affairs, *Soviet Diplomacy and Negotiating Behavior: Emerging New Context for US Diplomacy*, prepared by Joseph G Whelan, Congressional Research Service, Library of Congress (Washington, DC: US Government Printing Office, 1979), pp. 334–5.

44. TASS, as authorised by the Soviet Government, *Izvestiya*, 12 September 1962.
45. Hilsman, pp. 165–6 and Kennedy, p. 27.
46. Hilsman, p. 166.
47. *Ibid.*
48. *New York Times*, 23 October 1962.
49. Kennedy, pp. 105–6. Welch and Blight, pp. 19–20, note that while in restrospect many former ExComm members regarded the U-2 incident as a turning point of the crisis—and perhaps the moment when war seemed most imminent, this is not supported by the discussions recorded in the ExComm transcripts (pp. 66–72), wherein it appears to be only a passing distraction as discussion refocuses on means to resolve the crisis as a whole.
50. Hilsman, p. 224.
51. On this, see Garthoff, pp. 9, 11–12.
52. Anatolii Gromyko, 'Karibskii krizis,' *Voprosy istorii*. No. 7 (1971), pp. 135–44 and No. 8 (1971), pp. 121–9.
53. I V Lezhen (ed.), *Kratkiipoliticheskii slovar'* (Moscow, 1964), p. 5; see also *Bol'lshaia Sovetskaia entsiklopediia*, the entry for "*avantyura*."
54. *Pravda*, 28 October 1982.

CHAPTER 5

Deception and Surprise in Soviet Interventions

JIRI VALENTA[1]

. . . to stun the enemy by courage and daring, to catch him unawares when he is not prepared to repulse blows, to paralyse his will to resist, to deprive him of the possibility to take effective counter-measures.

Soviet Military Review[2]

Since the Second World War the Soviet Union has resorted to military intervention, direct or indirect, to resolve a number of crises within neighbouring or allied socialist countries. This chapter addresses the evolution of Soviet views and practices in regard to deception and strategic surprise in such intervention operations, which occur halfway between peace and war. The invasions of Hungary in 1956, Czechoslovakia in 1968, Afghanistan in 1979 and the crisis in Poland in 1980–81 provide examples from which some deductions may be drawn.

Generally speaking, military deception and surprise are aimed at misleading the potential victim over the instigator's military intentions, particularly when planning a strategic surprise attack, which is what an intervention operation amounts to. Political deception and surprise are aimed at influencing an antagonist's behaviour in a manner contrary to his interests. In the context of an intervention, deception and surprise at the political level might disarm, mislead or frighten governing elites, or they might create internal pressures, divisions or destabilisation within populations. Political action of these sorts is also known by the Soviets as *aktivnyye meropriatia*, 'active measures'.[3]

In this study of Soviet deception in interventions, there is a distinction to be made between deception to achieve surprise and deception to justify the act. If used to achieve surprise, as is usually but not always the case, deception may be aimed at the target country or at the non-Communist world. At the diplomatic level, it may take the form of public statements, 'official' government declarations, media reports, or negotiations intended to put the target off guard. At the military level, it could be overt military manoeuvres designed to de-sensitise or condition the intended victim, making him unsure of the specific intent or time of the deployment. Tactically, this type of deception could support manoeuvres or actions intended to disarm the victim's armed forces, such as seizing communications centres, airports, and the like.

Deception to justify the act is aimed at both the Communist and non-Communist world—the former to ensure that Party legitimacy and correctness of Politburo decision-making cannot be eroded, the latter to minimise damage to the reputation of the Soviet Union and the world Communist movement. This type of deception is usually implemented before the invasion begins by attempting to show the target as fascist or dupes of imperialism, aided by the CIA or other Western agencies. All four cases studied here reveal a predominant Soviet use of deception to achieve surprise; the Czech, Afghan and Polish cases also show some use of deception to obtain justification for Soviet intervention.

In examining Soviet deception before and during the interventions in Hungary, Czechoslovakia and Afghanistan, and during the Polish crisis of 1980–81, a lack of hard evidence has handicapped the author. This, of course, is a hazard of all deception research, but nevertheless this limiting factor deserves acknowledgement. What is particularly hard to establish is the motivations for various falsehoods: were they provocations, or spontaneous lies unrelated to wider aims; or were they parts of sophisticated deception operations? Until such questions can be answered, conclusions must necessarily be tentative. Before turning to these interventions, and the stratagems which supported them, it is appropriate to explore the place of deception and surprise in Soviet theory.

* * *

The Russian Tsars and their Bolshevik successors recognised surprise and deception as important weapons in political and military struggles. For a long time, deception and strategic surprise have been an integral part of Russian and Soviet military strategy. However, before the Second World War, Russian and Soviet leaders lacked confidence in the offensive capacity of their armed forces and this tended to minimise the usefulness of strategic deception and surprise. Perhaps the only exceptions were the Russian Marshals Suvorov and Kutuszov.

The history of the use of deception to achieve strategic surprise suggests that it is usually employed by the side that is on the offensive. This point was illustrated by Germany's infrequent use of deception during the latter part of the Second World War (1942–5), with the possible exception of the Battle of the Bulge, when, of course, the Nazis temporarily resumed the offensive. This too has been the case with Russia throughout most of its history. Many of Russia's major wars have been primarily of a defensive nature against other European powers. The best examples are Napoleon's invasion of Russia in 1812, and the German–Russian confrontations of 1914–18 and 1941–5. (The Russian experience in Central Asia was quite different.) By taking advantage of its vast lands and austere weather conditions, the Russian strategy was to immobilise the invaders slowly, and drive them out of the country, thereby transforming a defensive operation into an offensive. Traditionally, Russian defensive military strategy did not favour the use of strategic surprise and deception in large scale military operations.

The use of deception to justify intervention began in the first intervention operation launched under the Red Banner of Socialism. In 1921, before the Red Army's invasion of Outer Mongolia, the Soviets established, on Soviet soil, a shadow Mongolian government which requested military assistance. This was immediately supplied and the shadow regime was escorted into the annexed territory there to announce itself as the Mongolian People's Government. This technique of a puppet regime calling for 'fraternal aid' was repeated in the same year in what became the People's Republic of Tannu Tuva.[4] But when the Soviets tried the same technique in Finland in 1940, the scheme failed. The hastily concocted 'fraternal' government's call for help deceived no one because the Finns defended their country courageously and the Soviets failed to achieve a quick, successful military invasion.

One important reason for the tardy adoption of strategic deception and surprise by the Soviet Union was the influence of Josef Stalin on Soviet military thinking. Stalin's emphasis on the importance of 'permanently operating factors' in war (such as morale and industrial potential) inhibited proper considerations of strategic deception and surprise in conventional conflicts, as it would do later in the area of nuclear doctrine. Surprise was viewed as having only 'temporary' importance.[5] Stalin reserved to himself the role of chief military and political strategist of the Soviet Union. His complete control of the Soviet military and political systems, and his intimate involvement in the formulation of Soviet military theory, often inhibited debates among military theoreticians on the understanding and application of stratagem.

As a result, Stalin tended to downgrade the importance of

deception and strategic surprise prior to and during the Second World War, looking on them as transitory factors which could not significantly affect the ultimate outcome of the war. True, the Soviet Armed Forces occasionally achieved operational surprise during that war, but not at the beginning when the Soviet Forces were surprised by the Germans. It was only after a long defensive stance and the loss of significant territory, that surprise, combined with elements of deception, was employed by the Russians against Germany: first at Stalingrad (1942), then at Kursk (1943), and finally in Belorussia (1944). The attack against the Japanese in the Manchurian campaign of August 1945 is the only example of Stalin pulling off a successful strategic surprise.[6]

When Stalin discussed deception with Winston Churchill, the examples given by the Soviet leader were of a tactical nature, being limited mainly to the concealment of preparations for various operations—camouflage, dummy tanks and aircraft, and radio deception.[7] It took some time for Stalin and his subordinates to accept a minor role in Plan 'Bodyguard', the Anglo-American deception plan designed to support the Allied invasion of Normandy.[8] In this case, however, political considerations and suspicion of allies may have been major contributory factors.

After Stalin's death, and with the advent of nuclear weapons and the subsequent revolution in the conduct of military art, Soviet military thinkers had to reassess the role of strategic deception and surprise in military strategy. Shortly after Stalin's death, a debate was waged in Soviet military periodicals, such as the journal of the Soviet General Staff, *Voyennaya mysl'*, which led to a serious modification of Stalin's postulated doctrines of the 'permanently operating factors' of war. Perhaps the most significant turning point in this debate was the 1955 publication of Marshal P Rotmistrov's article in *Voyenaya mysl'* on the role of surprise in contemporary war. Rotmistrov set a new tone by arguing that the Soviet Armed Forces should not only prevent enemy surprise attacks, but also deal the enemy 'pre-emptive surprise blows'.[9]

From then onward, Soviet writers have upgraded their theories on strategic deception and surprise. A main theme of these discussions is that the Soviet Union should never again be caught unaware, such as in the German attack of 22 June 1941—the Soviet 'Pearl Harbor'. As elaborated by Leonid Brezhnev, 'We are taking into consideration the lessons of the past, and we are doing everything so that nobody takes us by surprise.'[10] Indeed, the Soviet Union has apparently learned from the German *blitzkrieg* strategy, the Japanese attack on Pearl Harbour, and the British practice of stratagem in the Second World War. Contrary to the view that prevailed in Stalin's day, surprise has come to be seen as 'one of the most important principles of military art . . . [consisting] in the choice of times, forms, and methods of combat operations which permit delivery of a strike when the enemy is least prepared to repulse it, thereby paralysing his will to offer an organised resistance'.[11] As pointed out by one Soviet writer, 'a more important condition for achieving victory than overall superiority in weapons and manpower is the ability to use concealment in preparing one's main forces for a major strike, and the element of surprise in launching an attack against important enemy targets'.[12]

The necessity of preventing a surprise enemy attack on the Soviet Union, and at the same time effecting a surprise attack on the enemy are central themes in the Soviet military literature of the 1960s and 1970s.[13] Soviet writers stress that surprise can be achieved at the tactical and operational level, but it is most important, and at the same time most difficult, at the strategic level. The significance of the surprise element has grown as a result of the deployment of nuclear weapons. Soviet military doctrine declares that both kinds of war—nuclear and conventional—would begin with the launching of a strategic surprise attack—an unexpected offensive. Thus it is often difficult to distinguish the kind of war under consideration in Soviet writings. The theme of surprise was particularly stressed in the important book by General S Ivanov, *Nachalnyy period voyny*. Using examples from military history such as Operation Barbarossa in 1941, in which Germany achieved strategic surprise, Ivanov illustrates how enemy offensives made it difficult for the Soviet Union to recapture the strategic initiative.[14]

Soviet writers recognise the difficulties in attaining complete surprise, yet they believe that most warning indicators can be eliminated by comprehensive efforts to maintain secrecy in the use of deception. In this effort, various forms of deception find wide employment.[15] These writers do not distinguish strictly between strategic and tactical deception. The Russian word for the kind of deception known as disinformation is *dezinformatsia*. The Russian word *maskirovka*, however, covers a range of meanings to include military practices such as camouflage and covert manoeuvres, and also political deception including *dezinformatsia* designed to protect the secrecy of military or political operations by propagating misleading information.[16]

Soviet military literature emphasises that the means of deception can vary. These include misleading the enemy as to one's intentions by camouflage, feigning actions, disseminating false information, lowering the enemy's morale through propaganda (leaflets and the media) and sabotage. Other means of achieving strategic surprise include neutralising and actively jamming the enemy's communication system, attacking at night when the enemy is exhausted, maintaining a continuously high state of troop combat readiness and concealing mobilisation.[17] Another recognised form of deception is the use of training exercises and manoeuvres as a cover for deploying military forces.[18]

After the Second World War, the growing Soviet theoretical appreciation of the elements of deception was followed by practical steps to facilitate its implementation by military and political intelligence services. During that war, and for some time thereafter, the Soviet intelligence services had difficulty in recovering from the purges of the 1930s. They were also hampered by Stalin's pervasive control and rigidity; this was particularly true of what is now the Committee on State Security (KGB). Military intelligence (GRU) was the only Soviet intelligence service that survived the great purge and so was able to provide high quality intelligence and practise some deception during the Second World War. This situation has gradually improved.

Since the late 1950s, the analytical function of Soviet political intelligence also has been perfected. The Information Centre, established by the Party's Central Committee in the late 1950s, co-ordinates intelligence gathering and provides more sophisticated assessments than were possible before. The 1959 centralisation of political deception planning in a special Disinformation Department of the KGB enabled the USSR to undertake co-ordinated political deception operations.[19] A KGB manual defines *dezinformatsia* thus:

> Strategic disinformation assists in the execution of state tasks, and is directed at misleading the enemy concerning the basic questions of the state policy, the military-economic status, and the scientific technical achievement of the Soviet Union; the policy of certain imperialist states with respect to each other and to other countries; and the specific counter-intelligence tasks of the organs of State Security. Tactical disinformation makes it possible to carry out the individual tasks of strategic deception and, in fact, comprises the principal deception work of the organs of State Security.[20]

The examples which follow provide an insight into Soviet application of these principles and capabilities. The first deals with Hungary in 1956, and may be seen as a learning experience for the Soviets; in the subsequent cases of direct intervention—Czechoslovakia in 1968 and Afghanistan in 1979—the lessons learned are applied with greater finesse. Finally, in the Polish crisis, where no direct intervention occurred in 1980–81, preparations for such a contingency nevertheless included elements of deception plans, and the 'internal' crackdown succeeded brilliantly in achieving surprise.

* * *

In Hungary, the first symptoms of open rebellion appeared in the Autumn of 1955 when prominent Hungarian Communist writers signed a paper denouncing the regime's cultural policies and forwarded it to the Hungarian Politburo. The First Secretary of the Hungarian Communist Party at that time was Mátyás Rákosi, a staunch and unpopular Stalinist. Two years earlier, in June 1953, the Soviets had forced the Hungarian leadership to appoint Imre Nagy as Premier, and Nagy quickly endeared himself to the Hungarian people by initiating 'new course' political policies. However, these liberal policies soon resulted in Nagy being labelled a right-wing deviant, and led to his dismissal from the premiership—and even from the party.

Rákosi continued as First Secretary until forced out by Khrushchev in July 1956 and replaced by Ernö Gerö—a choice hardly better than Rákosi. Hungarian society was becoming restless, responding to Khrushchev's de-Stalinisation programme, and on 13 October 1956 (just 10 days before the revolution began), the Hungarian Government decided to readmit Nagy to the Party, primarily as a compromise intended to quell some of the growing unrest.

The turmoil in Hungary came at a bad time for the Soviets, catching them off guard. The 1956 Suez Crisis, in which a Soviet client, Egypt, faced the combined intervention of Israel, Britain, and France was under way. Closer to home, Poland had gone through a rebellious period which was just beginning to diminish. These two situations left the Soviets somewhat distracted and ill-prepared for a third crisis. When the Hungarian Revolution flared up, it happened quickly, which meant that the Soviets did not have months, or even weeks, to prepare a response.[21] If the Soviets wished to intervene, military preparation had to be quick and relatively inconspicuous. The deception used in the crisis contributed to these qualities by lulling the Hungarians into a false sense of security, which in turn permitted the quiet occupation of key transporation and command and control nodes throughout the country.

The revolution broke out on 23 October 1956 when students began demonstrating for political reforms and democratisation. Demonstrations eventually turned violent when the police intervened. Ernö Gerö, the First Secretary of the Hungarian Communist Party, requested Soviet military assistance late the same day and the local Soviet commander obliged by bringing troops into

Budapest early the next morning. The presence of these and other Soviet troops throughout Hungary, quickly became a burning issue: at first with the revolutionaries, but soon with the beleaguered Hungarian Government. On 24 October, Gerö was replaced as the Hungarian leader by Imre Nagy, who was much more sympathetic to the revolutionaries' desires. Nagy attempted to deal with the situation, but gradually he began to incorporate the revolutionary demands into his governmental policies and public speeches—a situation that could only anger the Soviets.

On 27 October, when Nagy announced the formation of a new government to include some non-Communist members, the first step was taken that would eventually lead to the full-scale Soviet invasion.[22] Three days later, Nagy broadened his coalition government, a move the Soviets interpreted as the abolition of the one-party system of government.[23] The last straw was on 1 November, when Nagy announced Hungary's neutrality and demanded repeatedly that Soviet troops be withdrawn.[24] Against this backdrop, the official Soviet newspaper *Pravda* published a declaration of the Soviet Government on 30 October, which was conciliatory and accommodating in nature and which promised negotiations on the withdrawal of Soviet troops from Hungarian soil.

For the next few days, the activities of the Soviet military were confusing to Hungarian leaders. At times it seemed that some troops were being withdrawn; others appeared to be moving in large circles; and sometimes it was apparent that troops were entering Hungary.[25] Nevertheless, from 1–3 November, Soviet units quietly sealed off every major airport, railway station and railroad, while publicly announcing that this was to ensure the orderly withdrawal of troops. By this time, Nagy and other Hungarian leaders undoubtedly expected some sort of Soviet military action, but exactly when, or on what scale, was probably unknown.

Several noted analysts of the Hungarian revolution (especially Béla Király, Commander of the Hungarian National Guard at the time) contend that the *Pravda* declaration of 30 October was a deliberate Soviet attempt to deceive the Hungarian people while preparations for an invasion quietly continued.[26] Certainly, the most important statement within the declaration—the willingness to negotiate on the issue of troop withdrawals—was most likely a deliberate lie, designed to placate Hungarian fears while maintaining a Soviet military advantage; that is Soviet troops remaining in Hungary. The deliberate nature of this subtle deception became more apparent as the Soviets conducted negotiations with the Hungarian military about the supposed withdrawal, meanwhile moving troops into and around Hungary, claiming these movements were to ensure orderly withdrawal and prove their willingness to carry out that promise. The actual proof that this was a deception came on 4 November when the Soviets arrested the Hungarian officials with whom they were negotiating the withdrawal. About 120,000 Soviet troops, who were supposed to be ensuring an orderly withdrawal, formally executed the invasion.[27]

* * *

The invasion of Czechoslovakia in August 1968 was a clear example of Soviet achievement of strategic surprise using some elements of deception. Contingency planning for the invasion by the Soviet General Staff began several months before the event. The use of force in Czechoslovakia was considered to be an option of last resort. Military invasion was debated continuously and the debate reflected genuine ambivalence within the Soviet political leadership, who agonised over the decision for several months. After the invasion, Czechoslovak military intelligence officers estimated that preparations for a possible invasion had begun in February and March, at about the time the Czech-speaking Slavic specialists from Leningrad universities were said to have been mobilised. Some Soviet officials hinted at the time that they feared military action might prove necessary.[28]

This mobilisation may have been only a 'technical preparation' for the invasion in case of an emergency, some unexpected development such as an anti-Communist coup, or Czechoslovak withdrawal from the Warsaw Pact. Although the majority of Politburo members were slow to order a military action, sometime in May (as Brezhnev later admitted) they began to contemplate invasion as a viable option, but only as a worst-case scenario.[29] Consequently, the Soviet leadership decided to proceed apace with military deployments around Czecholovakia. The build-up served two purposes: first, it was intended as a form of psychological pressure and a warning to the reformists to keep events more tightly in hand; and second, it was a logistic preparation under the cover of military exercises, and a rehearsal for the worst case option—invasion. Thus the military build-up along the Czechoslovak border (primarily in Poland and East Germany) started in the early Spring of 1968, and continued for several months during the crisis. By late May and early June, the Soviet divisions in East Germany and Poland had moved from their regular garrison locations and camped on the Czechoslovak borders. In East Germany alone, 12 tank and mechanised divisions of the Soviet Army, and two East German divisions were stationed close to the borders.

Military manoeuvres, as recognised repeatedly by Soviet military writers, can serve as a useful form of deception. Thus, in a broad sense, the exercise lasting from May through August on the Czechoslovak borders, and on Czech territory, was used as a vehicle for deception—but deception of a unique kind. The massive exercises and military build-up were not only completely unconcealed, but at times it seemed that they were well-advertised—not surprising, perhaps, in the light of the apparent political purposes noted above. They took place in early May, June and July when it was anticipated that anti-Dubcek forces would attempt to slow down or reverse Dubcek's reforms. These nonstop military exercises might have served as a form of deception by de-sensitising Czech leaders to the possibility of a military invasion. In late July, the decision to use military force as a form of pressure was related to (and perhaps taken as) a decision to negotiate.

When it became known on 23 July that the Soviet Politburo had decided to negotiate with the Czechoslovak leadership at Cierna, the Soviet media announced the largest logistic exercise ever held by Soviet Ground Forces. Code-named 'Nemen', the exercise was under the command of the Commander-in-Chief of the Rear Services (Logistics), General S Mariakhin. Thousands of Soviet reservists were called up, and civilian transports were requisitioned. The manoeuvres took place in the western part of the Soviet Union, from the Baltic to the Black Sea. During the negotiations, they were extended to East Germany and Poland. In addition, before the Cierna Conference, a large scale exercise, code-named 'Sever', was begun in the Soviet North-east and Baltic Sea area, involving the Soviet Baltic fleets and the navies of East Germany and Poland.

Movements of the Warsaw Pact forces during the negotiations were aimed at intensifying psychological pressure, intimidating the Czechoslovak leadership, and giving the Soviet delegation at the Cierna Conference a strong bargaining position. At one time during the negotiations, the crews of tanks and armoured vehicles stationed on the Czechoslovak borders were noted to have started up their motors, and after a short while, switched them off again.[30] At this juncture, the movements were aimed at preparing for an actual military invasion in case agreement could not be reached.

Even after the negotiations, manoeuvres continued; in fact, they never really came to end. Thus 'Nemen' formally ended on 10 August, but a new major air defence exercise, code-named 'Sky Shield', began the next day, along with another exercise of 'communications troops' in the Western Ukraine, Poland and East Germany.[31] For the first time during the crisis, the manoeuvres were extended to Hungary on 16 August when Hungarian forces participated with Soviet armed forces.

Before the invasion, Soviet generals succeeded in lowering fuel and ammunition stocks of the Czechoslovak Armed Forces by transfer to East Germany, supposedly for an exercise. The manoueuvres were scheduled for 21 August, by which date the invasion would be in its second day, and were intended to divert the attention of Czechoslovak generals. Western electronic surveillance of the deployment was impaired by the Soviet use of electronic screens to cover the movement of troops, and by keeping the radio traffic of intervention units to a minimum.

Notwithstanding these dissimulations, American and NATO long range radar surveillance did pick up indications of Soviet preparations. Heavy concentrations of aircraft over Poland and refuelling by military transports at Leningrad were noted. However, these events were seen as continuations of the summer manoeuvres. No one knew that the transports carried a Soviet airborne division earmarked to spearhead the invasion by landing at Prague on the night of 20 August. What followed was intervention by airborne troops at the Czechoslovak nerve centre, backed up by tank and mechanised invasion across the land frontiers. Within six hours, all the important strategic points had been occupied; the entire country was occupied with three days. Surprise had been complete and no significant resistance was met.

Besides the Soviet Armed Forces, the KGB was a very important force influencing Soviet conduct during the Czechoslovak crisis. KGB officials who favoured an invasion tried to 'produce'—through various forms of political provocation, lies, and disinformation—proof of 'counter-revolution' to support arguments in favour of military intervention. Such proof consisted of caches of secret weapons, supposedly planted by imperialists, which were 'discovered' in July on the West German borders. Moreover, KGB interventionalists strove to create an impression of widespread opposition to Dubcek's supporters among 'healthy elements' in the party. Ladislav Bittman, a former Czechoslovak intelligence officer, wrote:

> The active role of the Soviet intelligence service in the events of 1968 and 1969 in Czechoslovakia centered on the systematic implementation of political provocation, disinformation, and propaganda campaigns aimed at influencing Czechoslovak public opinion, terrorising a selected group of liberals, and creating supportive arguments for the legitimisation of the Soviet invasion[32]

Other KGB deceptions included the falsification of documents to prove the existence of a Zionist conspiracy and involvement of the United States Central Intelligence Agency in the Czechoslovak affair. The target of some of these unsubtle ruses seemed to be the general public of the Warsaw Pact countries. At the same time, the KGB representatives in Czechoslovakia and the Soviet Ambassador to Czechoslovakia, S V Chervonenko, reportedly were engaged in providing biased information to the Soviet leadership, exaggerating negative developments in Czechoslovakia. This information may have influenced the perception of Soviet leaders.[33]

Deceptions of this kind can rather easily backfire on their instigators. In theory, any significant deception operation should be authorised at the highest level, whether military or civil. It must be conducted in total secrecy, which means that officials close to the scene will not be party to the ruse and may therefore accept the 'evidence' at face value. Of course experience of the system would create a measure of scepticism, but this might not provide complete protection. So an artificial event designed to impress public opinion in the Warsaw Pact might also affect the judgement and reporting of Soviet officials, thus distorting what is in any event a dark glass—the bureaucratic reporting channel. Add to this the problem of KGB agents exceeding their orders, either as overzealous individuals or as part of a service which feels entitled to use any means to enforce discipline, and you have the possibility of the tail wagging the dog, the deceivers influencing *their own* decision-makers.

Lies backfire in another way too. The stories told to Warsaw Pact countries to justify the invasion had to be repeated with greater zest to the soldiers who were about to make the invasion a reality. But when these liberators met passive resistance from crowds of working class Czechoslovaks, students and housewives, it soon dawned on the soldiers that it was they, the Warsaw Pact armies, and not some mythical conspirators, who were the aggressors. The impact on morale was serious.[34] Nor did the attempts at justification succeed outside the Soviet bloc: Western and Third World publics were disgusted by the aggression.

The most effective form of Soviet deception during the Czechoslovak crisis was undoubtedly a military one—the continuing series of military exercises. These illustrate perfectly what Barton Whaley calls dissimulation by dazzling—the blurring of distinctive patterns of invasion preparations by reducing certainty about the real nature of the thing. Since the deployment of huge Warsaw Pact forces could not be concealed, they were hidden, like a car in a traffic jam, within the turbulence of large scale manoeuvres. Within Donald Daniel's and Katherine Herbig's concept, the technique was one of 'ambiguity-increasing' to clutter the victim's sensors with a mass of contradictory indicators.[35] The technique gave the Soviet decision-makers complete flexibility. If they had decided not to intervene, no one need ever have known that preparations for such an event had been made; so long as the decision hung in the balance, the manoeuvres put psychological pressure on the Czechoslovak Government and maintained the viability of the military option; once the option was taken, they ensured surprise.

The Czech leaders failed to anticipate an invasion, as Dubcek later admitted:

> I declare on my honour as a Communist that I had no suspicion, no indication, that anyone would want to undertake such measures against us . . . that they should have done this to me after I have dedicated my whole life to the Soviet Union is the tragedy of my life.[36]

* * *

Another example of the Soviet Union's successful accomplishment of strategic surprise in an intervention was the invasion of Afghanistan in December 1979. As in Czechoslovakia, the preparations for a possible military operation must have begun months earlier. By September, the Soviet Union had undertaken the first serious attempt to influence developments in Afghanistan by military force. During Afghan President Noor Mohammed Taraki's visit to Moscow, *Pravda* reported on 13 September that he was assured that he could rely on the all-around support of the Soviet Union (which in Soviet language includes military support). During his visit, Taraki is reported to have been warned about the intrigues of his deputy, Hafizullah Amin, and (as some former Afghan Government officials have reported) the Soviets arranged a meeting between Taraki and Babrak Karmal.

Upon Taraki's return to Kabul, the Soviet Union appears to have organised or supported an anti-Amin coup on 14–15 September, the objective being to eliminate Amin and to establish a coalition government led by Taraki and Karmal. There was a reported attempt to assassinate Amin involving ambassador A Puzanov which has never been fully explained.

Concurrently, some Soviet units were deployed on the Soviet-Afghan border, and a 400-man airborne unit was sent to the important Bagram Air Base, 40 miles from Kabul. The coup was a failure; instead of Amin, Taraki was killed.[37] The Soviet Union accepted the *fait accompli*, at least for the time being, since there was no alternative. While preparing for his eventual overthrow, the game of deception with Amin continued. The plan for overthrowing Amin seems to have been finalised in late November when the final Soviet decision was made to intervene on a massive scale.[38] First Deputy Minister of the Interior, Lieutenant General

Viktor Paputin, was sent to Kabul with an official mission to 'help' Amin with police affairs and counter-insurgency, perhaps even to protect him; but this was only a cover. Seemingly, his real role was to organise the coup by mobilising Amin's opponents among supporters of Karmal and former supporters of Taraki.

Before the invasion, Soviet security officials and diplomats tried to turn factional struggles to their advantage. They secretly prepared a new pro-Soviet government that would make an orderly call for Soviet fraternal assistance. In Czechoslovakia the Soviet Union had tried to rally anti-reformist elements—mainly leaders who had objected to Dubcek's reforms, and were gradually being dismissed or feared dismissal. In Afghanistan, under the direction of Puzanov in September and Paputin in December, the Russians tried to overthrow Amin by rallying the supporters of his rivals.

The United States was not taken entirely by surprise by the invasion of Afghanistan. Twice the United States warned the Soviet Union against intervening, but they failed to anticipate the anti-Amin coup. United States intelligence detected the mobilisation of Soviet troops on Turkmenistan in late November, and in other areas along Afghan borders where local reserves were being called up. A steady military build-up on Afghan borders was noted in early December as various Soviet forces and tactical aircraft were shifted from the Soviet-Iranian frontier. On 8 and 9 December, airborne units of over 1,000 men, equipped with tanks and artillery, were airlifted to the Bagram Airfield to reinforce Soviet units deployed there in September. This enabled them to take effective control of the base. Their missions became clear in the wake of the invasion when they began handling incoming flights, and it transpired that they had cleared the Soviet-built highway between the Soviet border and Kabul from south to north. Concurrently, Moscow airlifted a number of small units into the Kabul airport.

The invasion was scheduled during the Christmas holidays when most Western officials would be on leave. Prior to the invasion, Soviet advisers actually disarmed two armoured divisions of the Afghan Armed Forces (one garrisoned in Kabul) using a simple but effective form of military deception. They convinced Afghan commanders that their ammunition and anti-tank weapons were due for inventory, that their batteries needed winterising, and that some tanks needed to go to repair depots for correction of defects.[39] Strategic surprise followed when 10,000 Soviet airborne troops were deployed in Kabul on 25–26 December, and two mechanised divisions crossed the northern border by land and advanced toward the capital and other vital centres.

Deception to justify the Soviet invasion was for the most part fabricated after the initial event. The principal illusion was that the Soviet troops had intervened in response to a request for 'fraternal assistance' from the Afghan Government. In fact, the 'request' was issued on the 27th, two days *after* the invasion had begun. Babrak Karmal, who was installed as leader in succession to Amin, has since claimed that he returned to Kabul secretly before the invasion, and arranged for the Afghan Communist Party leadership to depose Amin. But there is no evidence that he was indeed in Kabul at that time. He is generally thought to have been in the Soviet Union, arriving in Kabul only after the invasion; in any case, his own accounts of his movements and his knowledge of events have since proven contradictory. Moreover, his broadcast on the evening of 27th, in which he announced the overthrow of Amin and the request for Soviet assistance, was not transmitted by Radio Kabul, but by a Soviet station at Termez, in Uzbekistan, on Radio Kabul's wave-length. The broadcast preceded Amin's demise by several hours. It seems clear that Amin was killed during the assault on Darulaman Palace by Soviet airborne or 'Spetsnaz' troops, and that no Afghan forces were involved—except in his defence. Soviet and Afghan representatives later insisted however, that Amin had been sentenced by a 'revolutionary tribunal' and eliminated by Afghan action alone. In April 1980 *New Times* (Moscow) stated baldly:

> The fact that the removal of Amin took place concurrently with the beginning of the introduction of the Soviet contingent is a pure coincidence in time and there is no causal relationship between the two events. The Soviet units had nothing to do with the removal of Amin and his accomplices. That was the doing of the Afghans themselves.[40]

In 1981 *Novosti*, the official publishing house, produced a booklet which attempted to explain the action in terms of threats to socialism, 'the undeclared war of US imperialism and its allies against the democratic republic of Afghanistan—a glaring example of international terrorism' and Soviet duty to the international cause. Some rather poor photographs of rifles and other equipment of supposedly Western manufacture were included.[41] The book is so dull and unconvincing that one is tempted to conclude that the *Novosti* editors had little confidence in their work and were merely 'going through the motions' to reassure the worldwide party faithful.

As in the Czechoslovak case, troop briefings seem to have been based on the official deceptive propaganda, a dangerous procedure when the soldiers involved will soon be face to face with reality. It was not for several years that Soviet deserters began talking to Western journalists but when they did their stories frequently mentioned the contrast between their briefings and the facts. They had been told that 'our mission in Afghanistan was to defend the interests of the Afghan working class; that Afghanistan had been attacked from outside and that our soldiers were fighting against Pakistani and American mercenaries . . . I, of course, believed these tales, and everybody else believed them too.'[42] The contrasting reality tended to demoralise.[43]

* * *

The direct use of force in the Polish crisis has been considered by the Soviet leadership to be a worst-case option to be executed if all other means fail. The Soviet Union has followed the standard military track in an indirect way. Even more than in Czechoslovakia, military manoeuvres were mainly intended to exert pressure on Polish leaders, and to induce them to keep events under tighter control. This is suggested by the fact that some manoeuvres were well-advertised, including some that did not even take place.

Although the chief motive of the manoeuvres was to exert pressure on Poland to find a 'Polish solution', they also put the Soviet war-machine into gear for an actual invasion—if one were needed. The manoeuvres familiarised officers and troops (who were rotated as often as possible) with battle terrain, facilitated inspection of enemy military installations, and tested the reliability of the Polish Armed Forces. In addition, Soviet forces tested their deployment capabilities and their logistics and communications systems.

The Polish crisis caught the Soviet war-machine unprepared in the summer of 1980. Preparations for the first scheduled exercise, code-named 'Brotherhood-in-Arms', which took place in September in East Germany, close to the Polish border, and simultaneous manoeuvres in the western part of the Soviet Union were not very orderly; they even seemed chaotic in the Transcarpathian military district bordering Poland. Some senior officers apparently were dismissed as a result.

From late November to early December 1980, the Soviet military initiated a new, more comprehensive series of exercises intended to intimidate Solidarity (the Polish free trade union) and strengthen the position of Polish leaders to deal with it. Some features of these exercises suggest that again the Soviet Union may have been rehearsing for an invasion: direct radio communication was established with Moscow; fuel and ammunition dumps and medical facilities were built; Soviet air traffic over Poland increased; ships in the Baltic were unusually active; and Soviet marines were deployed on beaches near the city of Tallin. (A second motive for this particular feature of the manoeuvres may have been to intimidate the growing number of dissidents in the Baltic Republic of Estonia.)

In early December 1980, the United States detected the deployment of several Soviet airborne divisions in Poland. A series of diplomatic and political steps by the Carter administration, including a message from President Carter to Brezhnev, might have caused the Soviets to back off at this point. The Carter-Brzezinski team alerted Poland of the Soviet military build-up by publishing information that was not available to the Poles, and at the same time focused world attention on the Polish crisis. It was as though the United States had become more skilled in identifying invasion indicators among the dazzle of deceptive manoeuvres, and making them public. If the Soviets recognised this improved American capability, their confidence in being able to achieve strategic surprise would have been undermined, and this could have been a factor contributing to the Soviet decision not to intervene at that time.

Peter de Leon's research has explored the theoretical writings of Soviet authors on deception. Of particular interest is their attention to 'reflexive control', which is defined as 'influencing an opponent's will and mind when that opponent is making decisions in the course of preparing for and conducting combat operations'.[44] de Leon quotes V A Lefevr and G L Smolyan who contend that,

> Control of an opponent's decision, which in the end is a forcing of a certain behavioural strategy on him through reflexive interaction, is not achieved directly, not by blatant force, but by means of providing him with the grounds by which he is able to logically derive his own decision, but one that is predetermined by the other side.[45]

To some extent at least, Soviet deceptions prior to their earlier interventions had contributed to the achievement of surprise over their immediate opponents—the regimes whose countries were to be invaded, as well as over the outside world, particularly the United States. Decision-makers in Budapest, Prague and Kabul, as well as in Bonn, London, Paris and Washington had, to varying degrees, reached conclusions which they believed were logical but which had really been predetermined in Moscow. If the United States seemed more alert in 1980 to such deception, this would merely have borne out the warning of another Soviet writer:

> If one has succeeded in deceiving the enemy once, then he will not allow himself to be deceived a second time by the same technique. Therefore

> there is a continuous search for newer techniques and methods for achieving surprise.[46]

The new technique developed in the Polish case seemed to come in two parts: first, allow the situation to cool, so as to put the enemy off his guard; second, while the enemy's attention is focused on one threat (Soviet intervention), surprise him by some unexpected method.

The cool-headed preventive diplomacy initiated by the Carter Administration and pursued by the Reagan Administration may have deterred the Soviets from invading at earlier junctures of the crisis; at least it made it very difficult for the Soviet Union to achieve strategic surprise through direct intervention. But while the Reagan Administration continued its efforts to prevent such Soviet intervention, it failed to consider the possibility of intervention by proxy. Recall the confusion surrounding the unfortunate State Department statement (later clarified) which seemed to imply that if Polish forces alone were used to establish order, that would constitute an internal Polish matter and be of no concern to the West. Another even more self-injurious signal to the Soviet Union was the lifting of the grain embargo in the Spring of 1981. This made the Administration's rhetoric seem empty.[47]

By the Autumn, the Reagan Administration, like the media in the West, had become bored with the crisis and now paid little attention to it. On 13 December 1981, Moscow again surprised the West, this time by intervening by proxy. There can be no doubt that the Soviet Union had prior knowledge of the military takeover by General Jaruzelski, and helped to engineer it. Corroborating evidence includes the superb planning and execution of the coup itself; it was a sophisticated, even elegant, event. The timing was a complete surprise, not only for Solidarity, but also for Western intelligence. It happened, as prescribed by Soviet military doctrine, when least expected: shortly before Christmas (as in Afghanistan), on Sunday morning—when workers were absent from their factories and the West was generally at leisure. On the day of the crackdown, West German Chancellor Helmut Schmidt was negotiating with East German Party leader Erich Honecker in East Germany.

* * *

After the death of Stalin, deception and strategic surprise became central themes in Soviet military doctrine. The Soviet Union learned from its own experience in Finland in 1940, the German *blitzkrieg* strategy, particularly the German attack on the Soviet Union on 22 June 1941, the Japanese attack on Pearl Harbour and the British practice of stratagem during the Second World War. When it was confronted by crises calling for direct or indirect military intervention in Hungary, Czechoslovakia, Afghanistan and Poland, the Soviet Union used both military and political deception, mainly to achieve surprise and to claim justification but sometimes also for lesser purposes.

Endnotes

1. The author wishes to acknowledge the valuable contribution of David Hamilton in the research and writing of this chapter.
2. Colonel M Loginov, 'Surprise Actions', *Soviet Military Review*, No. 11, 1980, p. 25, quoted Jennie A Stevens and Henry S Marsh, 'Surprise and Deception in Soviet Military Thought', *Military Review*, June 1982, p. 5.
3. See Richard H Shultz and Roy Godson, *Dezinformatsia: Active Measures in Soviet Strategy* (New York, London: Pergamon-Brassey's 1984), pp. 2–3, 15–17, 187, 190, 193.
4. See Stephen T Hosmer and Thomas Wolfe, *Soviet Policy and Practice Toward Third World Conflicts* (Lexington, Massachusetts: Lexington Books, 1983), p. 185.
5. For a discussion see Herbert S Dinerstein, *War and the Soviet Union* (New York: Praeger, 1959), and Raymond L Garthoff, *The Soviet Image of Future War* (Washington, DC: Public Affairs Press, 1959).
6. For a detailed study of the Soviet strategy during the Manchurian campaign, see John Despres, Lilita Dzirkals and Barton Whaley, *Timely Lessons of History: The Manchurian Model for Soviet Strategy*, R-1825-NA (Santa Monica: The Rand Corporation, July 1976).
7. Barton Whaley, *Stratagem: Deception and Surprise in War* (Cambridge: MIT Center for International Studies, 1969) in note 2, p. 64; see also John R Deane, *The Strange Alliance: The Story of Our Efforts at Wartime Cooperation* (New York: Viking Press, 1947). General Dean was chief of the US military mission in Moscow in 1943–45.
8. Charles C Cruickshank, *Deception in World War II* (New York: Oxford, 1979), pp. 114–24.
9. Marshal P Rotmistrov, 'On the Role of Surprise in Contemporary War', *Voyennaya mysl'*, February 1955; also Dinerstein, *War and the Soviet Union*; and Garthoff, *op. cit.* in Note 5.
10. As quoted in Arthur D Nicholson Jr, *The Soviet Union and Strategic Nuclear War*, Technical Report 56-80-002 (Monterey: Naval Postgraduate School, June 1980). (Major Nicholson was the author's student at the Naval Postgraduate School: he was killed by a Soviet sentry while on legitimate duty in East Germany in 1985.)
11. M M Kiryan, 'Vnezapnost,' *Sovetskaya voyennaya entsiklopediya* (Soviet Military Encyclopedia), Vol. 2 (Moscow: Voenizdat, 1976), pp. 161–3.
12. Colonel A Postalov, 'Modelling the Combat Operations of the Ground Forces,' *Voyennaya mysl'*, No. 3, 1969.
13. For discussion see Peter H Vigor, 'Doubts and Difficulties Confronting a Would-be Soviet Attacker', *Journal of the Royal United Services Institute for Defence Studies* (June, 1980), pp. 32–8; and Joseph D Douglass Jr (i), and Amoretta M Hoeber, *Soviet Strategy for Nuclear War* (Standford: Hoover Institution Press, 1979); Douglass (ii), *Soviet Military Strategy in Europe* (New York: Pergamon, 1980).
14. General of the Army S P Ivanov, *Nachalnyy period voyny (The Initial Period of the War)* (Moscow: Voenizdat, 1974).
15. Douglass (ii), *op. cit.*
16. For discussion, see Roger Beaumont, *Maskirovka: Soviet Camouflage, Concealment and Deception*, (College Station, Texas: Center for Strategic Technology, Texas A & M, 1982), particularly pp. 1–3; Stevens and Marsh, *op. cit.*
17. Colonel L Kuleszynski, 'Some Problems of Surprise in Warfare', *Voyennaya mysl'*, No. 5, May 1971. Colonel Kuleszynski wrote this article originally for the Polish military journal *Mysl' wojskowa*, 1970.

18. Major General N Vasendin and Colonel N Kuznetxov, 'Modern Warfare and Surprise Attack', *Voyennaya mysl'*, June 1968.
19. See introductory note, 'Eastern Approaches'.
20. US Congress, House, Subcommittee on Oversight, Permanent Select Committee on Intelligence, 96th Congress, 2nd Session, *Soviet Covert Action (The Forgery Offensive)* (Washington: US Govt Printing Office, 1980), p. 63, Footnote 6.
21. Jiri Valenta(i), 'Soviet Policy Toward Hungary and Czechoslovakia,' Sally Terry, ed., *Soviet Policy in Eastern Europe* (New Haven & London: Yale University Press, 1984).
22. Imre Kovács, *Facts About Hungary* (New York: Hungarian Committee, 1958), p. 82.
23. Paul E Zinner, *Revolution in Hungary* (New York: Books for Libraries Press, 1972), p. 286.
24. *Ibid.*, pp. 326–28.
25. Ernest A Nagy, *Crisis Decision Setting and Response: The Hungarian Revolution* (Washington, DC: National Defense University, 1978), p. 16.
26. Interview with Béla Király, commander of the Hungarian National Guard during the revolution, New Jersey, Spring, 1980.
27. Valenta (i) in note 19; when Yuri Andropov became Chairman of the CPSU there was much speculation in Western media about his role in this deception, as he had been Soviet Ambassador to Hungary at the time. However, some analysts are inclined to attribute the main roles in this operation to the senior men brought in from Moscow to handle the crisis—Suslov and Mikoyan, and to Khrushchev himself, who retained hour-by-hour control from Moscow. Between them, these high officials also lured Imre Nagy to his death by a written guarantee of safe conduct signed by János Kádár. Nagy was killed without trial. See Martin Ebon, *The Andropov File* (New York: McGraw-Hill, 1983), pp. 64–74; Zhores Medvedev, *Andropov* (New York: W W Norton, 1983), pp. 32–40.
28. Jiri Valenta (ii), *Soviet Intervention in Czechoslovakia 1968: Anatomy of a Decision* (Baltimore: The Johns Hopkins University Press, 1979).
29. An interview with former Czechoslovak leader Z Mlynar, Vienna, 6 Ocober 1978.
30. *Ibid.*
31. Valenta (ii), in note 22, p. 113.
32. Bittman, p. 124.
33. Valenta (ii), pp. 123–8.
34. See Viktor Suvorov, *The Liberators* (London: Hamish Hamilton, 1981), pp. 161–97.
35. See Introduction.
36. Quoted Robert Littell, ed., *The Czech Black Book* (New York: Praeger, 1969), p. 17.
37. For a more detailed discussion, see Valenta (iii), 'The Soviet Invasion of Afghanistan: The Difficulty of Knowing Where to Stop', *Orbis* (Summer 1980), pp. 201–18; and Jiri Valenta (iv), 'From Prague to Kabul: The Mode of Soviet Invasions', *International Security*, Vol. 5 (Autumn, 1980), pp. 114–41.
38. For an analysis, see Valenta (iv).
39. *The Times* (London), 9 January 1980; and *Newsweek*, 21 January 1980, p. 115.
40. *New Times* quoted in Thomas T Hammond, *Red Flag Over Afghanistan: The Communist Coup, the Soviet Invasion and the Consequences* ((Boulder, Co: Westview Press, 1984), p. 100, Note 16; on related deception details, see Henry S Bradsher, *Afghanistan and the Soviet Union* (Durham, N.C.: Duke Press Policy Studies, 1983), pp. 173–86. Vladimir Kuzichkin, a KGB major who defected in Iran in June 1982, said in a *Time* magazine interview (22 November 1982) that Soviet commandos and KGB officers carried out the assault on the palace.
41. *The Truth about Afghanistan: Documents, Facts, Eyewitness Reports* (Moscow; Novosti, 1981).
42. Fatima Salkazanova, interview with Soviet Army deserter Rykov, broadcast on Russian Service of Radio Liberty on 1 and 2 March 1984. Radio Liberty background report RL 220/84.
43. Salkazanova interview with deserter Habib, broadcast 29 February 1984, RL 205/84.
44. Major General M Ionov, 'On the Methods of Influencing an Opponent's Decision', *Voyennaya mysl', (Military Thought)*, No. 12 (December 1971), p. 58, quoted Peter de Leon, *Soviet Views on Strategic Deception* P-6685 (Santa Monica: Rand Corporation, 1981.)

45. V A Lefevr and G L. Smolyan, *Algebra of Conflict* (Moscow: 1968 JPRS Translation 52700, March, 1971), pp. 33–34, quoted de Leon *op. cit.*
46. Colonel V Ye Savkin, *The Basic Principles of Operational Art and Tactics* (Washington, DC: Government Printing Office for the US Air Force, 1974), No. 4 in *Soviet Military Thought* series, p. 235, quoted de Leon, *op. cit.*
47. See Jiri Valenta (iv), 'Normalisation?: The Soviet Union and Poland', *Washington Quarterly*, Autumn, 1982; Valenta (v), 'Perspectives on Soviet Intervention: Soviet Use of Surprise and Deception', *Survival*, March-April, 1982; Valenta (vi), 'Soviet Options in Poland', *Survival*, March-April 1981.

CHAPTER 6

A 'Runaway Deception': Soviet Disinformation and the Six-Day War, 1967

MICHAEL I HANDEL

For secrets are edged
tools
And must be kept
from children and
fools.
Anon

Deception and intrigues have played a key role in the Middle and Near East since time immemorial. From biblical times through the rise of Islam, the Crusades, the Ottoman Empire, and the eras of Generals Allenby and Wavell, the use of stratagem has been a constant feature.

Historically, the political-strategic stakes in the region have also been high. The region is a strategic highway linking Eastern and Western civilisations, and North and South; it is the key for a warm water port for Russia and the trade route from Great Britain to India; there are strategic choke points such as the Suez Canal, the Bosphorus and Dardanelles, Bab-el-Mandeb and the Straits of Hormuz. In modern times, the region has become a crucial source of energy for the industrialised world. Recently, it has also become an arena for competition over the political and ideological support of the Islamic world and a large number of new independent countries. It is not surprising therefore that an area of such vital political and strategic importance should have generated competition between the Great Powers.

In the aftermath of the Second World War, the Soviet Union under Stalin concentrated on reconstruction at home and annexation and consolidation of its territorial gains in Eastern Europe. Yet, despite Stalin's conservative and cautious policy in the Middle

East at the time, the Soviet Union did not miss any opportunity to de-stabilise the region and undermine the position of the Western powers wherever and whenever possible. Direct Soviet intervention grew gradually, as the consolidation of Eastern Europe progressed. Following Stalin's death, by the late 1950s the Soviet Union had become one of the two major actors in the Middle East and its politics.[1] Deceptions and intrigues were among the tools used by the Soviets to penetrate the region.

Perhaps more than any other power, the Soviet Union has refined the art of supplying weapons as a means of improving her relative position *vis-à-vis* other powers. During the 1920s the Soviets clandestinely delivered weapons to the soldiers of Kemal Ataturk fighting the Greeks, to Persian rebels in the province of Gilan, and to the Kuomintang in China. In 1931, the Soviet Union gave military assistance to Chin Shu-jen, a provincial governor of Sinkiang. Later in the 1930s, Soviet weapons and military experts assisted the Spanish Republic as well as Chiang Kai-chek's Nationalist Chinese forces, and help was probably sent to Turkey and Yemen.[2] During the late 1940s and early 1950s, the Soviets used arms transfers as a primary means of penetrating the Middle East.

The first opportunity arose after Britain had handed the 'Palestine problem' to the United Nations to solve, opening the way to the creation of the State of Israel in part of the old mandated territory. Since the British and other Western powers had traditionally supplied military aid to the Arabs, who were militarily opposing the new state, the opportunity for the Russians was to step in as suppliers of Israel. This was an essentially pragmatic decision aimed at penetration: it was not based on pro-Israeli sentiment. Indeed, at the same moment, the Soviets made a similar, though smaller, arms deal with the Syrians.[3] Stalin, however, was anxious to avoid confrontation between the Soviet Union and the Western powers over the Palestine question: he wanted the benefits of penetration without the costs. An element of deception was consequently introduced into the arms deal which provided the necessary protection.[4]

Having achieved their short-range objectives soon after the Israeli War of Independence, the Soviets decided to change sides. Following Nasser's rise to power in Egypt, his aspirations to Arab leadership and Egyptian 'positive neutralism' gradually isolated him from the Western powers. Egypt's stridently anti-Israeli policies needed to be backed by large quantities of modern arms. Another opportunity beckoned to Moscow. Soviet and Egyptian interests converged, making a major arms transfer deal attractive to both.

The Egyptian arms build-up was one of the causes of the Suez campaign of 1956, in which Israel, France and Britain attacked Egypt. Some of the deceptions surrounding that event are described in the second part of this volume. It is sufficient to note that the Western powers failed in their bid to topple Nasser, but that Israel won a stunning victory in the Sinai.

* * *

By the middle of 1966 tensions once more began to mount in the Middle East. There were four main causes. First, the establishment and first operations of the Palestine Liberation Organisation, financed and supported by Syria, involved raids on Israeli towns and settlements across the Jordanian-Israeli border. A new cycle of Israeli reprisals and counter raids ensued, along the familiar lines of the early 1950s. Second, the coordinated Syrian, Jordanian and Egyptian plan to divert the waters of the Jordan River again led to Israeli military operations. Third, the dynamics of inter-Arab politics, in which each state tried to outbid all rivals in its anti-Israeli rhetoric, heightened tensions. Finally, the fading memories of the 1956 war and the gradual erosion of the Israeli deterrent posture lessened the disincentives to Arab adventurism.[5]

The situation created by this constellation of events proved to be too tempting for the Soviets to let pass. They cynically sought to exploit it in order to improve further their position in the region. This was done by a prolonged campaign of disinformation intended to capitalise on Arab fears of Israel by exaggerating Israeli aggressive intentions and, in fact, representing Israel as ready to launch a surprise attack on Syria. The Russians may have thought that, by destabilising the region, they stood to make substantial strategic gains, on the grounds that the greater the tension in the area the more the Arab states would come to depend on Soviet political and military support which in turn would permit Soviet penetration.[6]

Another factor favouring Soviet intervention in the Middle East was America's involvement in Vietnam. The United States, they thought, might be forced to choose between weakening its position in the Far East and denying military aid to its ally, Israel. Either way, the Soviets stood to gain.[7] According to one account, the Soviet leader Brezhnev met Ulbricht of East Germany and Gomulka of Poland in East Berlin on 22 April 1967 and spoke of 'inflicting one decisive blow' on the United States' position in the Middle East.[8] Soviet decision-making may have been further complicated by internal power play between the military and other institutions. Abraham Ben Zur argues that the Soviet military elite sought to increase military tension in the Middle East as a means of strengthening its case for a military rather than a civilian head for the Soviet defence ministry.[9]

Whatever their motives, the Soviets began to raise Middle Eastern tensions in May 1966, more than a year before their deception plan reached its climax.

On 25 May 1966 the Israeli Ambassador to Moscow, Katriel Katz, was called to report immediately to the Soviet Deputy Foreign Minister Vladimir Semyonov and was handed a note saying 'the Soviet Government holds in its hands proof regarding a recent concentration of Israeli troops along the borders with the Arab states. This concentration assumes a dangerous posture since it coincides with a hostile propaganda campaign conducted in Israel against Syria'.[10] Yet only five days earlier, on 20 May, a report by UN observers who patrolled the Syrian–Israeli border stated that there were no IDF concentrations in the area. The Soviet alarmist policy continued. On 6 June 1966 an article in *Mezhdunarodnaya Zhizn* stated

> . . . under pressure from the USA, the militarists of Israel have also intensified their provocative activities. The military units of Tel-Aviv and Jerusalem districts (a third of the entire military force of the country) have been transferred to the Syrian border . . . Newspapers write, not without grounds, that such provocations were intended to activate the reactionary forces within Syria as well . . .[11]

On 7 July 1966 an article in *International Affairs* (Moscow) stated that 'the Israeli Army has been put in a state of war preparedness and Israeli troops are concentrating on the border with Syria'.[12] Three months later, on 3 October, *Pravda* announced that Israel was planning to invade Syria in order to overturn the government. Between 11–12 October, Israeli Prime Minister Eshkol met twice with the Soviet Ambassador to Israel, Dimitry Chuvakhin, and through him asked the Soviet Government to restrain Syria. According to the Prime Minister, the Soviet Ambassador alleged at their first meeting (11 October) that:

> we [Israel] have mobilised our troops and that they are in a ready condition. Second that we plan to turn over [sic] the democratic government of Syria and that for this purpose we have concentrated on our northern borders thirteen brigades. The Ambassador further emphasized that we are being manipulated to act as proxies of a superpower that wants to rout the non-aligned government of Syria . . .[13]

On the 12th, Prime Minister Eshkol invited the Ambassador to take a car trip with him anywhere along the Syrian border to show him that the allegations of troop concentrations were false. The

Ambassador, presumably on orders from Moscow, declined the offer.[14] On that day the Soviet Government issued a new note to Israel stating:

> According to the information in our possession, concentrations of Israeli troops can again be discerned along the Syrian frontiers, and preparations are being made for an air attack on the areas bordering the Syrian frontier, so that, in its wake, Israeli troops may penetrate deep into Syria. The Chief of Staff of the Israel Defence Forces recently announced in public that Israel's military preparations will be directed first and foremost against the present regime in Syria.[15]

Following this note, a second one was issued on 9 November in which it was said

> . . . Prime Minister Eshkol, in his talk, tried to blame the Syrians for the worsening of the situation. But how can one speak of Syrian responsibility when Israel continues actively against a background of declarations designed to convince (the world) of her peaceful intentions, to mould public opinion towards preparations for a militant attack on Syria?[16]

After a two month period, during which the Soviets kept quiet, a commentary in *Pravda* on 18 January 1967 once more began to rekindle the issue. The commentary stated that '. . . it is evident that those who artificially created a troubled atmosphere on the Israeli–Syrian border wish to frighten Syria and to overthrow its progressive regime'.[17] On 3 February, an article in *Izvestia* further raised tension by announcing '. . . the Armed Forces of Israel are being put in readiness for battle. All military personnel have had their leaves cancelled and additional contingents of reservists have been called up. Large military concentrations have been concentrated on the northern border.'[18]

These largely overt simulations were meanwhile being reinforced by covert means. The Soviets were aware of the Arab nations' poor intelligence capabilities against Israel and manipulated this weakness to their own advantage. By 10 May 1967 President Nasser had received reports of Israeli troop concentrations along the Syrian border from four 'friendly' intelligence services—Egyptian, Syrian, Lebanese and Soviet. The multitude of apparently independent sources corroborating each other's warnings served to strengthen Nasser's belief in their reliability. He was unaware, of course, that all reports had in fact been fed into the various systems by the Soviets.[19]

On 12 May an intelligence officer from the Soviet Embassy in Cairo informed President Nasser that additional corroboration of earlier reports and warnings was now available. After the meeting,

Soviet Ambassador Podyedyeev sent a cable to the Ministry of Foreign Affairs in Moscow in which the penultimate sentence reportedly read 'today we passed on to the Egyptian authorities information concerning the massing of Israeli troops on the Northern frontiers for a surprise attack on Syria. We have advised the UAR Government [in this context, Egypt] to take the necessary steps'.[20] Meanwhile, in Israel on the same day, the Soviet ambassador to that country visited Arie Levavi, Director General of Israel's Foreign Office. The Ambassador was again offered a trip along the northern border to dispel the notion of an IDF concentration, but again this was declined.

The 'steps' referred to by Podyedyeev involved the ejection from the Sinai of the United Nations Emergency Force (UNEF), the mobilisation of the Egyptian Armed Forces, and the deployment of powerful infantry and armoured formations close to Israel's southern border, presumably to pose a threat that would neutralise Israel's supposed threat to Syria. On the next day, 13 May, Egyptian Vice President Anwar Sadat, while on a visit to Moscow, was informed by Soviet President Nikolai Podgorny that the Soviets had *actual information* that there were military concentrations against Syria.[21] The contents of this high-level warning were immediately cabled to Cairo. At the end of the meeting, a joint communiqué was issued in which both parties expressed their concern about an Israeli attack on Syria. Later, Soviet intelligence claimed that as many as 11 to 15 Israeli brigades were preparing for a swift and massive attack on Syria on 17 May, to begin at between 4 and 5 o'clock in the morning.[22]

In a speech to the officers and men of an advanced air force base in Sinai, President Nasser announced the closure of the Gulf of Aqaba to Israeli shipping. Then he went on to say:

> On 13 May we received definite information to the effect that Israel was concentrating huge forces of about 11 to 13 brigades on the Syrian frontier, and that these forces were divided into two fronts, one to the south and one to the north of Lake Tiberias. We also learned that the Israeli decision taken at this time was to carry out an attack on Syria starting on 17 May. On 14 May we took action, and discussed the matter and contacted our Syrian brothers. The Syrians also had this information. In the light of this, General Fawzi went to Syria to coordinate things. We told them that we had decided that if Syria was attacked we should enter the battle immediately. This was the situation on 14 May; forces started to move towards Sinai to take up their natural positions.[23]

In his 'resignation speech' on 9 June 1967, President Nasser reiterated his earlier claim, this time openly revealing the sources of the warnings:

> Syrian sources were quite definite on this point, and we ourselves had reliable information confirming it—*our friends in the Soviet Union even warned the parliamentary delegation that visited Moscow at the beginning of last month that there was a plot against Syria.*[24]

Six weeks later, in a speech commemorating the 15th anniversary of the 23 July Revolution, Nasser said it all again. Everyone, he claimed, knew that the war had been sparked off by Israel's attempt to attack Syria. Israel had not acted alone in this plot, 'but also on behalf of the forces which were at the end of their tether because of the Arab revolutionary movement'—a reference to Arab 'reactionaries', particularly within Syria. Nasser claimed his information had come from 'many sources', and that his staff had made a point of checking this information. 'Then there was our parliamentary mission . . . where our Soviet friends informed Mr A-Sadat that Israel was on the point of attacking Syria.'[25]

At the same time that Soviet deception was creating an imperative for Egyptian mobilisation and forward deployment, Nasser and his generals had been reassessing the relative strengths of Egypt and her Arab allies on the one hand and Israel on the other. They eventually succeeded in convincing themselves that their defeat in 1956 had been due more to Anglo-French intervention than to the prowess of the IDF. Their Soviet advisers, anxious to prove what a good job they had been doing, tended to overstate their success in reorganising and re-training the Egyptian Armed Forces, particularly its leadership qualities.[26] Intelligence reports on the IDF, mainly supplied by Moscow, underestimated Israel's military strength. In part this may have been due to the fact that recent German arms shipments to Israel had been kept secret, in part to poor intelligence and good Israeli security; in part it may have owed something to Israeli Prime Minister Eshkol's projected image of weakness and indecision.

It is unclear whether the Soviets deliberately encouraged Nasser and his aides to reach the conclusion that they did, or whether the Egyptians got there on their own, and Moscow simply accepted their conclusions. After the war, Nasser was to claim that 'accurate calculations' about relative strengths had provided the rationale for risk-taking.[27] The only source of such calculations available to Cairo would have been Soviet intelligence estimates. These may have undervalued qualities such as leadership and initiative. In any case, it is apparent that at some moment prior to the June 1967 war, probably within the last few months, Nasser and his staffs reversed their hitherto pessimistic assessment of their chances against Israel, and decided that the Arabs could win. In a speech by Nasser on the eve of war, 26 May 1967, the President said:

> One day, two years ago I stood up and said that we have no plan for liberating Palestine . . .
> But recently we felt ourselves so strong that if we shall engage in battle with Israel we could win with Allah's help, and on this we have decided practically to take real steps . . . the war will be general and our basic target would be the destruction of Israel. I could not perhaps say such things three or five years ago. If I would have said so, and I couldn't translate them into action, then my words would have been empty and worthless. But today, 11 years after 1956, I am saying those things because I feel confident. I know what Egypt and Syria have.[28]

This assurance rested on all the illusions discussed above and was reinforced by the belief that, if things went wrong, the Soviets would come to the Arabs' aid.

When the Egyptian defence minister Samseddin Badran left Moscow in mid-May after a visit, Marshal Gretchko who came to the airport to see him off advised: 'Stand up to them! The moment they attack you, or if the Americans make any move, you will find our troops on your side.'[29] Similar reassurances were given to Syria's Prime Minister Nureddin Atassi who arrived in Moscow on 29 May, just one week before the outbreak of the war.

On top of all this, Nasser was being urged on by an irresponsible propaganda campaign unleashed by Jordan which accused Egypt of cowardice and of hiding behind UNEF. In the Middle East, such gibes have consequences. In this case, the consequences were to be most dire for Jordan. The major factor leading to war, however, was Egypt's unjustified confidence. Until December 1966, Ben-Zur believes, the Egyptian general staff felt that their chances of winning a war with Israel were only 50:50. By May 1967 they calculated the odds in their favour as between 80 and 100 per cent.[30] The error in these figures is well known. The Arab armies suffered another humiliating defeat and Soviet influence in the region was undermined.

When the Arabs needed help, the Soviets were not to be seen. They did not even spread rumours about 'volunteers', as in 1956. On the last day of the war, when the IDF was moving towards Damascus, the Soviets began an energetic diplomatic campaign to save their Syrian clients from collapse. On 10 June the Kremlin sent a message to the President of the United States warning that unless the United States Government applied pressure on Israel to halt its advance, the Soviet Union would intervene directly on Syria's behalf. Moscow also broke off diplomatic relations with Israel, threatening sanctions unless military operations were stopped immediately. As early as the mid-1960s, Syria had become the Soviet's most reliable client state in the Middle East, hence the emphasis on saving the regime. Although it is questionable whether in 1967 the Soviet Union possessed the means of intervening effectively on Syria's behalf, nevertheless the pressure applied to the United States and Israel was effective. Moscow had been careful not to demand more than Washington and Jerusalem were willing to agree, and to demand nothing until the very last moment, when the danger of being drawn into the conflict was virtually nil.

The evidence pointing to deliberate deception about the supposed Israeli threat to Syria is overwhelming. What was the Soviet motive? As the campaign began long before the Egyptians were converted to the belief that they could win against Israel, it is unlikely that, at this stage, the purpose was to unleash a major Middle East war. Raising tensions, getting rid of UNEF, bringing the Egyptians back into the front line, putting Israel under intense pressure at a time when American support might be absent: all these were likely motives. At the last moment, after the Egyptian conversion to optimism, the motive might have changed. There is no evidence either way, although Brezhnev's April 1967 meeting and alleged remark about 'one decisive blow' would fit chronologically and rationally into such a widened agenda. If the eventual Soviet motive did include war, however, this must have been underwritten by a miscalculation of relative fighting qualities of quite staggering dimensions. If war was not on the Soviet agenda but only heightened tension, then the Soviet planners failed to anticipate how threatening Nasser's military deployment and closing of the Straits of Tiran would seem to Israel—justifying a preemptive offensive strike. Self-deception, brought on by concern over the Syrian client's insecurity, the reported strength and efficiency of Nasser's forces, and the supposed weakness of both Israel and America, almost certainly contributed to Moscow's failure. Above all, this example illustrates the difficulty of controlling a deception operation involving volatile and unstable allies and of knowing where it will lead. The 'threat to Syria' operation is the quintessential runaway deception operation. It also backfired on the instigator, creating results which were exactly the opposite of those desired.

Endnotes

1. The following are general works about the region: W B Fisher, *The Middle East. A Physical, Social and Regional Geography*. (Cambridge: Cambridge University Press, 1978), (7th edition Revised); George Lenczowski, *The Middle East in World Affairs* (Ithaca: Cornell University Press, 1980) (4th edition); Peter Mansfield, ed., *The Middle East: A Political and Economic Survey* (Oxford: Oxford University Press, 1980) (5th edition); Ronald M Devore *The Arab Israeli Conflict: A Historical Political Social and Military Bibliography* (Santa Barbara, California: Clio Books, 1976. The War/Peace Bibliography Series; Richard Dean Burns, ed.).
2. Uri Ra'anan, *The USSR Arms the Third World: Case Studies in Soviet Foreign Policy* (Cambridge, Mass: MIT Press, 1969), passim; Stephen S Kaplan *et al*, *Diplomacy of Power: Soviet Armed Forces as a Political Instrument* (Washington, DC: Brookings, 1981), pp. 27–64.
3. Ra'anan, *op. cit.*; Arnold Krammer, *The Forgotten Friendship: Israel and the Soviet Block, 1947–53* (Chicago: University of Illinois Press, 1974)—The arms for Syria were intercepted by Haganah agents and eventually finished up in Israeli hands.
4. For details see Michael Bar-Zohar (i), *Ben-Gurion: Biographia Politit* (A Political Biography) [Hebrew] 3 Vols (Tel Aviv: An Oved, 1977); Uri Bialer, 'The Czech-Israeli Army Deal Revisited' *The Journal of Strategic Studies* Vol. 8, September 1985, No. 3, pp. 307–15. Meir Mardor, *Shilkhut Aluma* (Secret Mission), (Tel Aviv: Ma'arachot, 6th edition, 1965), especially pp. 271–328; and Ezer Weizman, *Lekha Shamaim Lekha Ha'aretz* (To You the Sky to You the Land) [Hebrew]. (Tel Aviv: Ma'ariv, 1975). The abbreviated English translation is Ezer Weizman, *On Eagles' Wings* (Tel Aviv: Steimazky's, 1976); David J Bercuson, *The Secret Army* (Toronto: Lester and Orpen Dennys, 1983); Nadav Safran (i) *Israel, the Embattled Ally* (Cambridge: Belknap Press of Harvard, 1978); Krammer, cited.
5. See Michael I Handel (i), 'Intelligence and Deception,' *The Journal of Strategic Studies* Vol. 5, No. 1 (March 1982); and Handel (ii), 'Crisis and Surprise in Three Arab–Israel Wars', in Klaus Knorr and Patrick Morgan, eds., *Strategic Military Surprise. Incentives and Opportunities* (New Brunswick, New Jersey: Transaction Books, 1983).
6. Abraham Ben-Zur, *Gormim Sovyetim. U-Milkhemet Sheshet Ha-yamim* (Soviet Factors and the Six-Day War) [Hebrew], (Hakibutz Ha'artzi, 1975); Bar Zohar (i), *op. cit.*; Moshe Dayan (ii), *Avnay Derekh* (Story of my Life) [Hebrew] (Tel Aviv: Idanim, 1976); Spectator (Pseudonym, Major General Shlomo Gazit) 'Ma'agal Hatauyot shel Moskva' ('Moscow's Circle of Mistakes') [Hebrew] *Ma'arachot* No. 209 (August 1970); Yitzhak Rabin, *Pinkas Sherut* (Memoirs) [Hebrew], 2 Vols. (Tel Aviv: Ma'ariv, 1979); Weizman, *op. cit.*
7. Adam Ulam, *Expansion and Coexistence: the History of Soviet Foreign Policy 1917–1967* (New York: Praeger, 1968), pp. 741–7.
8. Moshe Gilboa. *Shesh Shanim, Shisha Yamim* (Six Years, Six Days) [Hebrew], (Tel Aviv: An Oved 1968).
9. Ben-Zur (i), *op. cit.*, relies on interviews as well as the Arab and Soviet Press. Some of his points suggest that he had access to intelligence services in Israel and the West, as well as Soviet defectors and immigrants.
10. *The USSR and Arab Belligerency* (Israel: Ministry for Foreign Affairs, Information Division, 1967), p. 13.
11. *Ibid.*, p. 19.
12. *Ibid.*
13. Gilboa, p. 61 and fn. 17.
14. *Ibid.*
15. *The USSR and Arab Belligerency*, p. 29.
16. *Ibid.*, p. 37.
17. *Ibid.*, p. 45.
18. *Ibid.*, p. 47.
19. Michael Bar-Zohar (ii). *Embassies in Crisis. Diplomats and Demagogues Behind the Six-Day War.* Translated from the French by Monro Stearns (Englewood Cliffs, New Jersey: Prentice Hall, 1970), p. 13.
20. *Ibid.*, p. 1.

21. This was confirmed by Nasser in his resignation speech of 9 June 1967. See *International Documents on Palestine, 1968*, ed. Fouad A Jabber (The University of Kuwait, Institute for Palestinian Studies: Beirut, 1970), Doc. No. 372, pp. 586–7, (9 June 1967). The fact that this was such a high level source added reliability to the information. Sadat cabled this immediately to President Nasser. See also Ben-Zur, *op. cit.*, p. 199,
22. Bar-Zohar (ii), p. 1; see also Lawrence L Whetten, 'The Arab–Israeli Dispute: Great Power Behaviour', *Adelphi Papers, No. 128* (London: IISS, 1977), pp. 3–5.
23. *International Documents on Palestine, op. cit.*, Doc. No. 318, p. 539 (22 May 1967).
24. *Ibid.*, Doc. No. 372, pp. 596–7 (9 June 1967) (emphasis added).
25. *Ibid.*, Doc. No. 393, pp. 620–21 (23 July 1967).
26. Ben-Zur, *op. cit.*, p. 212.
27. *Ibid.*, pp. 200, 209.
28. *Ibid.*, p. 208.
29. Mohammad Hassanin Heikal, *Sphinx and Commissar* (London: Collins, 1978), p. 28. In another place (*Ibid.*, p. 179), Heikal quotes Gretchko as saying 'Stand firm. Whatever you have to face you will find us with you. *Don't let yourself be blackmailed by the Americans or anyone else*' (emphasis added).
30. Ben-Zur, *op. cit.*, pp. 32–3.

CHAPTER 7

Deception and Revolutionary Warfare in Vietnam

GUENTER LEWY

To fight and conquer in all your battles is not supreme excellence; supreme excellence consists in breaking the enemy's resistance without fighting.

Sun T'zu[1]

The failure of the United States to achieve its key objective in South-East Asia—the creation of a free and independent South Vietnam—was due to many different factors. Despite much talk about 'winning the hearts and minds', the Americans never really learned to fight a counter-insurgency war and used force in largely traditional ways. Despite often heard charges that the South Vietnamese were American puppets, the fact is that the United States lacked or failed to use the leverage necessary to prevent its ally from making crucial mistakes in relation to military tactics, pacification and social policies generally. However, the decisive factor was that time ran out on the American effort in Vietnam. The capacity of people in a modern democracy to support a limited war is precarious at best. The mixture of propaganda and compulsion which a totalitarian regime can muster, in order to extract such support, is not available to the leaders of a democratic state. Hence, when a war for limited objectives drags on for a long time, it is bound to lose the popular backing which is essential for its successful pursuit. It may well be, as an American political scientist has concluded, that 'unless it is severely provoked or unless the war succeeds fast, a democracy cannot choose war as an instrument of policy'.[2]

That American public opinion, as Leslie Gelb has put it, was 'the essential domino' was, of course, recognised by both American policymakers and the Vietnamese Communists. Each geared his 'strategy—both the rhetoric and the conduct of the war—to this fact'.[3] And yet, given the limited leverage which the leaders of a democracy have on public opinion, the ability of American decision-makers to control this 'essential domino' was always precarious. For the Vietnamese Communists, on the other hand, ideological mobilisation at home and carrying the propaganda effort to the enemy were both relatively easy, and they worked at those two objectives relentlessly and with great success. Enormous effort, and huge quantities of manpower and money were devoted by them to the struggle for the sympathy and support of the outside world in general and the American people in particular. This struggle involved propaganda, designed to create the image of the Vietcong as freedom fighters, engaged in a just war against an imperialist aggressor and his corrupt puppets. It also utilised elaborate schemes of outright deception. It is the purpose of this chapter to describe some of these efforts at deception, to show why they were successful, and to assess their impact on the outcome of the war. Deception was used to project a false picture of the origins and character of the communist insurrection, to create a myth of systematic American war crimes, and to disguise the nature and goals of the North Vietnamese regime.

* * *

The 1954 Geneva Conference ended the French Indochina war, and brought about French withdrawal from the region. The Accords temporarily partitioned Vietnam into two sectors, north and south, which were to be reunified after elections in 1956. By July 1955, the end of the time period prescribed by the Accords for a change of residence, about one million persons had left the communist regime in the North in order to settle in the South, while about 80,000 to 100,000 Viet Minh troops and supporters had gone north. Many of the regrouped Viet Minh had contracted local marriages in order to establish family connections in the South. After further training in the North, these men were subsequently sent back as leaders of the southern insurgency. The communists also left behind several thousands of their best cadres as well as a large number of weapons caches. Five years later, these units were to become the nucleus of the developing Liberation Army, as it was to be called. According to documents captured later and the testimony of a high-ranking defector, the communists never expected the elections envisaged by the Geneva Conference to take place and from early on prepared for a strategy of armed struggle to reunify the country.[4]

The massive campaign of forceful oppression against the Viet Minh carried out by the Diem regime in South Vietnam in 1956–7 has been cited by both communists and non-communists as the real cause of the Southern insurgency. The spontaneity of the insurrection in South Vietnam was asserted in the 1960s by the publicity apparatus of the international communist movement[5] as well as by some Western scholars. According to this thesis, 'The people were literally driven by Diem to take up arms in self-defence'.[6] Eventually, in 1960, Southerners opposed to Diem formed the National Liberation Front (NLF), which

> gave political articulation and leadership to the widespread reaction against the harshness and heavy-handedness of Diem's government. It gained drive under the stimulus of Southern Vietminh veterans who felt betrayed by the Geneva Conference and abandoned by Hanoi . . . Insurrectionary activity against the Saigon government began in the South under Southern leadership not as a consequence of any dictate from Hanoi, but contrary to Hanoi's injunctions.[7]

Evidence available today—based on captured documents, the testimony of defectors familiar with internal party directives, and the boasting of North Vietnamese leaders since their victory in 1975—contradicts almost all this thesis. It is true that the repressiveness of the Diem regime in the years 1956–9 created pressure for armed action in the South, but the rest of the argument is false. The decision to begin the armed struggle in the South was made by the Central Committee of the Vietnamese Workers' (*Lao Dong*) Party (VWP), the Communist party of Vietnam, in Hanoi in 1959. 'The view that a co-ordinated policy of armed activity was initiated in the South by a militant group outside the Party, or by a militant southern faction breaking with the national leadership,' wrote the well-informed Jeffrey Race, a critic of the subsequent American intervention, 'is not supported by historical evidence—except that planted by the Party . . .' Two defectors separately interviewed by him, Race related, 'found very amusing several quotations from Western publications espousing this view. They both commented humorously that the Party had apparently been more successful than was expected in concealing its role.'[8] The NLF, the evidence clearly shows, was formed at the instigation of the party in Hanoi; it was established as a typical communist front organisation to hide the direction of the insurgency by the communists. 'The Central Committee', one defector stated, 'could hardly permit the International Control Commission to say that there was an invasion from the North, so it was necessary to have some name . . . to clothe these forces with some political organisation'.[9]

Why did the party wait until 1959 before launching the armed phase of the revolution? First, in the years immediately following the Geneva Conference, the communists in the North had severe problems with their own 'counter-revolutionaries'. In 1955–6, perhaps as many as 50,000 were executed in connection with the land reform law of 1953 and at least twice as many were arrested and sent to forced labour camps. A North Vietnamese exile put the number of victims at one-half million.[10] These domestic difficulties dictated a policy of waiting. The North Vietnamese leader, Ho Chi Minh, told the Fatherland Front Congress in 1955 that 'the North is the foundation, the root of our people's struggle . . . Only when the foundation is firm, does the house stand firm.'[11]

Secondly, the Central Committee felt that the revolutionary situation in the South was not yet 'ripe', the masses had not yet become convinced that armed struggle was really necessary. During the years 1956–9 the party contributed to the development of a revolutionary situation by assassinating the most effective local administrators, school teachers, medical personnel and social workers who tried to improve the lives of the peasants. This programme of systematic terror, known as 'the extermination of traitors', predictably goaded the Diem regime into stepping up its clumsily pursued and often brutal anti-terrorist campaign, creating an air of capricious lawlessness. A prominent defector recalls,

> the more the people were terrorised, the more they reacted in opposition, yet the more they reacted, the more violently they were terrorised. Continue this until the situation is truly ripe, and it will explode . . . we had to make the people suffer, suffer until they could no longer endure it. Only then would they carry out the Party's armed policy. That is why the Party waited until it did.[12]

By late 1958, as a result of Diem's anti-communist campaign, the party apparatus in the South had incurred severe losses, but the revolutionary potential was increasing. The southern branch of the Party increasingly now demanded a change in policy. The decision to form armed units throughout the South in order to smash the Saigon government was made by the Fifteenth Conference of the Central Committee, meeting in Hanoi in January 1959, though the new policy directive was not issued until May 1959.[13] July 1959 saw the beginning of large-scale infiltration of armed cadres trained to raise and lead insurgent forces; it is estimated that during 1959–60 some 4,000 Southerners who had gone north in 1954 returned to South Vietnam.[14] In January 1960 the North Vietnamese commander General Vo Nguyen Giap declared that 'the North has become a large rear echelon of our army. The North is the revolutionary base for the whole country.'[15]

The commitment of the Democratic Republic of Vietnam to support the southern insurgency was promulgated at the Third Congress of the VWP, which convened in Hanoi in September 1960. According to Secretary-General Le Duan, the party faced the task of promoting socialist construction in the North, making it 'an ever more solid base for the struggle for national reunification', as well as the task of liberating 'the South from the atrocious rule of the US imperialists and their henchmen'. To this end, the congress resolved that 'to ensure the complete success of the revolutionary struggle in South Vietnam, our people there must strive to . . . bring into being a broad National United Front directed against the US and Diem and based upon the worker-peasant alliance'.[16] The formation of the NLF was reported by the news media in Saigon in December 1960; nothing was known or said, of course, about the NLF's link to the Party leadership in Hanoi.

The 10-point programme of the NLF borrowed extensively from Le Duan's speech at the Third Party Congress, but otherwise North Vietnam made great efforts to show its non-involvement in the formation of the NLF, probably at least in part to deny the United States the excuse to expand the war to North Vietnam. Not until the end of January 1961 did Radio Hanoi announce that various forces opposed to the 'fascist' Diem regime on 20 December 1960 had formed the National Front for the Liberation of South Vietnam and had issued a manifesto and programme. Also designed to conceal the Party's role in the revolutionary movement in the South was a 'Declaration of the Veterans of the Resistance', issued in South Vietnam in March 1960, calling for armed struggle against Diem and the formation of a government of national union. As defectors told Jeffrey Race, the declaration 'was simply the product of a meeting called in accord with Central Committee policy, with the dual purpose of arousing internal support for the new phase of the revolution and of misleading public opinion about the true leadership of the revolution'.[17]

The NLF as well as the People's Liberation Armed Forces, established on 15 February 1961, undoubtedly included non-communists opposed to Diem's autocratic rule; however all key positions were in the hands of party members. For tactical reasons, NLF spokesmen occasionally took positions that differed slightly from the Hanoi line; from time to time, tensions may also have developed between what in effect was Hanoi's field command in the South and its parent headquarters in the North.[18] However, the ultimate control of the NLF by the VWP was never in doubt. 'Although welcoming support from all quarters', agree Kahin and Lewis, 'from its inception the organisation seems to have been dominated by the communists'.[19] Not every critic of American policy in Vietnam saw this domination so early; others tended to believe in an independent, essentially nationalist NLF.

The same tactical considerations which dictated the formation of the NLF led to the establishment of the People's Revolutionary Party in January 1962, allegedly an independent organisation without ties to the VWP in the North. In point of fact, the PRP was simply the southern branch of the VWP, set up to constitute the 'vanguard' of the NLF and to overcome the ideological isolation of communist cadres in the South. For the rural population in South Vietnam all these manoeuvres meant nothing new. As one defector told Jeffrey Race:

> The formation of the People's Revolutionary Party had no significance to the peasantry. They live in intimate contact with the Party and thus were aware that it was still the communists. The People's Revolutionary Party was useful only in dealing with city people, intellectuals, and foreigners.[20]

The perceived need to resist communist aggression was the key legitimising force behind American policy in Vietnam. The myth of an independent, nationalist struggle of David versus Goliath in South Vietnam, with America in the unpopular role, could undermine this legitimacy. Moreover, it was a myth that might readily be received in the West by those disposed to see the war in neo-colonial terms and by others who feared that a 'crusade against Communism' might eventually lead to nuclear war between super-powers. It offered an easy way out. Understandably, the propaganda apparatus of Hanoi and the NLF, Moscow and other communist states pushed the deception hard.

An organisation active in the international propaganda campaign was the Vietnam Support Committee, an offshoot of the Soviet-controlled Afro-Asian Solidarity Committee which, in its turn, was an instrument of the International Department of the Communist Party of the Soviet Union (ID, CPSU). The Afro-Asian Committee had some 90 African and Asian 'Solidarity Committees' and 'national liberation movements' affiliated, in addition to associate members in East and West Europe, the United States and Canada.[21] This network was turned over to the Vietnam cause, each unit developing or encouraging apparently indigenous local support for the NLF.

In the Autumn of 1967, for instance, Sol Stern was one of 41 'assorted anti-war Americans and delegations from the NLF and from North Vietnam' who attended a conference at Bratislava, Czechoslovakia expressly designed, it would seem, to propagate the 'independence' deception. The effectiveness of this form of transmission can be judged by Stern's subsequent writing:

> One of the great American delusions about Vietnam, shared even by many of official Washington's critics, is that the key to peace lies in Hanoi . . .

> [but] it is the Front [the NLF], alive and doing quite well on its own, which is confronting the Pentagon . . . The Front seems to be embarked on a course of trying to create a broad working coalition secretly and quietly while fighting is still going on . . .[22]

However, the impotence of the NLF was to be revealed after the collapse of South Vietnam in 1975, when all important positions in the temporary administration of the South and later in the government of the united Vietnam were given to Northerners. In August 1976, Jean Lacouture finally conceded that he had overestimated the autonomy of the NLF which, he now acknowledged, was 'piloted, directed and inspired by the political bureau of the Lao Dong Party, whose chief was and remains in Hanoi'.[23] In a French television documentary, broadcast on 16 February 1983, the North Communist military commanders, General Giap and his colleague General Vo Bam, confirmed that the NLF had been organised as a result of a decision taken by the North Vietnamese Communist Party. They proudly described the building of the Ho Chi Minh Trail, which was used from 1959 onwards to infiltrate in 'absolute secrecy' men and supplies into the South. The London *Economist*, which reported on this programme on 26 February 1983, commented: 'A quarter of a century after the original events, Hanoi's men evidently feel that the West's desire to forget all about the war is strong enough to allow them to come clean at last.'

Unfortunately, all this came too late. Between 1965 and 1967, the communists succeeded in projecting the illusion of the NLF as an independent and indigenous southern political entity with a policy of its own, fighting for the cause of freedom, independence and justice. This effort complemented and coalesced with the independent judgements of those in the West who blamed the war on Diem and were predisposed to think well of the insurgents. The combined effect was an erosion of domestic and international support for the policy of the United States in Vietnam, because the legitimacy of that policy was in doubt.

* * *

Since the victory of their cause in 1975, the communist leaders of Vietnam have spoken with pride of the success of their diplomacy and international activities in support of the war against the United States. In an article entitled 'Thirty Years of Diplomatic Struggle', Vice-Premier and Foreign Minister Nguyen Duy Trinh described how, in order to isolate and weaken the United States and to win support from the people of the world, the North Vietnamese concentrated 'on the denunciation and condemnation of the unjust and barbarous character of the US war of aggression and US war crimes'. This campaign succeeded in moving 'the conscience of humanity and put US imperialism in the pillory'.[24]

The campaign consisted of straightforward communist propaganda backed up and made to seem credible by media events in neutral or Western countries. On 13 November 1965, *Izvestia* wrote that the Pentagon had trained thousands of professional murderers. 'They are conscious murderers. They kill with pleasure, they kill on orders and of their own will . . . Sometimes prisoners are interrogated aboard airplanes, before an open door. If the prisoner refuses to answer the questions, he is thrown out of the plane.' And earlier, on 26 August of the same year, *Pravda* had described in lurid detail American air attacks against North Vietnam, concluding: 'This is true genocide.'

The first of the many public fora set up to publicise and denounce such alleged American war crimes was the International War Crimes Tribunal organised by Bertrand Russell. The idea of organising a tribunal to pass judgement on American war crimes in Vietnam had been suggested to Russell by M S Arnoni, the editor of the American journal *Minority of One*, in 1965. Russell was sceptical at first, but he soon changed his mind, probably under the prodding of his young American confidant, Ralph Schoenman.[25] In early 1966 Schoenman was sent to North Vietnam to collect evidence. Foreign Minister Nguyen Duy Trinh later spoke with satisfaction of the Party's wisdom of 'promoting' the Russell Tribunal.[26] The members of this Tribunal, beginning with Jean-Paul Sartre, the executive president, were all well-known partisans of North Vietnam such as Vladimir Dedijer, Simone de Beauvoir, Stokeley Carmichael, Dave Dellinger, Carl Oglesby, Isaac Deutscher; indeed, the tribunal was described by its secretary-general, Schoenman, as 'a partial body of committed men'. President Johnson had been invited to appear in his own defence to answer charges that the United States was waging a 'war like that waged by fascist Japan and Nazi Germany in South-East Asia and Eastern Europe, respectively', but he declined to participate in a proceeding characterised by an assumption of guilt before trial, a tribunal, as Russell put it, convened 'in order to expose . . . barbarous crimes . . . reported daily from Vietnam'.[27]

Many potential supporters were disillusioned by this abandonment of any pretence of impartiality. One letter of protest asked whether it was indeed the position of the tribunal that 'when a little child is killed by American napalm it is clearly a crime, but that if that same child were killed by an NLF terrorist it would be no crime at all?'[28] Others, on the other hand, were so pleased with the contribution of the tribunal to the anti-war cause that they swallowed their reservations. Professor Richard A Falk, who in 1967 had called the Russell tribunal 'a juridical farce', by 1968 argued that, despite a 'one-sided adjudicative machinery and procedure', the tribunal 'did turn up a good deal of evidence about the manner in which the war was conducted and developed persuasively some of the legal implications it seems reasonable to draw from that war'.[29] In 1971 Falk wrote that the proceedings of the Tribunal 'stand up well under the test of time and independent scrutiny'.[30]

Acting as an accuser, juror and judge all at once, the tribunal at times found it difficult to make its witnesses perform as expected. For example, Donald Duncan, a member of the American Special Forces who had served in Vietnam until September 1965, testified about some instances of torture during his tour of duty early in the war. He was then asked by one of the Judges, the Pakistani lawyer Muhamud Ali Kasuri: 'Now, if I were to conclude now that there are more numbers of American troops in Vietnam, the evidence which has been given that there are more cases of direct barbarities by American troops should be believed, would you have anything to say against that?' Duncan replied that he was really familiar only with events up to and including September 1965. He had read and heard about developments after that time, 'but I have no facts and figures and I certainly have no firsthand knowledge of it'. Kasuri continued to press him: 'But you wouldn't be in a position to say that this could not be true?' To which Duncan answered: 'Oh, no.' Such testimony helped the Tribunal reach the conclusion that the torture of prisoners 'is practiced daily'.[31]

Communist propaganda against the American bombing of North Vietnam was massive and ceaseless, and the Russell Tribunal did its part in publicising allegations of 'barbarous crimes.' In May 1967 witnesses before the Tribunal charged that American flyers systematically and intentionally bombed North Vietnamese medical facilities. Hospitals, it was stated, 'are shown *on the maps of targets* in the hands of the US pilots who have been shot down . . .'. (emphasis added). This testimony, of course, lent itself to a different interpretation: the pilots had to have good maps and hospitals would be marked in order to avoid hitting them. To present such a construction, the United States Committee of Concerned Asian Scholars reported the testimony in this way: 'Maps with hospitals *marked as targets* on them have been found in the possession of US pilots shot down over North Vietnam' (emphasis added).[33]

One member of an investigating team sent to North Vietnam by the Russell tribunal, a Scottish professor, was bothered by the question of intent. He told Harrison E Salisbury:

> It is easy enough to report that there has been enormous destruction of civilian property and non-military objectives. Anyone who travels in North Vietnam can see this. But is it intentional? This is the difficult question. How can we read the mind of the pilot who dropped the bombs or loosed the rockets? How can we know, for certain, what his orders were? How can we know from the ground what the airman high in the sky really thought he was doing?[34]

But the tribunal preferred to ignore such scruples and did not call the professor as a witness. On 10 May 1967 it concluded unanimously that 'the government and armed forces of the United States are guilty of the deliberate, systematic and large-scale bombardment of civilian targets, including civilian populations, dwellings, villages, dams, dykes, medical establishments, leper colonies, schools, churches, pagodas, historical and cultural monuments'. By subjecting the civilian population of the DRVN to such bombardment the United States had committed a war crime.[35]

Going a step further, the Tribunal next convicted the United States of genocide. At its second session, in 1967, it adopted a statement formulated by Jean-Paul Sartre according to which the United States Government was engaged in 'wiping out a whole people and imposing the Pax Americana on an uninhabited Vietnam'. In the South, specifically, American forces were conducting the 'massive extermination' of the people of South Vietnam, killing men, women and children merely because they were Vietnamese, and this represented 'genocide in the strictest sense'.[36] Whilst such damning findings were doubtless impressive in a propaganda sense, they strained the sympathy of some more serious critics of American policy in Vietnam. The American philosopher Hugo A Bedeau noted, for instance, that although obliteration bombing, free-fire zones and search-and-destroy missions 'tended towards genocidal results', these tactics were not employed with the *intent* to destroy the Vietnamese people *as such*. Thus while the charge of genocide, according to Bedeau, had 'undeniable rhetorical appropriateness', in actual fact the United States had not committed the crime of genocide.[37] Indeed, population figures compiled by the United Nations show that the populations of North and South Vietnam increased steadily during and despite the war—at annual rates of change roughly double that of the United States.[38] The use of the word genocide by the Tribunal was politically motivated. By association, all American military actions in Vietnam might come to be seen in these terms.

The theme was resurrected during the American bombing of the Hanoi-Haiphong complexes in December 1972. This bombing campaign, Hanoi charged, surpassed the atrocities committed by the Hitlerite fascists and represented an 'escalation of genocide to an all-time high'. In only 12 days 'the Nixon administration wrought innumerable Oradours, Lydicies, Guernicas, Coventrys . . .'.[39] Anti-war groups and commentators in America jumped on the bandwagon. A *New York Times* editorial of 22 December 1972 headlined 'Terror from the Skies', protested what it called the 'indiscriminate use of the United States' overwhelming aerial might'. Columnist Anthony Lewis on 23 December called the bombing a crime against humanity.'[40] One scholar spoke of 'unabashed terror tactics' and a 'flaunting of established norms of civilised conduct'.[41]

By any rational assessment, both Hanoi's outburst and its echo in America were deceptive propaganda. The Hanoi death toll, wrote the London *Economist* 'is smaller than the number of civilians killed by the North Vietnamese in their artillery bombardment of An Loc in April or the toll of refugees ambushed when trying to escape from Quang Tri at the begining of May. That is what makes the denunciation of Mr Nixon as another Hitler sound so unreal.'[42]

Among the civilian facilities hit was the Bach Mai Hospital in Hanoi. The North Vietnamese cited the extensive damage to this hospital as proof of American criminal intentions, and the charge of deliberate attacks on civilian targets was accepted as true by Dale S De Haan, counsel to the Kennedy committee on refugees, who visited Hanoi in March 1973, and by Senator Kennedy himself.[43] However, other observers offered a different explanation. The hospital unfortunately was located about 1,000 yards from the Bach Mai airstrip and its military barracks, which were heavily bombed. The attack, wrote Telford Taylor after visiting the site in January 1973, 'was probably directed at the airfield and nearby barracks and oil-storage units'.[44] Aerial photographs released by the Defence Department in May 1973 further confirmed that the hospital was hit by bombs escaping the normal bomb train.[45]

Clearly, communist deceptive propaganda in this instance was not acccepted uncritically by the Western media. The piece by the *Economist* illustrated this reluctance to be deceived, and over the Bach Mai Hospital, Murray Marder of the *Washington Post* and Peter Ward of the *Baltimore Sun* concurred in Telford Taylor's view. Moreover, the bombings may have created one occasion when the North Vietnamese policy of encouraging media and other visitors backfired. Malcolm W Browne of the *New York Times* was greatly surprised by the condition in which he found Hanoi and wrote that 'the damage caused by American bombing was grossly overstated by North Vietnamese propaganda . . .'. 'Hanoi has certainly been damaged', noted Peter Ward of the *Baltimore*

Sun on 25 March 1973 after a visit, 'but evidence on the ground disproves charges of indiscriminate bombing. Several bomb loads obviously went astray into civilian residential areas, but damage there is minor, compared to the total destruction of selected targets'.[46] Stanley Karnow, who visited Vietnam in early 1981, concluded that 'Hanoi and Haiphong are almost completely unscathed', and that the B-52s, programmed to spare civilians, had been able to pinpoint their targets 'with extraordinary precision'.[47]

However, a failure on Hanoi's part to deceive the West's elite media would not necessarily have implied failure against all audiences. Propaganda that is picked up and replayed by the world's newspapers may be influential, even if the leader-writers repudiate it. When audiences have been conditioned by years of ceaseless atrocity stories, and when words such as genocide have been placed in circulation within a particular context, subliminal reactions may be more significant than rational analysis.

American critics of the war seem to have followed the example set by the Vietnamese communists and used self-styled war crime tribunals to whip up anti-war sentiment. In early 1970 three young American anti-war activists, including a West Point graduate disillusioned after service in Vietnam, dubbed themselves the National Committee for a Citizens' Commission of Inquiry on US War Crimes in Vietnam (CCI) and organised a series of hearings in various locations across the country at which veterans testified about their personal experience in Vietnam. In the first week of December 1970 they convened a large hearing, lasting three days, at the DuPont Plaza Hotel in the national capital.[48]

Some 40 veterans testified in Washington during the first three days of December 1970. Among them was a former military intelligence officer and alleged CIA operative, Kenneth Burton Osborn, whom one reporter called the inquiry's uncontested superstar. Osborn testified about instances of torture, but he refused to give the names of the individuals involved on the grounds that he had signed an agreement with the CIA not to reveal the specifics of secret operations. Such agreements, he explained, 'are an attempt on the organisation's part to cover their ass', but he had decided to adhere to it 'to avoid endangering intelligence operatives who are still active'.[49] Several years later, Osborn became an official of the Organizing Committee for a Fifth Estate, the publishers of the magazine *Counterspy* which freely and frequently exposed CIA operatives. In 1970, he gave another reason for not revealing names: 'There's no reason to identify them. The thing to do is to attack the thing at its source, which is at the policy-making level'.[50] This was in line with the official position of the CCI which, as one of its spokesmen stated at the end of the Washington hearing, sought to show 'that war crimes in Vietnam are not isolated, aberrant acts; that war crimes are a way of life in Vietnam; and that they are a logical consequence of our war policies'.[51] The generalised, sweeping condemnation, which could not be tested by the rules of evidence, was preferred to the detailed exposure of facts, which could.

The refusal of men like Osborn to provide crucial details and factual information to back up their atrocity stories prevented the military and other authorities from investigating most of these allegations. Nevertheless, the accusers received generous publicity for their sensational charges. Both NBC and ABC network news covered the hearings. The *New York Times* and *Washington Post* ran stories under headlines such as 'Slides said to show US Torture in War', 'Random Firing in Viet Village', and, most damning, 'War Veterans at Inquiry feel "Atrocities" are Result of Policy'. This last allegation gradually became part of the conventional wisdom about the war. War crimes, concluded a paper issued by the American Friends Service Committee, were 'a considered aspect of US policy in Vietnam'.[52]

Another organisation active in airing charges of American atrocities in Vietnam was the Vietnam Veterans Against the War (VVAW), which was founded in 1967; by 1970 it was said to have 600 members. From 31 January to 2 February 1971, the VVAW, with financial backing from actress Jane Fonda,[53] convened a hearing, known as the Winter Soldier Investigation, in the City of Detroit. More than 100 veterans and 16 civilians testified at this hearing about 'war crimes which they either committed or witnessed'; some of them had given similar testimony at the CCI inquiry in Washington. The allegations included using prisoners for target practice and subjecting them to a variety of grisly tortures to extract information, cutting off the ears of dead VCs, throwing VC suspects out of helicopters, burning villages, gang rapes of women, and packing the vagina of a North Vietnamese nurse full of grease with a grease gun.[54]

Among the persons assisting the VVAW in organising and preparing this hearing was Mark Lane, author of a book attacking the Warren Commission probe of the Kennedy assassination and more recently of *Conversations with Americans*, a book of interviews with Vietnam veterans about war crimes. On 22 December 1970, Lane's book received a highly critical review in the *New York Times Book Review* by Neil Sheehan, who was able to show that some of the alleged 'witnesses' to crimes alleged by Lane had never even served in Vietnam, while others had not been in the combat situations they described in horrid detail. Writing in the *Saturday Review* a few days later, James Reston, Jr, called *Conversations with Americans* a 'hodgepodge of hearsay' which ignored 'a soldier's talent for embellishment', and a 'disreputable book'.[55] To prevent the Detroit hearing from being tainted by such irregularities, all the veterans testifying fully identified the units in which they had served and provided geographical descriptions of where the alleged atrocities had taken place.

Yet the appearance of exactitude was deceptive. Senator Mark O Hatfield of Oregon was impressed by the charges made by the veterans and inserted the transcript of the Detroit hearing into the *Congressional Record*. Furthermore, he asked the Commandant of the Marine Corps to investigate the numerous allegations of wrongdoings made against the marines in particular. The results of this investigation, carried out by the Naval Investigative Service, revealed that many of the veterans, though assured that they would not be questioned about atrocities they might have committed personally, refused to be interviewed. One of the active members of the VVAW told investigators that the VVAW leadership had directed the entire membership not to co-operate with the military authorities.[56] A black marine who agreed to be interviewed was unable to prove details of the outrages he had described at the hearing, but he called the Vietnam war 'one huge atrocity' and 'a racist plot'. He admitted that the question of atrocities had not occurred to him while he was in Vietnam and that he had been assisted in preparation of his testimony by a member of the Nation of Islam. But the most damaging findings consisted of the sworn statements of several veterans, corroborated by witnesses, that they had in fact not attended the hearing in Detroit. One of them had never been to Detroit in all his life. He did not know, he stated, who might have used his name.[57]

Careful investigations, such as that conducted by the Naval Investigative Service, help set the record straight for those historians and others who take the trouble to research. But coming as they did months or years after the initial allegations, they were almost worthless in the war of ideas that was so important to the outcome of the war in Vietnam. It is the sensational headline or television story that registers in the public consciousness, not the dry repudiation which may follow.

One of the stories told and retold by the media was that of prisoners being pushed out of helicopters in order to scare others into giving information. The Russell Tribunal in 1967 convicted the United States of this practice just as *Izvestia* had two years earlier. On 2 December 1970, Kenneth B Osborn testified at the CCI war crimes hearing in Washington that he had witnessed two Vietcong suspects being thrown to their death. On the following day the *Washington Post* ran this atrocity story under the headline 'Forced Plunge from Copter to Coerce V C is Described.' Osborn dramatically recalled the details of one of these incidents: 'He screamed on the way down.'[58]

War is a brutal business. Even when disciplined troops act within the laws and usages of warfare and in accordance with orders, the reality of battle includes burnt villages, dead civilians and terrible suffering. Counter-revolutionary warfare multiplies the brutality because the rebels reject law and conventions and deliberately place civilians in the front line. There is evidence that under the strains imposed by combat in Vietnam, discipline in the United States Armed Forces suffered and instances of torture, mutilation of corpses, and other individual acts of misconduct occurred. It is entirely possible that some American interrogators engaged in the criminal practice of throwing prisoners from helicopters, although no instance of this particular offence has ever been confirmed, while one alleged instance was unmasked as a grim hoax.[59]

Insofar as real atrocities were committed, it was appropriate that the media should have reported them. The propaganda disadvantage for the United States of such disclosures would have been justified, provided the media also reported the constant, massive and organised atrocities of the enemy. But once the American media had turned against the war in Vietnam, that is against the ends of government policy, it seemed to view all the means of accomplishing that policy as inherently evil too. This may account for a tendency to absorb the deceptive propaganda theme that American crimes were not merely individual acts of indiscipline but represented 'a way of life' approved by authority. From this belief it was but a short step for much of the media and an influential section of the American public to view the war as an atrocity writ large. Many intellectuals dedicated their talents to such national self-flagellation.[60] Newspapers carried advertisements written and paid for by guilt-ridden citizens whose capacity for objective thought seemed to have deserted them. In time, such attitudes inevitably affected young Americans drafted into army service in Vietnam, to the detriment of discipline and motivation.[61]

* * *

The tendency among Western publics to believe the worst about the conduct of the American forces in Vietnam was often accompanied by a naive acceptance of propaganda and deception from Hanoi about the benevolent character of the North Vietnamese regime, its promise of reconciliation with the South, and about the humane treatment extended to Americans captured in Vietnam. The dishonesty of these assertions with regard to the American captives was revealed after the release of the American prisoners following the 1973 Paris Agreement. By 1 April 1973, 566 United States military personnel and 25 civilians had returned alive; the remains of 23 prisoners said to have died in captivity in North Vietnam were also returned. That left 1,284 men who remained unaccounted for and were officially reported as missing. Some 50 of these had been listed by the North Vietnamese news agency as captured alive, others had been seen alive after the capture. The subsequent fate of most of them remained unknown.[62] In view of what the returned prisoners related about the conditions of their captivity, it is likely that an undeterminable number were killed some time after capture or died as a result of mistreatment.

A relatively small number of Americans were captured in South Vietnam. The majority of these prisoners died of malnutrition, lack of medical care or were executed. Four Special Forces sergeants who had been captured in November 1963 were interviewed by the Australian pro-communist journalist Wilfred G Burchett. The sergeants had assured him, he related in a book published shortly thereafter, that they were being well treated.[63] An American NLF sympathiser, writing in 1970, admitted that the Viet Cong took few prisoners and sometimes killed those that were taken. But on the subject of mistreatment, he simply stated that 'torture of prisoners by the NLF is very rarely reported'.[64] Reported or not, maltreatment was the common lot of those who were not killed, and it is an academic point whether this amounted to 'torture' or not. Those with a stomach for the clinical details of slow death by malnutrition and disease may want to read the grim tale told by two US officials who survived, published in *Nutrition Today* in May/June 1973.[65]

Just like the NLF in the South, Hanoi assured the world that, despite the terrible crimes that the American pilots had committed, they were being treated decently and indeed better than they deserved. Ever willing to be of service, Wilfred G Burchett, after an interview with North Vietnamese officials in the Spring of 1966, wrote: 'In view of what is going on in the South and in the North, it is indeed a tribute to the discipline and truly civilized outlook of the Vietnamese people that pilots have been humanely treated from the moment of their capture'.[66] Between 1967 and 1972 the North Vietnamese released 15 prisoners in batches of three to various American anti-war groups. Each release provided the North Vietnamese and the anti-war organisations receiving the prisoners with extensive newspaper, television and radio publicity and each was heralded as proof of Hanoi's conciliatory attitude. If the purpose had been truly humanitarian, the North Vietnamese would, of course, have released the seriously wounded and sick prisoners as required by Article 109 of the Geneva convention. Instead, they picked prisoners who had been in captivity a relatively short time and were in more or less presentable physical condition.[67]

On 5 September 1969 at a news conference held at Bethesda Navy Hospital, Navy Lieutenant Robert F Frishman and Seaman Douglas B Hegdahl, released the preceding month to a group of American anti-war activists led by Rennie Davis, described the brutality and torture they had experienced and witnessed. Frishman had lost 45 pounds, Hegdahl 60 pounds. At a press conference held upon arrival of the group at Kennedy International Airport in New York, Rennie Davis and his friends had praised the humane treatment of the prisoners. According to Davis, the prisoners had been protected within the very villages they had bombed and they had received better food than that provided for their guards.[68] A soft-spoken rebuttal now came from Frishman:

> I don't think solitary confinement, forced statements, living in a cage for three years, being put in straps, not being allowed to sleep or eat, removal of finger nails, being hung from the ceiling, having an infected arm almost lost without medical care, being dragged along the ground with a broken leg and not allowing exchange of mail for prisoners are humane.[69]

Pulitzer prize-winning reporter Seymour Hersh commented on Frishman's recital of horror in February 1971: 'There is evidence in the public record that Frishman seriously distorted and misrepresented the prison conditions inside North Vietnam.'[70]

By late 1969, the extensive publicity and growing concern over the mistreatment of the American captives had produced some improvement in prison conditions in North Vietnam, though the torture of prisoners evidently continued intermittently. The wish to ignore or disbelieve the unpleasant news about prison conditions in North Vietnam persisted, however, and in December 1970 Dave Dellinger denounced what he called the 'Prisoner of War Hoax—the Nixon Administration's creation of the myth of the innocent and mistreated American prisoners held in North Vietnam'. 'The only verified torture associated with American prisoners held by the North Vietnamese', he continued, 'is the torture of the prisoners' families by the State Department, the Pentagon and the White House.'[71] Back in 1967, wrote Dellinger, he had personally interviewed Commander Richard Stratton in Hanoi, who had assured him that he was being well-treated. Stratton, it was revealed in 1973,[72] following his capture on 5 January 1967, had been brutally tortured until he had signed a much-publicised confession just two months before Dellinger's arrival: 'Since my capture, I have been led to see the full and true nature of my criminal acts against the people of Vietnam in terms of injury, death and destruction. I sincerely acknowledge my crimes and repent at having committed them.'[73] Stratton's deep bow before his captors, we now know, was meant to show the American audience, which, he thought, might view his filmed appearance, that he was not in control of the situation, but this signal was lost on Dellinger.[74]

For Richard A Falk, writing in *The Progressive* in March 1971, the new concern about the mistreatment of American prisoners in North Vietnam was 'the result of a deliberate and cynical effort on the part of the Nixon administration to exploit the plight of the POW's' in order to rally support for the government's policy of 'prolonging and expanding the war'.[75] Falk referred to reports by American, French and Japanese peace groups who had visited North Vietnam and had determined that no systematic torture or brainwashing was taking place. The most informed and reliable study, Falk wrote, was that of a former State Department official, Jon M van Dyke, who had concluded that there was no evidence that North Vietnam was pursuing a policy of torture and whose 'careful analysis of the most publicised claims of torture by some released prisoners casts considerable doubt on the authenticity of their allegations'.[76] Richard J Barnet of the Institute of Policy Studies told the House Foreign Affairs Committee on 31 March 1971, that 'the evidence of mistreatment is itself highly suspect,'[77] and Steward Meacham, peace secretary of the American Friends Service Committee, informed the same committee that his own first-hand observations in Hanoi had convinced him that 'the interrogation process, unlike the situation in the prisons of the Republic of Vietnam (Saigon), is not accompanied by torture'.[78]

After the last American prisoners had returned from Vietnam in April 1973, the testimony of numerous senior officers, identical in practically every detail, finally put to rest the myth of 'humane' treatment. Testifying before Congress, at news conferences and in many books and articles, these men described how they were tortured into confessions of criminal conduct and, in preparation for meeting delegations from Eastern Europe, North Korea, Japan and America, were rehearsed with tape recorders in the testimony to be given.[79] Lieutenant Commander David W Hoffman told how he was 'persuaded' to meet with former Attorney General Ramsey

Clark in August 1972 by being hung by his broken arm. Upon his return from Hanoi, Clark assured a committee of Congress chaired by Senator Kennedy that the prisoners he had met were well-treated, had exercise, got all they wanted to eat and generally were 'good, strong Americans . . . I think when we say those men are "brainwashed", and they don't know what they are doing, we do a terrible disservice to them.'[80]

The value of the prisoner deception lay in its intended soothing effect on American public opinion. The North Vietnamese needed to be seen as injured innocents, unflinching in defence but magnanimous in victory. Such chivalry would obviously govern the North's conduct in the South, in the event that America abandoned her imperialist, atrocious war policy and withdrew her troops. This theme of deception became predominant during the concluding phase of the war, when a guilty, frustrated America was looking for a way out of its predicament. Naturally, it was not confined to prisoners of war. It dealt specifically with the happy condition that all of Vietnam would share once the Americans were gone. For many Vietnamese, it was a terminal deception.

All through the war, the NLF and the Provisional Revolutionary Government (PRG), formed in June 1969, assured the world that they were fighting not to establish Communism in South Vietnam, but rather to build a government that would be neutral and democratic. The reunification of the country would take a very long time, would proceed step by step and would be settled between the NLF and the North on a basis of equality. Reunification, Premier Pham Van Dong assured the *New York Times*' Harrison Salisbury in early 1967, would be worked out 'as between brothers'.[81]

These deceptive promises were confirmed formally in Chapter five of the Paris Agreement of 27 January 1973 that was to end the war, and helped foster the image of a genuine movement of national liberation. Within South Vietnam there were many who hoped that a victorious North Vietnam would honour the independence of the South. Among those who were rudely disabused of such illusions was one of the founders of the NLF and Minister of Justice in the PRG, Truong Nhu Tang, who finally escaped from Vietnam in November 1979. All too late, he wrote in an article entitled 'The Myth of a Liberation', 'We discovered that the North Vietnamese communists had engaged in a deliberate deception to achieve what had been their true goal from the start, the destruction of South Vietnam as a political or social entity in any way separate from the North.'[82] Soon after the fall of Saigon, all important positions in the South were given to North Vietnamese cadres and the PRG was dissolved. 'The idea that the country was reunified by some process of negotiation between northern and southern delegations followed by the election of 26 April 1976', writes another Vietnamese observer, 'was just a propaganda stunt.'[83]

The promise of a peaceful reconciliation between the communists and their opponents also helped to weaken the anti-communist opposition. Vietnamese Buddhists and other so-called 'Third Force' elements were repeatedly assured that the communists had no desire to seek revenge against their enemies. Pro-Hanoi propaganda centres in the United States, like the Washington-based Indochina Resource Center promoted the same message. The Center's director, Gareth Porter, assured the readers of the *New York Times* on 1 November 1975 that the victorious Vietnamese communists have shown 'wisdom and compassion', and that there had been no reprisals. The 're-education' of former foes, Porter explained, was not a 'punitive process'; 'the re-education courses, consisting of classes, discussion groups, farm labour and entertainment, represents an effort to counteract the monopoly of political education that the United States and its client regime had in the Saigon zone of control for more than two decades'. A brutal reality soon overtook these claims of benevolence. 'With other liberals', wrote the former PRG Minister of Justice, Truong Nhu Tang, 'I shared the romantic notion that those who had fought so persistently against oppression would not themselves become oppressors.' Yet, in fact, the communists have imposed a system of oppression 'unparalleled in Vietnam's history'.[84]

The question whether the number of political executions carried out by the victorious communists should be called a 'bloodbath' is still unresolved. It is uncontested that some 10 years after the communist victory many thousands of former Saigon officials as well as members of the Third Force remain incarcerated in jungle prisons known as 're-education camps'—uncharged, untried and unsure whether they will ever be released. Conditions in these camps are extremely harsh. The prisoners are subjected to hard labour; food and medical attention are inadequate and the mortality rate is high. A delegation of Amnesty International, which visited three such camps in December 1979, sharply criticised the continuing detention without trial of thousands of former members of the pre-1975 administration and armed forces, in clear violation of the Paris Agreement of 27 January, 1973 which prohibited such reprisals. Compulsory detention for purposes of 're-education'—without a regular review procedure, without inspection of the facilities by an independent body like the International Committee of the Red Cross, and without adequate safeguards against mistreatment and torture—Amnesty International concluded, violated 'basic principles of justice'.[85]

The horrors of the Vietnamese Gulag, standing in such sharp contrast to earlier promises of reconciliation, were denounced by a group of American former anti-war activists, led by Joan Baez. In an open letter to the Socialist Republic of Vietnam, published in the *New York Times* of 30 May 1979, the signers charged that 'instead of bringing hope and reconciliation to war-torn Vietnam, your government has created a painful nightmare'—thousands of innocent persons have become victims 'to the totalitarian policies of your government'. The United States, which by 1984 had admitted more than one half million Vietnamese refugees, including many thousands of 'boat people,' offered to take all of the political prisoners held by the communist regime.

* * *

Although only one factor among many, North Vietnamese deceptions made an important contribution to Hanoi's eventual triumph. They included: the downplaying of the communist character of the North Vietnamese regime and its revolutionary goals and the promotion in its place of a nationalist liberation myth; the concealment of northern leadership and invasion; the creation of belief in a possible compromise settlement; the denial of communist atrocities and the propagation in their place of unfounded allegations of American genocide or systematic violations of the rules of war—'guilt transfer' to American shoulders of all the blame for the horrors of conflict.

Unlike the experiences of recent major wars, where deception operations have usually been aimed covertly at the opposing leader to distort his vision of reality and thus undermine his judgement, the North Vietnamese more often addressed deception overtly to mass audiences. In South Vietnam, the principal instrument as well as victim was the National Liberation Front. In the West, especially in America, deception began with the political left and quickly spread to the liberal establishment who, in due course, gained influence over mainstream opinion. The choice of target illustrated how well the communists understood the vulnerabilities of a democracy engaged in a protracted conflict of apparent peripheral importance: the 'essential domino'—American public opinion—was recognised as the key to victory in the field because once this domino was knocked down, the United States Government was powerless to continue the fight.

The 'transmission belts' for these deceptive messages were ubiquitous, but the main ones were diplomatic, the global propaganda network controlled by the International Department,

CPSU, the fronts set up in South Vietnam and in the West to promote North Vietnam's interests and, through them, and through professional agents of influence such as Burchett and the 'innocents' who journeyed to Hanoi, the international news media. The *New York Times*'s acceptance of Indochina Resource Center material was a classic, if relatively unimportant and routine, example of the transmission belt in action.

The character of counter-revolutionary warfare, the 'imperialist' connotation of American involvement and the war's protracted and highly political nature, rather easily stimulated traditional liberal guilt over the use of force, particularly in the Third World. As the conflict wore on without prospect of early victory, this latent guilt may have created a susceptibility to the themes of American genocide and lawlessness. Certainly, once the anti-war movement was in motion, even activists who were not communist sympathisers might have felt subconsciously that the greater good of ending the war justified the lesser evil of uncritical acceptance of horror stories of doubtful veracity which might nevertheless be politically effective. Taken as a whole, Hanoi's deception operations were relatively easy because they delivered messages their intended victims wanted to hear. But the complex organisation and immense perseverance necessary to penetrate the targets were remarkable: the war may have been unique for the sheer scale of its deception.

In a wider sense, these deceptions may have affected the East-West balance of power—the correlation of forces as the Soviets call it. The loss of a war has inevitable consequences on the confidence of friends and the respect of potential adversaries. But in the case of Vietnam, the endlessly repeated accusations of American criminality and gross immorality have also left America with a strong sense of guilt which affects national self-confidence, and undermines the moral basis of America's leadership in the Western Alliance. The United States' capacity to act in world affairs has been severely curtailed.

The record does not bear out charges of genocide and wholesale violation of the laws of war. A detailed examination of battlefield practices reveals that the loss of civilian life in Vietnam was less great than in the Second World War or Korea and that concern with minimising the ravages of war was strong. It is perhaps natural that Americans should have turned their backs on that war, closing their minds to further argument. But it is important that Vietnamese deceptions be refuted not only for the sake of historical truth[86] and to prevent the teaching of falsehoods to future generations, but also to recover moral authority in the world and to alert democratic publics to the dangers of such disinformation.

Endnotes

1. James Clavell, ed., *Sun T'zu, The Art of War* (New York: Delacorte Press, 1983), p. 15.
2. Ithiel de Sola Pool in Richard M Pfeffer, ed., *No More Vietnams? The War and the Future of American Foreign Policy* (New York: Harper and Row, 1968), p. 206.
3. Leslie H Gelb, 'The Essential Domino: American Politics and Vietnam,' *Foreign Affairs* L (1972), p. 459.
4. Jeffrey Race, *War Comes to Long An: Revolutionary Conflict in a Vietnamese Province* (Berkeley: University of California Press, 1972), pp. 34–35; the defector quoted was Vo van An, a Party member for 19 years and one of only four provincial committee members to accept amnesty from the Saigon Government.
5. For instance, *Izvestia*, 31 July, 1965, p. 1 referred to the 'independent southern insurrection'; see also *Pravda* editorial 25 December 1965.
6. Philippe Devillers, 'The Struggle for the Unification of Vietnam', in P J Honey, ed., *North Vietnam Today* (New York: Praeger, 1962), p. 42. This thesis was also supported by Jean Lacouture in France and George McTurnan Kahin and John W Lewis, in the US.
7. George McTurnan Kahin and John W Lewis, *The United States in Vietnam* (New York: Dial Press, 1967), pp. 119–200; writing in the July 1965 issue of *Ramparts*, p. 130, Robert Scheer referred to the NLF's resort to violence: 'By all indications this was done without the approval of the government in Hanoi. They have since fought almost entirely on their own, painstakingly securing arms and building up combat units with little more than moral support from Hanoi and Peking.'
8. Race, p. 107, No. 5. The defectors were Vo van An and Le van Chan, former deputy secretary of the Interprovince Committee for Western Nam Bo, who had been in the Party since 1947.
9. Bernard B Fall, *The Two Viet-Nams: A Political and Military Analysis*, 2nd rev. ed. (New York: Praeger, 1967), p. 156.
10. Race, p. 122. Hoang Van Chi, *From Colonialism to Communism: A Case History of North Vietnam* (New York: Praeger, 1964), p. 72. Attempts by Gareth Porter to deny the scope of this terror remain unconvincing. See his *The Myth of the Bloodbath: North Vietnam's Land Reform Reconsidered* (Ithaca, N Y: Cornell University Press, 1972), and US Senate Committee on the Judiciary, Subcommittee to Investigate the Administration of the Internal Security Act and other Internal Security Laws, *The Human Cost of Communism in Vietnam—II: The Myth of No Bloodbath*, Hearing, 93rd Congress 1st session, 5 January, 1973.
11. Ho Chi Minh, *Selected Works*, 4:128, cited by Robert F Turner, *Vietnamese Communism: Its Origins and Development* (Stanford, Ca: Hoover Institution Press, 1975), pp. 168–9.
12. Race, p. 112, quoting Vo van An.
13. King C Chen, 'Hanoi's Three Decisions and the Escalation of the Vietnam War', *Political Science Quarterly* XC (1975), p. 247, n. 27.
14. *The Pentagon Papers: The Defense Department History of United States Decision-making on Vietnam*, Senator Gravel edition (Boston: Beacon Press, 1971) Vol. I, p. 264.
15. *Hoc Tap*, January, 1960, cited by Douglas Pike, *Viet Cong: The Organization and Techniques of the National Liberation Front of South Vietnam* (Cambridge, Mass.: MIT Press, 1966), p. 78.
16. Cited by Turner, pp. 203–4.
17. Race, pp. 107, n. 5; pp. 120–21.
18. George A Carver, Jr., 'The Faceless Viet Cong,' *Foreign Affairs* XLIV (1966), p. 372.
19. Kahin and Lewis, p. 132.
20. Race, p. 123; Pike, pp. 137–42.
21. Clive Rose, *Campaigns Against Western Defence: NATO's Adversaries and Critics* (London: Macmillan, 1985), pp. 254–5; see also testimony of Stanislav Levchenko, former major in the KGB, 14 July, 1982, in US House, Permanent Select Committee on Intelligence, *Soviet Active Measures*, Hearings, 97th Congress, 2nd Session, 13–14 July, 1982, pp. 139–40; see also Richard H Shultz and Roy Godson, *Dezinformatsia: Active Measures in Soviet Strategy* (Washington, DC, London: Pergamon-Brassey's, 1984), pp. 64–6, 124–6, 181–2.

22. Sol Stern, 'A Talk with the Front', *Ramparts*, November, 1967, pp. 31–2; seven months later Tom Hayden wrote of the NLF 'broadening its political programme to enlist new support from the Vietnamese middle and upper classes . . .', *Ramparts*, 29 June 1968.
23. Jean Lacouture, "A Bittersweet Journey to Vietnam', *New York Times*, 23 August, 1976.
24. *HocTap*, October 1975, excerpted in *Vietnam Courier*, December 1975, p. 20.
25. Ronald William Clark, *The Life of Bertrand Russell* (New York: Knopf, 1975), p. 624.
26. *Vietnam Courier*, December, 1975, p. 21.
27. Russell to Johnson, 25 August 1966, in John Duffett, ed., *Against the Crime of Silence: Proceedings of the International War Crimes Tribunal* (New York: Simon and Schuster, 1970), pp. 18-19, 49-50.
28. Clark, p. 625.
29. Richard A Falk (i), ed., *The Vietnam War and International Law*, Vol. I (Princeton, N J: Princeton University Press, 1968), p. 451, n. 12, and Vol. II (1969), p. 252.
30. Richard A Falk (ii), 'The American POWs: Pawns in Power Politics', *The Progressive*, XXXV, No. 3 (March 1971), p. 16.
31. Duffett, p. 506.
32. *Ibid.*, p. 189.
33. Committee of Concerned Asian Scholars, *The Indochina Story: A Fully Documented Account* (New York: Bantam, 1970), p. 126.
34. Harrison E Salisbury, *Behind the Lines—Hanoi: December 23, 1966–January 7 1967* (New York: Harper & Row, 1967), p. 58.
35. Duffett, pp. 308–9.
36. *Ibid.*, p. 620.
37. Hugo A Bedeau, 'Genocide in Vietnam?.' in Virginia Held *et al.*, eds., *Philosophy, Morality and Public Affairs* (New York: Oxford University Press, 1974), p. 43.
38. *United Nations Demographic Yearbook 1974* (New York: United Nations, 1975), p. 130.
39. DRVN Commission for Investigation of the US Imperialists' War Crimes in Viet Nam, *The Late December 1972 US Blitz on North Viet Nam* (No place; no date), pp. 7, 25.
40. *New York Times*, 22, 23 December 1972.
41. Peter A Poole, *The United States and Indochina: From FDR to Nixon* (Hinsdale, Ill: Dryden Press, 1973), pp. 226–8.
42. 'Use of Air Power', *Economist*, 13 January, 1973, p. 14.
43. US Senate, Committee on the Judiciary, Subcommittee to Investigate Problems Connected with Refugees and Escapees, *Relief and Rehabilitation of War Victims in Indochina*, Part III: *North Vietnam and Laos*, Hearing, 93rd Congress, 1st Session, 31 July, 1973, p. 72.
44. *New York Times*, 7 January, 1973.
45. Quoted in US Senate, *North Vietnam and Laos Hearing 1973*, pp. 78, 88; Drew Middleton, 'Hanoi Films Show No "Carpet Bombing" ', *New York Times*, 2 May 1973.
46. Quoted in US Senate, *North Vietnam and Laos Hearing 1973*, pp. 74, 78.
47. Stanley Karnow, *Vietnam: A History* (New York: Viking, 1983), pp. 41, 653.
48. James Simon Kunen, *Standard Operating Procedure: Notes of a Draft-Age American* (New York: Avon, 1971), pp. 23–5. This book reproduces the verbatim text of large portions of the CCI hearing in Washington.
49. *Ibid.*, p. 232.
50. *Ibid.*
51. *Ibid.*, p. 369.
52. American Friends Service Committee, *Indochina 1971* (Philadelphia: American Friends Service Committee, 1970), p. 18.
53. Vietnam Veterans Against the War, *The Winter Soldier Investigation: An Inquiry into American War Crimes* (Boston: Beacon, 1972), p. xv.
54. *Ibid.*, p. xiv.
55. *Saturday Review*, 9 January, 1971, p. 26.
56. Office of the Director, Judge Advocate Division, Headquarters USMC, Winter Soldier Investigation files.
57. *Ibid.*
58. Kunen, *op. cit.*; *Washington Post*, 3 December 1970.
59. See US Department of the army, Office of the Judge Advocate General, International Affairs Division, files of atrocity allegations. (On 29–30 November 1969 a photograph

appeared in the *Chicago Sun-Times* and *Washington Post* of a body falling from a helicopter. Accompanying stories of interrogation under duress generated considerable public interest. An official investigation established that the helicopter crew had picked up a dead NVA soldier, staged the incident, and taken the photo. One soldier had invented the interrogation story and mailed this with the photograph to his girlfriend through whom it eventually reached the *Sun-Times*.)

60. See on this generally Sandy Vogelgesang, *The Long Dark Night of the Soul: The American Intellectual Left and the Vietnam War* (New York: Harper & Row, 1974); Eric F Goldman, *The Tragedy of Lyndon Johnson* (New York: Knopf, 1969); Arnold Beichman, *Nine Lies about America* (New York: Library Press, 1972).
61. See William L Hauser, *America's Army in Crisis: A Study in Civil-Military Relations* (Baltimore: Johns Hopkins, 1973), pp. 98–104.
62. US House, Committee on Foreign Affairs, Subcommittee on National Security Policy and Scientific Developments, *American Prisoners of War and Missing in Action in South-East Asia*, 1973, Hearings, 93rd Congress, 1st Session, 23–31 May 1973, pp. 41, 59.
63. Wilfred G Burchett (i), *Vietnam: Inside Story of the Guerilla War*, 2nd ed. (New York: International Publications, 1965), p. 103. The story of the interview is also told by one of the captives. Cf. George E Smith, *POW: Two Years with the Vietcong* (Berkeley, Ca: Ramparts Press, 1971), pp. 158–9.
64. Edward S Herman, *Atrocities in Vietnam: Myths and Realities* Boston: Pilgrim Press, 1970), p. 32.
65. Reprinted in US House, *American Prisoners of War Hearings 1973*, pp. 215–233.
66. Wilfred G Burchett (ii), *Vietnam North* (London: Lawrence and Wishart, 1966) p. 53.
67. Cf. Howard S Levie in Falk (i), Vol. IV (1976), p. 356.
68. US House, Committee on Armed Services, *Problems of Prisoners of War and Their Families*, Hearing, 91st Congress, 2nd Session, 6 March 1970, p. 6086.
69. *Ibid.*, p. 6052.
70. Seymour M Hersh, 'The Prisoners of War', reprinted in US House, Committee on Foreign Affairs, Subcommittee on National Security Policy and Scientific Developments, *American Prisoners of War in South-East Asia, 1971*, Hearings, 92nd Congress, 1st Session, 23 March–20 April 1971, p. 501.
71. Dave Dellinger (i), "The Prisoner of War Hoax', *Liberation* XV, No. 10 (December 1970), pp. 8–9. See also his earlier article (i), 'Indomitable Vietnam—A Fresh Look', *Ibid.*, XII, No. 3 (May–June 1967), pp. 14–24.
72. See the account of Lieutenant Commander John S McCain III, 'How the POWs Fought Back', *US News and World Report*, 14 May 1973, p. 52.
73. 'Confession by Richard Allen Stratton', *Vietnam Courier*, Nos. 101–2, 13–20 March 1967, p. 7.
74. See McCain, *op. cit.*
75. *The Progressive*, March 1971, p. 16.
76. *Ibid.*
77. US House, *American Prisoners of War Hearings 1971*, p. 218.
78. *Ibid.*, p. 229
79. In addition to sources cited above see John A Dramesi, *Code of Honor* (New York: Norton, 1975); Ralph Gaither and Steve Henry, *With God in a POW Camp* (Nashville, Tenn: Broadman, 1973); John G Hubbell, *POW* (New York: Reader's Digest Press, 1976); and the collective account of eight senior officers, 'Torture . . . Solitary . . . Starvation: POWs Tell the Inside Story', *US News and World Report*, 3 April 1973. See also the earlier book by James N Rowe, *Five Years to Freedom* (Boston: Little, Brown, 1971).
80. US Senate, Committee on the Judiciary, Subcommittee to Investigate Problems Connected with Refugees and Escapees, *Problems of War Victims in Indochina. Part III: North Vietnam*, Hearings, 92nd Congress, 2nd session, 16–17 August 1972, p. 16.
81. Salisbury, p. 151.
82. Truong Nhu Tang, 'The Myth of a Liberation,' *New York Review of Books*, 21 October 1982, p. 33.
83. Nguyen Van Canh, *Vietnam Under Communism: 1975–1982* (Stanford, Ca: Hoover Institution Press, 1983), p. 13.

84. Truong Nhu Tang, p. 35.
85. Amnesty International, *Report of an Amnesty International Mission to the Socialist Republic of Vietnam: 10–21 December 1979* (London: Amnesty International, 1981), p. 7.
86. See also Chapter 16 in this volume.

CHAPTER 8

Red Herring: FM 30-31B and the Murder of Aldo Moro

DAVID A CHARTERS

A lie can be half-way round the world before truth has got its boots on.

*James Callaghan**

Shortly after 9.00 am on 16 March 1978, Aldo Moro, President of Italy's Christian Democratic Party and former Prime Minister, settled into the back seat of his blue Fiat 130 for the short trip from his parish church to Parliament.[1] On that day, a historic vote of confidence was to bring the Italian Communist Party (PCI)—since the 1976 elections, Italy's second largest elected party—into the governing coalition for the first time. The controversial arrangement, which fell far short of the PCI's objectives, but represented the most that conservative elements in the Christian Democratic Party were prepared to tolerate, was the product of Moro's handiwork. Indeed, that Moro had been able to arrange a workable compromise at all was a tribute to his consummate skill as a politician.[2] By the vote that morning, Moro was to earn a lasting place in history as *the* elder statesman of modern Italian politics.

Aldo Moro never reached Parliament that day, or ever again. As his two-car motorcade turned into Rome's Via Mario Fani, it drove into a carefully staged ambush. Moro's driver and four bodyguards were killed in a hail of machine gun fire. Moro himself was abducted by his assailants and held for 54 days while the security forces searched in vain. Then, on 9 May, his captors shot him and dumped his body in the boot of a car left in the centre of Rome.[3]

The abduction and assassination of Aldo Moro was claimed by the Brigate Rosse (Red Brigades),[4] the largest and most active of the many leftist terrorist groups that had been spawned during the political and economic crisis of 1968–69. Red Brigades actions, together with those of the other groups, pushed the number of terrorist incidents in Italy to over 2,000 in 1978 alone, and to more than 12,000 for the decade as a whole. By the latter half of the 1970s then, leftist terrorism had surpassed that of the right as the principal threat to the security and stability of the Italian state. But groups such as the Red Brigades shared with the neo-fascists a love of political violence and a virulent hostility towards the institutions of Italian parliamentary democracy.[5] A special hatred was reserved for the PCI. The younger generation of leftists (represented in the Red Brigades and other groups) believed the PCI had betrayed its revolutionary principles, had 'sold out' to become part of the bourgeois establishment; the PCI's admission to the governing coalition confirmed this suspicion beyond all doubt.[6] The Red Brigades set out to sabotage the arrangement. One observer has commented that,

> In killing Moro, the Red Brigades destroyed the key man who could have paved the way for the entry of the PCI in the government. The target had been chosen with great political accuracy . . . by kidnapping Moro and murdering him, the Red Brigades effectively killed any hope the PCI had of entering the government.[7]

At the time of the Moro operation it was noted that Renato Curcio, founder of the Red Brigades, had spent time in Czechoslovakia and that Moro's abductors had used Czech-made CZ61 Skorpion automatic pistols. A 'liaison centre' for international terrorists was thought to exist at Karlovy Vary, a Czech resort town.[8] Subsequently, largely through information provided both by Czech intelligence service defector general Jan Sejna and by captured Red Brigades members, it was learned that Curcio was only one of many Italian leftist revolutionaries to have received political-military training at special camps in Czechoslovakia, and that such training continued through the 1970s. Czechoslovakia was also one of several sources of weapons.[9]

This is not to suggest that the Soviets or their allies either initiated or controlled Italian leftist terrorism; they merely assisted it. Given the potential diplomatic repercussions resulting from the exposure of such assistance, their motives for doing so remain unclear. Interpretations include the destabilisation of Italy's democratic system (a key link in the NATO chain), an attempt to make the PCI appear more moderate by contrast with extreme elements of the Left or, conversely, an effort to hinder the PCI's accommodation with the centrist parties.[10] With regard to the latter thesis, historian Norman Kogan has observed that, 'The Czech Communist government hated the PCI intensely; the hatred was reciprocated.'[11] In any case, it did not take long for Italian political leaders of differing persuasions to start hinting darkly about the involvement of sinister unnamed 'foreign' forces in the Moro operation and in the terrorist assault on the Italian state generally.[12] The Soviet Government clearly was not going to wait for suspicions to focus on itself; it immediately launched a disinformation campaign to place the blame for foreign involvement in the Moro affair squarely on the United States Government.

Unfortunately for the United States, there was no shortage of material in the public domain from which the Soviet propagandists could construct their conspiracy theories. First, during the previous four years, official investigations by the United States Congress, together with unofficial memoirs and exposés, had provided details of American attempts to influence the political process in Italy during the late 1940s, the 1950s and as recently as the 1972 elections, by means of covert funding and other activity conducted by the United States Central Intelligence Agency.[13] Secondly, the United States Government, through the State Department and the American Ambassador, Richard Gardner, had made it clear in January 1978 that it did not favour PCI participation in the Italian Government. Both the PCI and some Christian Democrats interpreted the American position as constituting 'undue interference' in Italy's internal affairs.[14]

Furthermore, prevailing political attitudes—deeply engrained suspicion of corruption and political manipulation by the powerful, conservative Christian Democratic Party, and the genuine and not unrelated fears of violence from the extreme right—provided fertile ground for Soviet deception. To the Italian far left, the idea of 'state terrorism' in relation to the Moro case was not incredible at all. They did not need an American 'connection'; they could elaborate their views fully on the basis of what they perceived as evidence of indigenous, institutionalised Italian 'fascism'.[15] It goes without saying, however, that if such attitudes were pervasive, they would provide a useful starting point for Soviet propagandists to construct a more elaborate and deceptive conspiracy theory. In this regard, a senior CIA official testified in 1980 that an Italian colleague had told him that many people in Italy believed the Red Brigades were controlled and manipulatd by right-wing extremists and funded by the CIA; it remained only for the Soviets to provide documentary proof.[16]

As if this were not enough, several Italian politicians, undoubtedly unware that they were making the task of Soviet propagandists that much easier, made public comments which could provide useful 'independent' confirmation of Soviet propaganda themes, although the Soviets rarely quoted them directly. On 17 March, *L'Unita*, the PCI party newspaper, quoted from a PCI 'communiqué' to the effect that, 'The conspiracy is on a large scale and utilises Nazi-fascist methods and groups disguised under various names as its perpetrators.'[17] Two days later in *Le Monde*, Ugo Pecchioli, PCI 'shadow interior minister', elaborated and embellished this theme. He expressed the view that the kidnapping must have involved several dozen people, and then suggested that the terrorists possibly had been assisted by 'professional gangsters or that they had external support . . . the intervention of foreign secret services can neither be ruled out nor confirmed'.[18] But it was a former senior member of the Socialist Party, Giacomo Mancini, who accused the United States directly. 'The responsibilities for the Moro kidnapping are perhaps to be found in the United States', he said, in an interview for the Italian magazine *Panorama*, in which he drew attention to the United States Government's warning to the Christian Democrats concerning the PCI. Saying he believed President Carter's claims of non-interference in Italy's internal affairs, Mancini nonetheless added, 'The American secret services have many heads.'[19]

Given all of this, it would not have required a propagandist of above-average imagination and skill to weave together these strands into a convincing web of conspiracy. But the Soviets did not wait for the Italian politicians to provide them with propaganda ammunition. They launched their campaign immediately, and relied on Italian commentary for credible support later. The Soviets ran two campaigns: one overt, through recognised official Soviet channels, and one covert, through 'independent' publications, to support the ongoing overt campaign. Both campaigns were deceptive in content. The covert campaign contained the added deceptions of concealed Soviet instigation and the surfacing and replaying of a forged document.

The overt campaign began immediately. Several hours after the kidnapping a Radio Moscow English-language broadcast described the incident as 'a crime of reaction . . . one of several attempts by a right-wing force to aggravate the situation in Italy'.[20] The Radio Moscow commentary broadcast to Italy the following day was more circumspect, but on the 18th it embellished the original theme by referring to 'a plan scrupulously prepared by internal and international reactionary forces'.[21] It supported this assertion with a vague allusion to 'widespread opinion in Italy', and by linking the Moro case to incidents in other countries where outbreaks of violence were associated with periods of political change.[22] The CIA was not mentioned in international broadcasts until 23 March. Even then TASS (The Soviet international news agency),[23] broadcasting in English, did not directly link the CIA with the kidnapping. Quoting from the French Communist Party newspaper *L'Humanité*, TASS merely made three cleverly-related points that could serve as 'planted axioms' for later commentary. First, it picked up the theme, advanced initially by Pecchioli, that the kidnapping would have taken so much organisation that outside assistance might have been necessary. Second, it stated that it was 'nobody's secret' that the CIA was operating in the Appenines Peninsula [Italy]. Finally, it claimed that earlier disorders in Bologna had been attributed by the Italian press to 'American Agents'.[24] These three themes, originating from the same source (*L'Humanité*), were then broadcast to Italy on 28 March by Radio Moscow. On this occasion, however, a further embellishment was added. The broadcast stated that three main political parties (the Christian Democrats, the PCI, and the Socialists) had no doubts about the interference and role of foreign secret services.[25]

Subsequently, English-language commentary by TASS sought to expand upon these themes by 'clarifying' the links beween the CIA and Italian terrorism. On 31 March, TASS elaborated in detail on the earlier theme of CIA activity in Italy. It described extensive CIA penetration of Italian military and economic circles and manipulation of the political process. The CIA's main goal—on Washington's behalf—was said to be the prevention of PCI participation in government. To this end the CIA was accused of financing the preparation of *putsches* and rendering material support to right-wing subversive groups.[26] Clearly uncomfortable with the 'leftist' ideological stance of the Red Brigades, the Soviets sought to discredit their left-wing credentials and to link them to the United States. TASS commentary of 17 April first noted the Brigades' 'Maoist' orientation, then stated that the 'crimes committed by the so-called "left-extremists" ' played into the hands of extreme reactionaries and neo-fascists. Further, TASS said that 'ultra-reactionaries' had infiltrated the ranks of 'leftists' and had carried out many of the crimes perpetrated by 'ultra-revolutionaries'. Finally, it connected the extreme left, the extreme right, and the CIA by saying the Italian press surmised, 'that neo-fascists and representatives of imperialist intelligence services are involved in Moro's kidnapping'.[27] The blurring of distinctions between Italian right and left extremism and American/CIA interference thus was complete, and the blame for the Moro affair placed squarely in the ideologically correct place. TASS concluded: 'The interests of extreme reaction and the so-called "left" extremists are interwoven in a complex knot.'[28] Three weeks later, citing Italian publications, TASS accused the CIA directly of setting up the Red Brigades, recruiting terrorists and hiring professional killers.[29]

In commentary on the murder of Moro, the Soviet media initially stopped short of naming the CIA as the instigator of the assassination. Instead, with its East European counterparts, it fell back on euphemisms—'reactionary forces' inside and outside Italy.[30] By the end of May, however, the CIA was again being linked to the Red Brigades and the Moro case.[31] Thereafter, the overt campaign lay fallow until 1981 when, in response to American accusations of Soviet involvement in international terrorism, the Soviet media accused the CIA of masterminding Moro's abduction and murder.[32] At that time, the Soviet media launched an all-out campaign to discredit the CIA, accusing it of committing a vast array of crimes worldwide. This campaign, which was intensified as a consequence of Western speculation about a possible Soviet role in the assassination attempt on Pope John Paul II, included direct charges that the CIA had directed the kidnapping and murder of Moro.[33] Up to this time, the Soviet propagandists had relied almost exclusively on their own speculations and on those appearing in either the independent Italian press or newspapers controlled by Western communist parties. Moreover, a campaign carried out through recognised official Soviet channels such as TASS, Radio Moscow, and Soviet publications was going to convince only that audience already pre-disposed to favour the Soviet position. But in an obvious effort to broaden their appeal, in 1981 they revived a covert campaign that had begun in the Autumn of 1978 with the surfacing of a skilfully fabricated piece of 'documentary evidence'.

The 'evidence' was a 'Supplement B' to the United States Army *Field Manual 30–31*, entitled 'Stability Operations Intelligence—Special Fields', dated 18 March 1970, and distributed under the signature of General William C Westmoreland, then United States Army Chief of Staff. The manual purported to contain operational guidelines for United States Army intelligence personnel with regard to exerting influence on the internal security affairs of friendly countries where American forces are based and whose internal security is threatened by leftist and communist forces. In dire circumstances, the manual advised, United States intelligence personnel should penetrate existing local leftist organisations, form 'special action groups' within them, and then carry out 'special operations' of a violent (or non-violent) nature to convince the host country of the seriousness of the threat and the need for counter-measures.[34] There was, in fact, a genuine *FM 30-31*, but

'Supplement B' did not exist. In this sense, it was not so much a distorted version of a genuine document as a complete invention. At first glance it appeared to be authentic, but a number of format errors, among them an exessively high classification (Top Secret), exposed it as a fake.[35]

FM 30-31B had a curious history, completely consistent with other 'Active Measures' forgery campaigns conducted by Service A of the KGB.[36] On 24 March 1975, *Baris*, a left-wing Turkish newspaper, contained an article about *FM 30-31* and mentioned that a 'Supplement B' would be discussed in future issues. Neither the original manual nor the supplement were mentioned again in *Baris*, and the reporter who had written the article disappeared. Then, in September 1976, a photocopy of the fake *FM 30-31B* was left on the bulletin board of the Philippine Embassy in Bangkok, Thailand, accompanied by an anonymous note from the sender. At the time the manual was exposed as a fake and apparently attracted little attention.[37]

Then, in September 1978, four months after Moro's assassination, *FM 30-31B* was reprinted in two left-wing Spanish newspapers: *El Pais* (18 September), and *El Triunfo* (23 September). The *El Triunfo* reprint was accompanied by an article by Fernando Gonzalez, a member of the Spanish Communist Party who, the United States Government alleges, maintained a close relationship with the Soviet Embassy in Madrid, in particular with an official who had been involved with the KGB. Gonzalez cited the bogus manual as proof to support his allegations that the United States was involved with various Western European terrorist groups, the Italian Red Brigades in particular. Copies of the document and the Gonzalez article had been offered to *El Triunfo* and other Spanish publications by Luis Verdecia, a Cuban embassy official and known member of the DGI, the Cuban intelligence service.[38]

The *El Triunfo* article was quickly picked up and replayed in a number of European newspapers, including the Italian weekly *L'Europeo*, which appeals to the intellectual left. Ultimately, it appeared in the press in more than 20 countries, including the United States. The American surfacing of *FM 30-31B* occurred in the January 1979 issue of *Covert Action Information Bulletin*, the radical anti-CIA magazine. A leading article by William Schaap, co-editor of the *Bulletin*, preceded the document itself, testifying to its authenticity.[39] Consequently, when, in 1981, the Soviets wanted proof of American involvement in terrorism, they had only to quote from non-Soviet publications which had reprinted or discussed the manual. This they did with considerable enthusiasm and embellishment. A TASS commentary in September 1981 linked

FM 30-31 to the CIA, the Red Brigades, Aldo Moro's murder, and the mysterious P-2 'Masonic' Lodge – then under investigation in Italy for suspected corruption and criminal activity. The Grand Master, Licio Gelli, who by that time had become a fugitive, was notorious for his fascist views and was later shown to have links to right-wing terrorists in Italy. TASS said that his daughter (who had been arrested at Rome airport in July) was carrying a copy of *FM 30-31* in her bag.[40] These charges were amplified in a Soviet propaganda pamphlet published the following year. This particularly clumsy effort cited as a source a work by an alleged former CIA agent. In it, the title of the Field Manual was incorrect, as was the quote from the portion of the manual dealing with penetration and manipulation of insurgent groups. The chronology of events was also badly distorted; the pamphlet had the Italian Government falling in May on the basis of documents found in July.[41] The Soviets did not repeat these mistakes. A 1983 pamphlet, *International Terrorism and the CIA*, which extended the P2 conspiracy to the Reagan Administration itself, gave the correct title for the document, and quoted the relevant portion accurately from the copy published in the *Covert Action Information Bulletin*. In what appears now as an ironic twist, the pamphlet noted that the kind of technique described was not new; that in a similar manner Hitler's supporters had framed the communists and eliminated the opposition by staging the Reichstag fire.[42] The CIA link to the Moro assassination was repeated yet again in Soviet pamphlets published over the next two years.[43]

What did the Soviets gain from this not inconsiderable effort, sustained over several years? By mid-Summer 1978, Italian attitudes did not appear to have changed a great deal. The main political parties shared the conviction that Moro was abducted and killed solely to prevent the participation of the PCI in the governing coalition. Beyond that, there was little consensus about the origins of the perpetrators. The Christian Democrats were still talking vaguely of a 'destabilising design of much greater scope', possibly with international ramifications, and *Panorama* was continuing to circulate suggestions of American involvement, but of no greater substance than those in Mancini's interview in March. Still, the American Embassy felt moved to describe *Panorama*'s insinuations as false and despicable.[44] *Panorama*'s persistence was appreciated by the Soviets. They cited the magazine frequently as a source for allegations of American and neo-fascist links to the Moro case,[45] although the script writers were careful never to cite the publication date or page, presumably to hamper efforts to verify the accuracy of quotes and paraphrasing.

They had somewhat greater success with *FM 30-31B*. On 19 September 1978, one day after it appeared in the Spanish newspaper *El Pais*, the US embassy in Rome repeated its denial of allegations appearing in the Soviet and Italian media of CIA involvement in the Moro affair.[46] The fact that the document was reprinted or reported in many publications could be regarded by Soviet propagandists as a genuine success. Its reprinting in *Covert Action Information Bulletin*, together with a leading article verifying its authenticity and disputing official claims of its fraudulent origins, was a significant achievement, given that the *Bulletin's* main readership is within the American radical community. Nonetheless, that feature article did not connect the Manual and the Moro case. The most recent replay in a Western publication of the CIA–Moro connection is a veiled hint in the book *The Rise and Fall of the Bulgarian Connection*, by Edward S Herman and Frank Brodhead. After acknowledging that the Red Brigades killed Moro, they state that 'the ultimate source of his death is in dispute'. Then, noting that Moro had been the target of an earlier right-wing conspiracy, they add that 'a variety of political interests', including the CIA, had made contact with the Red Brigades. No specifics were given.[47] A similar suggestion was included in Herman's earlier book, *The Real Terror Network*.[48] It is interesting to note, however, that in neither book did Herman—who has written for the *Bulletin*—cite the *FM 30-31B* story as a source. Moreover, serious scholars of Italian terrorism are inclined to argue that on the basis of available evidence, Moro's abduction and murder cannot be attributed to the CIA.[49] All of which suggests something less than the 'significant victory for the Kremlin' with which Ladislav Bittman credits the *FM 30-31B* campaign.[50]

The Soviet disinformation campaign concerning the Moro case was a gratuitous attempt to score points off the Americans in a situation which posed virtually no risk to the Soviets themselves. There was no Soviet 'smoking gun' to be concealed in respect of Aldo Moro. Although Soviet bloc support for the Red Brigades was suspected at the time and subsequently verified, no one then or since has accused them of engineering or sponsoring Moro's abduction and murder. This made the Soviet task that much easier; it allowed them to confine their effort to simulation—showing the false. The overt campaign involved decoying. It drew attention away from the observable facts of the case (that Moro had been kidnapped and killed by an Italian leftist terrorist group), and focused it on an alternative hypothesis—that the incident had been carried out by right-wing extremists posing as leftists, with direct assistance from the CIA. The covert campaign required invention, the creation of 'evidence' (*FM 30-31B)* with which to sustain and enhance the overt campaign.

That the covert campaign probably had more impact than the overt one should hardly be surprising. So long as disinformation came from official Soviet sources, it would have credibility only amongst the already converted (or the incredibly naîve) and could be exposed and dismissed easily. Certainly by the Summer of 1978, the overt Soviet campaign did not appear to have shifted mainstream Italian editorial opinion from the positions taken at the outset of the crisis. Furthermore, even when the disinformation surfaces in a manner which is 'plausibly deniable', it may have only a limited useful lifespan before being exposed. In the meantime, of course, it can be effective and damaging. *FM 30-31B* was surfaced in an effective manner; it acquired a limited constituency of believers, and temporarily put the United States Government on the defensive, forcing it to respond to the allegations. Over the long term, however, the deception campaign appears to have yielded at best a modest harvest of credibility.

Endnotes

* House of Commons, 1 November 1976, *Hansard*, Vol. 918, No. 176, colm. 976.

1. 'Striking at Italy's Heart', *Newsweek* (UK edition), 27 March 1978, p. 6.
2. On Aldo Moro, the PCI, and the politics of the 'historic compromise', see Norman Kogan, *A Political History of Post War Italy: From the Old to the New Center Left* (New York: Praeger, 1981), p. 124; Peter Lange, 'Crisis and Consent, Change and Compromise: Dilemmas of Italian Communism in the 1970s', in Peter Lange and Sidney Tarrow, eds., *Italy in Transition: Conflict and Consensus* (London: Frank Cass, 1980), p. 110; James Ruscoe, *On the Threshold of Government: The Italian Communist Party, 1976–81* (New York: St Martin's Press, 1982), pp. 32, 111, 142, 153, 160–61.
3. Robin Erica Wagner-Pacifici, *The Moro Morality Play: Terrorism as Social Drama* (Chicago: University of Chicago Press, 1986), p. 62. This scholarly volume takes a sociological and literary criticism approach to the way the event was presented and interpreted through a variety of media; see also 'Italy's Trial by Terror', *Newsweek* (UK edition), 3 April 1978, pp. 5–6; *The Times* (London), 17 March, 10 May 1978.
4. *The Times*, 17 March 1980; 'Moro: No Exit?', *Newsweek* (UK edition) 15 May 1980, p. 16.
5. The literature in the English language on the origins and development of the Red Brigades is extensive, although the quality is uneven. Scholarly accounts include: Alessandro Silj, *Never Again Without a Rifle: the Origins of Italian Terrorism* (New York: Karz Publishers, 1979); Daniela Salvioni and Anders Stephanson, 'Reflections on the Red Brigades', *Orbis*, Vol. 29, No. 3 (Autumn, 1985), pp. 489–506; Richard Drake, 'The Red Brigades and the Italian Political Tradition', in Yonah Alexander and Kenneth A Myers, *Terrorism in Europe* (New York: St Martin's Press, 1982), pp. 102–40; other relevant works include; 'Special Issue: Terrorism in Italy', *Terrorism: an International Journal*, Vol. 2, Nos 3 & 4 (1979); Thomas Sheehan, 'Italy: Behind the Ski Mask', *New York Review of Books*, 16 August 1979, pp. 20–26; Vittorfranco S Pisano, *Conflict Studies No. 120 The Red Brigades: A Challenge to Italian Democracy* (London: Institution for the Study of Conflict, 1980), pp. 3–11; Paolo Stoppa, 'Revolutionary Culture Italian-Style', *Washington Quarterly*, Vol. 4, No. 1 (Spring, 1981), pp. 100–113.
6. Kogan, p. 124; Sidney Tarrow, 'Italy; Crisis, Crises, or Transition?', in Lange and

Tarrow, *Italy in Transition*, p. 178; Stoppa, pp. 101, 103, 105; Drake, pp. 105–6, 113–14, 117–18, 122–3; Claire Sterling, Italian Terrorists—Life and Death in a Violent Generation', *Encounter* (July 1981), p. 23.

7. Donald Sassoon, *The Strategy of the Italian Communist Party: From the Resistance to the Historic Compromise* (London: Frances Pinter, 1981), p. 230.
8. *The Times*, 10 May 1978; 'Terror International', *Newsweek* (UK edition), 22 May 1978, pp. 18–19.
9. Pisano, p. 17; Roberta Goren, *The Soviet Union and Terrorism*, edited by Jillian Becket (London: George Allen and Unwin, 1984), pp. 156–7, 160–61; Shlomi Elad and Ariel Merari, *The Soviet Bloc and World Terrorism*, Jaffee Center for Strategic Studies, Paper No. 26 (Tel Aviv: Tel Aviv University, 1984) p. 14; Sterling, 'Italian Terrorists . . .' pp. 27–8; see also quote by Italian leftist Rossellini in *Le Matin*, reprinted in *The Times*, 5 October 1978.
10. Elad and Merari, p. 50; Goren, pp. 161–5.
11. Kogan, p. 107. Wagner-Pacifici, p. 75 notes that in 1977 PCI leader Berlinger had hinted at Soviet bloc covert action to hinder the PCI's move toward Eurocommunism.
12. See comments by the Christian Democrats, and by Enrico Berlinguer, PCI leader, in *The Times*, 18, 21 March 1978. Comments by other Italian politicians and groups are cited in Wagner-Pacifici, pp. 72–5.
13. United States, Senate, Select Committee to Study Government Operations with Respect to Intelligence Activities, *Book 1: Foreign and Military Intelligence* (Washington, DC: United States Government Printing Office, 1976), p. 49; see also Victor Marchetti and John D. Marks, *The CIA and the Cult of Intelligence* (New York: Dell, 1976; repr. 1980), pp. 39, 144; Ray S Cline, *Secrets, Spies and Scholars: Blueprints of the Essential CIA* (Washington, DC: Acropolis Books, 1976), pp. 99–102; William Colby and Peter Forbath, *Honourable Men: My Life in the CIA* (New York: Simon and Shuster, 1978), pp. 108–40; Salvioni and Stephanson, p. 495, n. 14. See also this volume, Chapter 12.
14. The State Department statement which created the furor was read to news correspondents by a departmental spokesman, 12 January 1978: text reproduced in full in *Department of State Bulletin*, Vol. 78, No. 2011 (February 1978), p. 32; for reaction in Italy, see *The Times*, 14 January 1978; on its impact in the Italian government crisis, see Kogan, p. 124; the foreign policy context of the Carter Administration's attitude towards the PCI is discussed in Robert J Lieber and Nancy I Lieber, 'Eurosocialism, and US Foreign Policy', in Kenneth A Oye, *et al*, eds., *Eagle Entangled: US Foreign Policy in a Complex World* (New York: Longman, 1979), pp. 274–87; Zbigniew Brzezinski, *Power and Principle: Memoirs of the National Security Adviser 1977–1981* (New York: Farrar Straus Giroux, 1983), p. 312 provides an insider's view of the shaping of the Administration's attitude towards PCI participation in government. Ruscoe, p. 153, has observed since that some in the PCI suspected that Moro himself may have inspired American intervention on the issue, both to placate his own party's right wing and to force the PCI to moderate their demands. Apparently there was nothing in Moro's letters written in captivity to suggest that he was particularly displeased with Ambassador Gardner's approach to US–Italian relations: *New York Times*, 19 October 1978.
15. For a far left viewpoint, see Gianfranco Sanguinetti, *On Terrorism and the State: The Theory and Practice of Terrorism Divulged for the First Time* (Italy, 1979; repr. London: Aldgate Press, 1982).
16. A viewpoint probably closer to that of the mainstream left might be represented by that of Giorgio Bocca, quoted in *La Repubblica* (6 April 1979) to the effect that terrorism in Genoa was widely believed to be financed by the right: cited in Jon Fraser, *Italy: Society in Crisis/Society in Transition* (London: Routledge and Kegan Paul, 1981), p. 143, n. 10, p. 276; for the comments of the CIA official, see, United States Congress, House of Representatives, Select Committee on Intelligence, Hearings Before the Subcommittee on Oversight, *Soviet Covert Action (The Forgery Offensive)*, 96th Congress, 2nd Session (Washington, DC: USGOP, February 1980), p. 16.
17. *L'Unita* (Rome), 17 March 1978, in Foreign Broadcast Information Service (FBIS) Daily Reports, Western Europe Series, 21 March 1978, p. L2.
18. *Le Monde* (Paris), 19–20 March 1978, FBIS, 22 March 1978, pp. L3–4; see also Alberto Ronchey, 'Guns and Gray Matter: Terrorism in Italy', *Foreign Affairs*, Vol. 57, No. 4

(Spring 1979), pp. 934–9, which suggests that attitudes prevailing within and about Italian political culture provided fallow ground for suspicions of foreign involvement in the event. Wagner–Pacifici, pp. 73–4, 149–54 suggests that, fearing repression, the Italian 'ultra left' was anxious to put some distance between themselves and the Red Brigades. Pointing to 'foreign' sources fulfilled both survival and ideological imperatives.

19. *Corriere Della Sera* (Milan), 28 March 1978, FBIS, 31 March 1978, p. L5. Based on an unattributed report of interview granted by Mancini to *Panorama*.
20. Quoted in Vittorfranco S Pisano, 'Communist Bloc Covert Action: the Italian Case', *Clandestine Tactics and Technology* (Gaithersburg, Maryland: International Association of Chiefs of Police, 1981), p. 15.
21. Radio Moscow broadcast to Italy in Italian, 18 March 1978, FBIS, USSR series, 20 March 1978, p. E5.
22. *Ibid.*, pp. E4–5.
23. On the interlocking news agency and propaganda roles of TASS, see Baruch A Hazan, *Soviet Propaganda: a Case Study of the Middle East Conflict* (Jerusalem: Keter, 1976), pp. 37–43; Gayle Durham Hannah, *Soviet Information Networks* (Washington, DC: Center for Strategic and International Studies, Georgetown University, 1977), pp. 14–15; Lilita Dzirkals, 'Media Direction and Control in the USSR', in Jane Leftwich Curry and Joan R Dassin, eds., *Press Control Around the World* (New York: Praeger, 1982), pp. 87, 91–3; Paul Lendvai, *The Bureaucracy of Truth: How Communist Governments Manage the News* (London: André Deutsch, 1981), pp. 30–31, 35, 130–32.
24. TASS in English, 23 March 1978, FBIS, 24 March 1978, p. E2. By contrast, for the domestic Russian audience the involvement of 'foreign intelligence services' associated with the NATO [US] Base in Naples, was mentioned in *Pravda*, 18 March. FBIS, 21 March 1978, p. E3.
25. Radio Moscow, in Italian, 28 March 1978, FBIS, 29 March 1978, p. EI.
26. TASS in English, 31 March 1978, FBIS, 5 April 1978, p. E7.
27. TASS in English 17 April 1978, FBIS, 21 April 1978, pp. E3–4.
28. *Ibid.*
29. TASS in English, 7 May 1978, FBIS, 9 May 1978, pp. E13–14. TASS attributed its main accusations to the Italian publication *Giorni-Vie Nuove*. Wagner–Pacifici, pp. 84–5, observes however, that allegations linking the Red Brigades to the fascist right were common to many Italian leftist publications at the time.
30. Radio Moscow, in Polish, 10 May 1978, FBIS, 11 May 1978, p. E3; Soviet and East European media commentary also digested in *The Times* (London), 10 May 1978; *Manchester Guardian*, 11 May 1978.
31. Radio Moscow, in Italian, 30 May 1978, FBIS, 31 May 1978, p. E7.
32. 'The Enemy Without', *Economist*, 7 February 1981. In the interval, articles on the Moro case in several Soviet bloc publications tended to mention only 'reactionary forces' and 'ultra-rightists'. See *World Marxist Review (Problems of Peace and Socialism)* (Prague), Vol. 21, No. 6 (June 1978), p. 14, No. 9 (September 1978), p. 81; *International Affairs* (Moscow), (May 1979), p. 67.
33. See, for eg., *World Marxist Review*, Vol. 24, No. 4 (April 1981), p. 63, No. 7 (July 1981), p. 80, Vol. 25, No. 7 (July 1982), p. 65; *International Affairs* (Moscow) (May, 1981), pp. 93–102 (June 1981), pp. 67–73 (October 1982), pp. 102–8, 138; Radio Moscow, World Service in English, 11 January 1983, FBIS, 12 January 1983, p. A6.
34. *Soviet Covert Action*, pp. 66, 86, and Annex A-I, pp. 89–101.
35. *Ibid.*, pp. 16, 66, 86.
36. *Ibid.*, pp. 61–70; see also, US Congress, House, Select Committee on Intelligence Hearings, *Soviet Active Measures*, 97th Congress, 2nd Session (Washington, DC: USGPO, July 1982), pp. 35, 37–8, 47, 52–4.
37. *Soviet Covert Action*, pp. 66, 86; see also, 'The Mysterious Supplement B; Sticking it to the Host Country', *Covert Action Information Bulletin*, No. 3 (January 1979), p. 9. *CAIB* says that the *Baris* article appeared in April, and that 'Supplement B' surfaced in a number of North African capitals during the next few years. It does not mention the supplement's appearance in Thailand or its exposure as a fabrication at that time.
38. *Soviet Covert Action*, pp. 12, 66–7.

39. *Ibid.*, pp. 66–67, 86; *CAIB*, No. 3 (1979), pp. 9–121, 14–18.
40. TASS in English, 11 September 1981, FBIS, 15 September 1981. A Reuters report, 6 July 1981, said that Gelli's daughter was arrested while trying to smuggle into Italy documents relevant to the investigation. In July 1984, the Italian Government released the report of a parliamentary investigation of the P-2 affair, which implicated senior political, military and security officials in subversive activities.
41. Andrei Grachev, *In the Grip of Terror* (Moscow: Progress Publishers, 1982), pp. 93–4.
42. *International Terrorism and the CIA: Documents, Eyewitness Reports, Facts* (Moscow: Progress Publishers, 1983), pp. 221–58. The Reichstag fire deceptions are discussed by Christopher Andrew and Harold James in Chapter 1.
43. Afansi Veselitski, *Murderers: the Strategy of Destabilization and the Tactics of Terror in the Appenines* (Moscow, 1985); E Kovalev and V Malyshev, *Terror: the Inspirators and the Hatchetmen* (Moscow, 1984), cited in Walter Laqueur, *The Age of Terrorism* (Boston: Little, Brown, 1987), p. 271n.
44. *Ansa* (Rome) in English, 3 August 1978, FBIS, Western Europe, 3 August 1978, p. L1. See also *The Times*, 28, 29 July 1978.
45. See, for eg., TASS in English, 7 May 1978, 11 September, 1981; Grachev, pp. 94–5; *International Terrorism and the CIA*, p. 237.
46. *The Times*, 20 September 1978.
47. Edward S Herman and Frank Brodhead, *The Rise and Fall of the Bulgarian Connection* (New York: Sheridan Square Publications, 1986), p. 66n. The source cited in support is an Italian publication.
48. Edward S Herman, *The Real Terror Network: Terrorism in Fact and Propaganda* (Boston: South End Press, 1982; repr. 1983), p. 56.
49. See, for example, Gianfranco Pasquino and Donatella Della Porta, 'Interpretations of Italian Left-Wing Terrorism', in Peter H. Merkl, ed., *Political Violence and Terror; Motifs and Motivations* (Berkeley: University of California Press, 1986), p. 171.
50. Ladislav Bittman, *The KGB and Soviet Disinformation: an Insider's View* (McLean, Virginia: Pergamon-Brassey's International Defense Publishers, 1985), p. 106.

CHAPTER 9

'A True Picture of Reality': The Case of Korean Airlines Flight Number 007

MAURICE A J TUGWELL

The truthfulness of our propaganda springs from its genuinely scientific character and its party spirit . . . Our propaganda gives people a true picture of reality, explaining it from working-class positions, from the correct class positions.

Kommunist[1]

On 2 September 1983, newspapers and radio commentators throughout the non-communist world were united in condemnation of a grave event. Headlines such as 'Massacre in the Sky', '269 passagers assassinés', and '269 Tote: Sowjets schossen südkoreanischen Jumbo', were typical.[2] Leader writers saw 'no conceivable excuse'[3] for a 'cold-blooded, aggressive act that deserves the condemnation of the world'.[4]

The event referred to was the shooting down of Korean Air Lines Flight 007 during the night of 31 August–1 September 1983. Evidently the airliner had departed from its planned route on a flight from Anchorage, Alaska, to Seoul, entering Soviet airspace. When a Soviet fighter aircraft fired two air-to-air missiles at it, the plane plunged into the sea. All 240 passengers and 29 crew were killed. The available facts about the incident have been carefully recorded and analysed and it is not the intention to make a detailed reconstruction here.[5] This examination of the deceptions arising out of the shootdown accepts the propositions that Flight 007 was off course for some reason unknown, presumably pilot error, and that the plane was shot down by a Soviet fighter whose pilot may or may not have been aware of its civilian identity. Six years after the event, no convincing evidence has come to light in support of alternative theories, although there has been no shortage of these. On the other hand, the careful investigations of responsible authorities and journalists have done much to strengthen the pilot error explanation.

For the Soviet authorities, the initial world reaction obviously represented a serious propaganda set back. They were being portrayed as the ruthless assassins of innocent travellers and held entirely responsible for an unjustifiable action. In the light of the enormous Soviet effort to claim leadership of a world peace movement, the event and its interpretation must have seemed especially serious.[6] Moreover, the state propaganda apparatus had been taken by surprise. By the time officials were aware of it, the main facts of the incident were already public throughout the non-communist world and would be seeping into the Soviet Union and Bloc countries through foreign radio broadcasts, there to circulate as rumours.[7] Time was of the essence, but circumstances also dictated caution.

Within the Soviet Union before *glasnost*, news, like history, was contingent upon the needs of the rulers. 'Truth', as the quotation from *Kommunist* makes clear, is a manifestation of party spirit, that is to say the requirements of the ruling elite in the Communist Party. Any disastrous consequences of party policy cannot be 'news': they were non-events, to be ignored, denied, covered up, or explained in some way which exempted the party from blame.

When an embarrassing event could not be completely concealed, delays were often imposed to gain time to prepare, first the propaganda apparatus, and then the Soviet public, for news which eventually has to be released. Such delays occurred during the Cuban crisis of 1962, the Czechoslovak invasion in 1968, the revolution in Iran in 1978–79, and the Soviet attack on Afghanistan in 1979.[8] Now, on 2 September 1983, Soviet leaders and their news managers were denied the option of hushing-up the event, and had to fashion their response under pressure of time.

The bare facts would have been available to these leaders from the Soviet air defence command, *Voyska Protivovozdushnoy Oborony* (PVO). An unidentified aircraft had overflown Soviet air-space in a militarily sensitive area. Efforts to intercept it over Kamchatka Peninsula had failed. However, when it overflew Sakhalin Island, a fighter plane made contact and, on instructions from ground control, shot it down. Soviet leaders on the morning of the 2nd would also have been aware that the downed plane was a Korean airliner carrying civilian passengers. But whether they had this information from PVO or from Western sources no one in the West knows.

Policy guidance on the handling of the incident in the United Nations, at a diplomatic level, and in propaganda would undoubtedly have come from the Politburo.[9] The options open, broadly speaking, were to deny any role in the incident; to accept responsibility for the shootdown, perhaps blaming it on the airline pilot's navigational error, but nevertheless apologising for the loss of life and possibly making restitution; and to acknowledge the shootdown while justifying it in a way that placed the blame squarely on another's shoulders, thus obviating any need to apologise or offer compensation. The last option would not exclude an expression of regret, provided this was unapologetic, pointing an accusing finger elsewhere. The first option was a non-starter, given what the West already knew from the Soviet radio traffic between ground controller and pilot. In the choice between honesty linked to an apology and deceit preserving party infallibility, Marxist-Leninist ideology and practice were presumably important factors influencing the outcome. In addition, the Russian cultural heritage may also have played a role.

* * *

In the early hours of 18 October 1904, the Russian Baltic Fleet was settling into its long journey to reinforce the Russian Far Eastern squadrons, which had suffered severely in the war with Japan. The Fleet's course took it over the Dogger Bank, where the 'gamecock fleet', as the North Sea fishing boats were known in Britain, operated. As soon as the English boats were spotted, Russian gunners, fearing a Japanese torpedo attack, opened fire. Their shells sunk the *Crane* and killed or wounded crewmen in other fishing boats. The battleship *Orel* alone fired 17 six-inch shells and 500 of smaller calibre. Five of her shells hit another Russian ship, *Aurora*. *The Times* wondered how officers wearing the uniform of any civilised power 'could suspect that they had been butchering poor fishermen with the guns of a great fleet and then steam away without endeavourng to rescue the victims of their unpardonable mistake'.[10]

Privately, the Tsar expressed 'regret' to King Edward VII but would not apologise. Publicly, Russia was completely unrepentant. Admiral Rozhdestvenski and the officers of the Baltic Fleet persuaded themselves either that the fishing smacks had really been disguised Japanese torpedo boats, or that torpedo boats had mingled with the gamecock fleet, using the trawlers as cover. A Captain Klado told an *Echo de Paris* correspondent: 'I have been in the Navy for the last 26 years. I know what a torpedo boat is. I know what a fishing boat is. I know what a torpedo boat disguised like a fishing boat is, too . . .'[11] Whatever the cause of the incident, so far as the Russians were concerned, it certainly was neither an accident nor a mistake.

Nathan Leites has cited Dostoevsky and Chekhov to illustrate a trait in Russian character which insists on the determinate nature of major negative events.[12] '. . . nothing is more mortifying and insufferable than to be ruined by an accident', one of Dostoevsky's characters remarks. 'For an intelligent being it is humiliating.'[13] This generalised tendency, Leites explains, was sharpened and drafted into Bolshevik doctrine by Lenin and his successors, becoming part of their operating code. If an unwelcome event is to be viewed as 'accidental', the party would feel dominated by an unintelligible outer force—a feeling particularly dreaded by Bolsheviks, for the alternative to controlling is being controlled. Much better that such an event be predictable by the Party, or at least explained after it has happened, because this feat of intellectual mastery reasserts a measure of Party control. The Party's responsibility for an undesirable event is thus reduced by denying the 'accidental' character of the incident.[14]

The order for the Imperial Navy to open fire on the trawlers, like the order for the Soviet fighter to destroy 007, was deliberate. Both, however, arose out of chance events—the unanticipated meeting of fleets on the Dogger Bank, and the flight of an off-course civil airliner over sensitive Soviet bases. Seemingly, these events were accidents, unplanned and unwanted, but they gave rise to deliberate reactions. If the Soviet leaders were to admit that their air force had responded blindly to an accidental overflight, killing 269 people simply because they happened to be off course, this would imply a loss of control, domination by fate, fallibility. If, in 1904, such an admission had been out of the question, this was even more so in the Soviet Russia of 1983. The Politburo apparently opted for the third course: guilt would be transferred.

In making this decision, Soviet leaders evidently decided that guilt for the incident was to be transferred to American, not Korean, shoulders. The United States had led the world condemnation of the shootdown and had most to gain politically and psychologically from Soviet discomfiture. Therefore this 'main enemy' would be made to suffer.

Given the Soviet world outlook, which tends to view all foreigners with suspicion and fear, it is entirely possible that political and military leaders jumped to the assumption that, if the plane had been civilian, this must simply have been a ruse by hostile foreigners to disguise some wicked mission, such as spying. Or, given the propaganda advantage the event had created for the West, these leaders might have seen the whole affair as a provocation. Within its operating code, the party must never permit itself to be provoked, just as it must not provoke the enemy into actions which are unwanted at the moment. To 'yield' to provocation means to be controlled by the enemy who has succeeded in his attempt to induce in the party or its agents precisely the action which he thinks is in his interest.[15] In the case of 007, the PVO could be seen as the victim of a provocation that caused it to kill 269 civilians. Yet, the Command had done no more than obey standing orders, to defend Soviet airspace against hostile intrusion. Clearly this was, this must have been, 'sophisticated provocation', and these were indeed the words used by Party Chairman Yuri Andropov in his first statement about 007.[16]

* * *

As explained in the introduction, the first three steps to be taken in any deception operation relate to defining the mission. If the Politburo ordered an operation to clear the party of guilt for the incident, the first phase would have been to define the wider goal: the transfer of guilt to the Americans. Next, the deception planners would consider how they wanted their target audiences to react. These audiences would be the politically conscious world public, not the enemy leadership. As such, desired reactions from each audience would have been different. Judging by subsequent messages, furious indignation was the reaction desired from the all-important Soviet domestic and East European audiences. For the non-communist audience, particularly the West, a more modest reaction seemed to be required; nagging uncertainty and latent guilt sufficient to stifle criticism of the Soviet Union. There would of course be shadings and cross-overs between categories: Western and Third World communists and Soviet sympathisers might share the fury required from the domestic faithful, while from the cynical, silent Russian majority, a pretence at indignation might be the best that could be hoped for.

By analysing subsequent Soviet statements about the incident, it is possible to surmise the Soviet planners' probable answer to phase three: what the target audiences were to be made to think about the event, what they should perceive. Here the planning process was probably affected by the inverted nature of the operation. Instead of holding the initiative and creating an illusion that would cause the victim to act to his own disadvantage in the future, planners were reacting to recent history, trying to conjure up an illusion about what lay behind a past event which would cause target audiences to react to the disadvantage of the Reagan Administration. Because the planners lacked the initiative, they evidently decided that it was impracticable to construct one solitary illusion and defend it through thick and thin. Instead, they advanced on a broad front of perceptions, probing for weaknesses in the enemy's position, reinforcing success, abandoning ideas which proved impossible to sell. This method protected them against unforeseen revelations which might destroy the credibility of one essential argument, and it enabled them to utilise scraps of information or, more importantly, conspiracy theories, which might surface in the non-communist world.

The choice of perceptions offered by the Soviet planners divided into three principal illusions. First, the Korean airliner was spying on behalf of American intelligence agencies, specifically the CIA; it was not off course accidentally but in order to gather intelligence about the important Soviet military installations over which it flew. The second concerned provocation. This alleged that the plane had been sent into dangerous airspace above sensitive military bases with the deliberate intention of provoking just the Soviet response that was forthcoming. By forcing the PVO to shoot 007 down, with high civilian casualties, the American government hoped to smear the Soviet Union as a ruthless, dangerous power, to discredit the peace movement, and justify an 'arms race'. Since on the face of it this was incompatible with the spying story, a means had to be found of linking them.

Finally, there was the 'monitoring' scenario. This presented an imaginative script in which spectators and controllers of American military radar and other surveillance devices followed the flight of the doomed 007, saw it was off course and liable to interception by Soviet missiles or fighters, rubbed their hands in gleeful anticipation at what might might happen, but issued no warning. This illusion would reinforce either the spying or the provocation scenarios or, if these had been rejected by target audiences, it could stand alone.

The fourth phase for any deception planner is to decide which facts have to be hidden and what simulations have to be displayed in their stead. In the case of KAL 007, Soviet planners had limited options. For the domestic audience, which in the early days seemed to enjoy priority, the hiding of the truth could only be temporary and partial, given the penetration by foreign broadcasters. The time was used to prepare audiences. Only one essential fact remained concealed throughout so far as official media were concerned: the size and composition of the casualty list. But many other 'smaller' truths were concealed from domestic publics, including every fact and opinion that contradicted the Soviet version being presented. For non-communist audiences, dissimulation was different. Everyone knew that a Soviet fighter had shot 007 down killing 269: nothing could hide this fact. But fortunately for the Soviet planners, no one knew why the plane had gone off course: with the crew dead and the black box lost, there was no 'truth' about why 007 was over Soviet territory to be hidden. As for the motives of those who, in the provocation story, had deliberately sent innocents to their death, these are always invisible; and since the details of military surveillance equipment are secret, the truth of what the radar operators knew, and when, was also hidden from public view and needed no dissimulation by Moscow. Given that they could not deny the shootdown itself, the planners had little need for hiding facts and could concentrate instead on simulating alternative pictures of reality.

For spying, simulation was needed to offer evidence of motive, opportunity and execution, with America being the prime mover. Concerning provocation, a picture had to be presented which showed the callous political and propaganda exploitation of a cynically planned event. For the monitoring illusion, simulation had to show technical capability to monitor 007's flight in 'real time' and a deliberate decision not to warn the plane or other authorities who might have saved it.

The resources at the disposal of the Soviet deception planners were formidable. The Soviet media machine in the pre-'*glasnost*' period was monolithic and obedient to Party control. In 1983 there were three principal agencies for foreign deception: the International Information Department responsible for overt means, TASS and *Novosti*, international radio broadcasting, Party publications such as *Pravda*, periodicals and books like *New Times*, and embassy information; the International Department controlling fronts; and Service A of the 1st Chief Directorate of the KGB, with responsibility for covert propaganda, forgeries, disinformation, influence operations, foreign media manipulation and whatever violent actions might be needed to support such operations. Service A, in effect, worked as the servant of the International Information and International Departments, bringing all the resources of the KGB to the aid of deception.[17]

One such resource was *Ethnos*, an Athens tabloid newspaper created in 1981 out of an unholy alliance of capitalist avarice and Leninist cynicism. Although nominally and apparently independent, *Ethnos* followed the Moscow line: it was a psychological 'Trojan Horse'. On 5 September 1983, it wrote that the 007 incident was a 'prepared provocation by the CIA and definite espionage'.[18]

In pluralist societies, every news item is liable to generate a hundred critical reactions and every crisis, half a dozen conspiracy theories. Sometimes these can be useful to the Soviets. A high proportion of TASS and *Novosti* staff abroad are engaged not in writing news but in collecting summaries of important articles from the local press and writing brief evaluations. This open source material becomes highly classified the moment it is transmitted to Moscow and is used for political, technological, military and other types of analysis. Material of use in deception operations is retrieved from this trawling operation, which includes careful monitoring of Western television and radio programmes.[19] Judging by what followed, the Soviets used virtually all their agencies in carrying their deceptive messages to target audiences. However it was the last, the trawling operation, which proved most useful.

* * *

TASS's first, brief announcement on 2 September 1983 merely talked of an aircraft of unestablished identity twice violating Soviet airspace, flying without lights, failing either to respond to inquiries or to answer radio calls. After failing to respond to signals from Soviet fighters, the plane 'continued its flight towards the Sea of Japan'. This release seems to have been a play for time, possibly because the policy line was still under discussion. Caught unprepared, the Soviet propaganda apparatus seemed slow and clumsy in the early stages. For domestic audiences, this statement might have acted as a warning that more was to come, the start of the preparation phase. For non-communists, it was a disaster, being seen as totally evasive and burdened with guilt. Nevertheless, besides the massive dissimulation of not mentioning the plane's fate, the first probable simulations were floated: flying without lights, failing to respond.

Deception became more apparent next day, when TASS stated: 'It is noteworthy that in the very first report about this [in the West], reference was made to the US Central Intelligence Agency.' The link binding the shootdown to American intrigue was being forged. The same statement now laid the foundation stones for the three illusions.

> Further reports from the United States provide more and more reason to believe that the itinerary and nature of the flight was not accidental . . . the airspace intrusion by the aforementioned plane cannot be regarded as anything but a pre-planned act. The obvious hope was that special intelligence objectives could be obtained . . .

TASS continued:

> There is reason to think that those who organised this provocation were deliberately trying for a further exacerbation of the international situation, striving to smear the Soviet Union, to sow feelings of hostility towards it, and to cast aspersions on peace-loving Soviet policy.

On the subject of monitoring, the statement added, 'The American side . . . cites data from which it is evident that the relevant American services kept a very close watch on the flight throughout its duration.' Thus the full range of desired perceptions of what had led up to the shootdown of 007 was launched into the deception channels, within 60 hours of the shootdown.

On 4 September, *Pravda* published what may have appeared to Western readers as just another mendacious statement: 'US journalists are also putting these questions to the American administration . . . Who, and for what purposes, sent this airplane into Soviet airspace?' In fact, this was oblique language, referred to in a study of Soviet propaganda[20] as the 'indicative-imperative' tense. Communists who read such statements recognise them as orders. The order was: 'find evidence'.

For Soviet domestic audiences, propaganda channels compounded warnings of an intrusion and of spying by whipping up resentment and anger against an American campaign of 'frenzied hatred and malice for the Soviet State, for Socialism . . . torrents of vicious abuses'.[21] This audience was being prepared for the 7 September announcement in the same paper that their airmen had 'fulfilled the command station's order to stop the flight'. Presumably such euphemism was understood by readers long schooled in Orwellian language. First, Soviet publics were made to feel angry, threatened and insulted: then they were notified that the source of these torments had been destroyed. Later, the shootdown was described as 'brilliant'. [22] The West has no reliable means of measuring the effectiveness of Soviet deception of its domestic audience. Citizens who harboured doubts had no means of expressing them publicly. It may be presumed, however, that this treatment of a negative event was adequate.[23]

As the Soviets developed their simulations, they used the trawling operation to provide facts, statements and theories surfacing in the West to strengthen their case. One of the most useful came from the White House itself. On 4 September, Larry Speakes, the official spokesman, reported that an RC-135 reconnaissance plane had been flying in the region and that Soviet radar operators initially might have confused the two airplanes.[24] The Soviets had not yet mentioned the RC-135, but once the Americans had drawn their attention to it, they made it into a central part of their case. This was presented in a developed form by Marshal Ogarkov, the Chief of Staff, at a well-organised Moscow news conference on 9 September. Besides weaving the 135 into the simulation, Ogarkov deftly removed the contradiction between the spying and the provocation illusions: 'It has been proved irrefutably that the intrusion of a South Korean airline's plane into Soviet airspace was a deliberate, carefully planned intelligence operation . . . disregarding, or possibly, counting on the loss of human lives'.[25] The supposed American scheme, according to this interpretation, was one of 'heads I win, tails you lose'.

Eleven days later, Marshal Kirsanov, a senior air force officer, wrote a lengthy account. This introduced one new theory, that the overflying of Soviet territory by 007 was coordinated with the revolutions of an American Ferret 'D' reconnaissance satellite.[26] By this stage almost all the simulations were in place. They can be summarised as follows:

Desired Perception: 007 was spying

Simulations	*Comments*
• 'According to a report from an American radio station on the island of Guam, at 9.10 pm on 31 August the Tokyo flight Control Centre received a radio message from the plane: "We have safely passed over southern Kamchatka. The plane is proceeding normally".' (*Pravda*, 6 September.)	No foundation in fact. Not followed up.
• 'From time to time Soviet radio control services picked up short coded radio signals, which are usually used in transmitting intelligence information.' (*Pravda*, 7 September.	No foundation in fact. Not followed up with faked evidence.
• *Izvestia*, 8 September, reported *New York Times* as saying: 'This type of aircraft is controlled by electronic apparatus which has never let down the pilots.'	Deceptive selection. Omitted the *Times'* subsequent comment that incorrect data programmed into the plane's computers could account for the pilot being unaware of his deviation off course.[27]
• Marshal Ogarkov insisted that 'The termination of the flight was *not an accident or an error*.' (*Pravda*, 10 September.)	See earlier discussion on Soviet attitudes to accidents.
• Alexander Dallin summarises the contradiction inherent in another part of the Marshal's statement: '(1) We did not know it was a commercial plane with civilians aboard; (2) if need be, we will do the same thing again.'[28]	'ideology . . . can take precedence over an opportunist desire to influence the Western public'.[29] In this case, the right of the Party to do whatever it thinks necessary could not be questioned.

- 007 was disguised as an RC-135, according to *Krasnaya Zvezda*, 13 September, 'while the people who were passengers on the plane—if there really were any—were no more than hairs in those pasted-on, painted-on moustaches'.

  For domestic consumption. Presumably intended to cast doubt on the plane's civilian status. Compare with remarks of Captain Klado, 1904.

- *Krasnaya Zvezda* (16 September) also launched the story of longstanding ties between the CIA and Korean Air Lines and of how 007's pilot, Chun Byung, had boasted of his previous intelligence missions and shown his friends the espionage equipment aboard his plane.

  No foundation in fact. Pilot dead, so could not deny report. Pilots do not 'have' planes, but fly whatever machine is scheduled.

- Marshal Kirsanov (*Pravda*, 20 September) provided stories that: 'the South Korean plane not only was performing an intelligence assignment but also was one link in an overall system of major intelligence actions'. Details of United States ship and plane deployments were given.

  United States deployments were probable, given an expected Soviet missile test. Security considerations might have inhibited American responses.

- The 40-minute delayed take-off by 007 was due to the need to 'strictly synchronise the time of the plane's approach to the shores of Kamchatka and Sakhalin' with the satellite. (*Ibid.*)

  The 27-minute delay was to compensate for tail winds and to prevent 007 arriving at Seoul airport before 6 am, when it opened to traffic.[30]

- The extra crew members aboard 007 were 'specialists operating the reconnaissance equipment installed on the plane, of course'. (*Ibid.*)

  They were airline employees being repositioned at Seoul.[31]

- Since 007 'had special reconnaissance equipment on board, it did not, of course, want to land on a Soviet airfield, because it would be caught red-handed'. (*Ibid*)

  No foundation in fact that there was reconnaissance equipment, and ample evidence that there was not.[32]

- Richard Nixon had been booked to travel on 007's fatal flight but, at the last moment, had cancelled. (Soviet story reported by AP from Moscow.[33])

  Without foundation. Not followed up.

Desired Perception: 007 was Provoking

- According to *Pravda*, 6 September, the United States planned to exploit the incident in order to turn Europe into 'The Pentagon's nuclear-missile testing ground, for military actions in Central America and in Lebanon . . .'

  All these simulations advance from the planted axiom that 007 was sacrificed as a deliberate American action; this had no factual basis.

- On the next day *Izvestia* viewed the provocation as helpful to the Reagan administration in 'the implementation of a nuclear first-strike strategy', and the paper's Washington correspondent noted that: 'If the incident had not happened, it would have been necessary to invent it . . . The act of provocation concocted in the Washington corridors of power was, so to speak, a "cluster" provocation . . .'

- At a conference in Madrid, Soviet Foreign Minister Gromyko alleged that 'This major provocation' was used by its instigators 'in the interests of their militaristic policy and of the inflation of military psychosis'. (quoted *New York Times*, 8 September.)

- On 8 September, *Izvestia* explained that the purpose behind the incident was to kill 'all hopes engendered in the international community by "the Soviets' Peace Offensive" '.

From this false assumption, simulations are created as to motive. Any response to such extravagant allegations tends to increase their credibility and news value.

- Kirsanov (*Pravda*, 20 September) echoed Ogarkov in saying that 'The US special services undertook the implementation of a major intelligence operation, or in case it was terminated, the transformation of that operation into a large-scale provocation.'

Desired Perception: 007's flight was monitored, but no warning was issued.

- *Pravda*, 6 September, published what it claimed were details of American air, naval and military deployments in the Pacific at the material time—1000 servicemen on Hokkaido, radio-intercept stations galore, the Cobra radar system, warship and airplane positions.

Based, no doubt, on Western open sources, and embellished with invented detail that the United States could deny only at the risk of compromising military secrets.

- On 8 September, *Izvestia* wrote: 'In monitoring the Soviet pilots' conversations, American and Japanese intelligence services were not concerned about the fate of the passengers on the South Korean airliner . . .'

The Japanese monitoring of these conversations was done mechanically. Human analysis was some while after the event.[34]

- Kirsanov (*Pravda*, 20 September): 'The flight . . . took place not only in the zone of air traffic control radio services but also in the working area of the American Loran-C radio navigation system . . .'

Air traffic radar covered only the first and the last sections of the flight, as 007 departed Anchorage and approached Japan, leaving a 900 nautical mile gap uncovered. If United States military radar covered any of the gap it had no air traffic mandate or responsibility.[35]

Western newspapers published summaries of Soviet statements about the incident. The Ogarkov press conference was televised abroad but was not broadcast in the Soviet Union, presumably because foreign journalists might have spoken out of turn.[36] The various Soviet simulations intended to demonstrate American guilt for the shootdown probably were not accepted at face value by audiences of which the majority were non-communist. Moscow may, however, have succeeded in placing in circulation in the West a number of questions suggesting that there might be explanations for the shootdown other than those put out by White House spokesmen. One American defence journal republished a long extract from *Pravda*'s 5 November summary version of the Moscow deceptions, concluding with the comment '*Hmm*', which seemed to imply some measure of acceptance, or at least doubt about earlier explanations.[37] Perhaps it would be accurate to conclude that Soviet deception, unassisted, penetrated the target audience's 'absorption screen'; it attracted attention. But it failed to penetrate the 'personality screen' and therefore made little impression on audience values and behaviour.[38] But help was on its way.

* * *

Soviet explanations aimed at transferring guilt for the fate of 007 fell into the category of deceptions deliberately induced by another as discussed in the Introduction. When examining the plethora of explanations which surfaced in the West during the year following the shootdown, it is probably safe to assume that most belong to other categories, such as unintentional misrepresentation and, more especially, self-deception where preconceptions or preconditioning have been too strong to permit the individual to see the truth.

According to a Canadian journalist, sometime in the Autumn of 1983 'it came to be generally understood that the earth was about to be blown up'.[39] Fear of nuclear war, discussed in Chapter 10, created in many hearts a need for accommodation between East and West. The need for accommodation may have led in turn to a desire to believe in Soviet goodwill and to discount any notion of political or military aggressiveness. A theory of two morally symmetrical superpowers, each equally responsible for the 'arms race', had already gained some acceptance.[40]

Joseph Stalin is supposed to have said that 'one man's death is a tragedy; 10,000 deaths is merely a statistic',[41] and Arthur Koestler added his famous observation that 'Statistics don't bleed; it is the detail which counts.'[42] In a world seemingly unmoved by the pain of millions—the victims of famine, war in Afghanistan, the Persian Gulf and Cambodia—an event such as the 007 shootdown produced a remarkably sharp response. Presumably, the circumstances were dramatic and the numbers slain were within the grasp of people's comprehension. Moreover, the event was newsworthy by almost every evaluation. The deaths were tragedies; the victims did bleed. Their destruction, and the early Soviet media reporting, seemed threatening to the notions of goodwill and brotherhood which underpinned hopes of accommodation. It was suddenly difficult to maintain the belief that the United States was every bit as bad and as dangerous as the Soviet Union.

In ancient times, bearers of ill tidings were put to death. In Toronto's Queen's Park a mourner for one of the 007 Canadian victims carried her picture through the ranks of a peace rally. 'There, like a leper ringing his bell, he made his way through the masses of anti-nuke demonstrators until he was stopped by a young man who yelled, "Why don't you go back across the street with the rest of the assholes?" '[43] The shootdown and its implications were bad news for such people. Evidence that the United States had really been responsible would have been good news, re-establishing the moral equilibrium. The search for such evidence seemingly became, for some, a re-enactment of some romantic medieval quest.

Not every knight who joined this quest belonged to the category seeking to re-established East–West equilibrium. Hardened social critics and radical activists had their own reasons for blaming the United States. And there were others whose perceptions of United States Government morality had been so influenced by revelations of CIA activities, the Watergate trauma or other experiences, that their reflexes were conditioned to reject all official statements. The common quality among the questing knights was a desire to establish a *particular* truth—to demonstrate American guilt.

News editors on the other hand sought out the 'story behind the story', whatever it might turn out to be. Generally speaking, the media was not committed to the quest, only to 'news'. However, the first rule of news editing is that man bites dog is news: dog bites man is not. Consequently, once the early furore died down, stories or evidence backing the original version of the shootdown could not be news, and generally speaking did not get much secondary publicity, whereas conspiracy theories did.

The theories which surfaced in the West covered a wide spectrum. The quest produced material blaming America for the loss of 007. At least two writers from the political right alleged that the Soviets lured planes into Soviet airspace, where they could be attacked.[44] Several investigators found no evidence of conspiracy, and said so, although in one case, at least, there was a suspicion that 007's pilots might have decided to fly by the most direct route to save fuel, even though this meant accepting the risks involved in violating Soviet airspace.[45] Ernest Volkman, described by Canadian radio and newspapers as an editor for *Defense Science* magazine, but in reality a writer for *Penthouse*, was reported as saying that Korean Air Lines planes 'overfly Soviet territory whenever they can manage it'. Volkman later retracted, but the story was launched.[46]

Early knights of the quest included two ex-Vietnam War RC-135 crewmen, T Edward Eskelson and Tom Bernard. They felt certain that the 135 which crossed paths with 007 must have been replaced on station by another, and that the replacement aircraft would surely have used its equipment to monitor 007's flight. Having made these two questionable assumptions, they built another upon them, to the effect that 'the US Government possessed the capability to intercede during the entire sweep of events culminating in the annihilation of Korean Air Lines Flight 007'.[47] The Eskelson–Bernard theme was expanded by James Bamford in *Washington Post Magazine*, 8 January 1984. Author of a book on the United States National Security Agency's signals intelligence operations, Bamford suggested that various American agencies were probably monitoring 007's entire flight, but he thought it unlikely that they notified Washington of the intrusion. Bamford was not joining the quest.

John Keppel was a former foreign service officer who, in 1960, had been offended by the presentation of a flawed cover story after Francis Gary Powers' U-2 surveillance aircraft had been shot down over the Soviet Union. This experience apparently convinced him that 007 must have been on a similar mission, and that everything said this time by American officials must also be a false cover story.[48] He earned no thanks from the Soviets for sharing and promulgating their view of the incident. On 25 August 1984, Radio Moscow claimed that he, Keppel, had asserted that 007 was destroyed by an American bomb remotely detonated. Keppel had mentioned this possibility merely as the least probable explanation.[49] In his search for evidence to support his convictions, Keppel became an ardent knight of the quest, urging the US Congress to open an inquiry into the affair.

R W Johnson, a Fellow of Magdalen College, Oxford, wrote about the incident in the *Guardian* in December, 1983.[50] He began by rehearsing arguments against NATO's deployment of cruise and Pershing II missiles in Europe and stressed the importance of Soviet military bases in the area overflown by 007. He quoted Eskelson and Bernard. Then he argued that, prior to the shootdown, the CIA had two problems—to gather better intelligence about Soviet bases at Okhotsk, and to undermine the European peace movement's opposition to cruise and Pershing II deployment. He mused: 'What would be really convenient would be one of the periodic displays of ruthless brutality the Russians are so well capable of.'[51] The same solution was found to both problems. What if, he asked, a surveillance mission was mounted by a *civilian* airliner? Heads I win, and reap an intelligence coup; tails you lose, because by shooting down a civilian plane you, the Soviets, are seen as barbarians. The reader was asked to share the agony of the Soviet pilots, ground controllers, air force commanders, defence minister and Andropov himself, torn between humanitarian concerns and 'the security complex below, on which the country's whole strategic future depends'.[52] Reluctantly, it is decided to shoot 007 down. At the end of the article, Johnson decided that 'the US bears the major responsibility for putting the KAL 007 at risk and thus for the deaths of its passengers'.[53]

'P Q Mann' was the *nom de plume* chosen by a British public relations executive, Tony Devereux, when he wrote an article 'Reassessing the Sakhalin Incident' published in the London defence journal *Defence Attaché* in June 1984.[54] The approach here was technological rather than political, concentrating on the supposed co-ordination of 007's intrusions with the Ferret satellite's gyrations, and comparing the recent incident with various shootdowns in the 1960s. But Mann raised a new proposal, that the United States Space Shuttle mission STS-8 was also part of the American plot, being sent aloft to monitor events over Sakhalin Island. The article ended with the original suggestion that the Americans and Russians agreed not to expose America's wicked conspiracy and in return the United States would put 'an end to this species of intelligence gathering operation'. Mann concluded with a rousing call for investigative journalists of the world, who had 'not pursued the enquiry with anything like the vigour that might be expected',[55] to unite. The tone of the piece had switched from pseudo-scientific to polemical.[56]

David Pearson, a PhD candidate in Sociology at Yale University, who had studied American global military communications, devoted nearly a year to developing his conspiracy charge. 'KAL 007: What the US Knew and When We Knew It' appeared in *The Nation* in August, 1984, on the eve of the first anniversary of the shootdown.[57] Pearson acknowledged assistance from John Keppel. The burden of Pearson's argument was that the US Air Force, the National Security Agency, the CIA, the North American Aerospace Defense Command and the Pentagon 'had to have known' that 007 was off course, and those agencies 'had to have known' that the plane was in grave danger and that the agencies had the time and means to correct 007's course, but did not. Additional assertions were that Soviet radars in the region were probably jammed and that the White House and the Secretary of Defense presumably knew of the events as they transpired, well in advance of the shootdown. They had time to have 007's course corrected, but did nothing.

Pearson traced the history of earlier shootdowns by the Soviets, concluding that 'the United States, at least in its intelligence-gathering activities, places a very high priority on penetrating Soviet airspace, sometimes at the expense of human life. That is the context in which the tragedy of Flight 007 must be understood'. He claimed that 007's delayed departure from Anchorage had 'never been adequately explained'. He quoted Mann, but the main part of his diatribe concerned America's surveillance devices in the region (a further development of Bamford's article) which, he claimed, were monitoring the flight in real time. While suggesting that 'all the evidence points in that direction', Pearson, in the *Nation* article, cautiously avoided reaching the firm conclusion that the plane was on a spying mission. In a later interview, however, he was quoted as saying, 'You could bet the store' that KAL 007 was spying. The important point for Pearson was that 'it can be presumed' that the highest in the land knew about the intrusion 'well in advance of the shootdown'. Americans should, therefore, 'take responsibility for our contribution to this tragedy and the tensions, fears and international disorder that it has promoted'.[58]

All three articles transferred guilt for the loss of life and for the worsened East–West climate from the Soviet Union to the United States. All three supported the 'spying' theme, but whereas Johnson portrayed 007 as the direct instrument of surveillance, Mann and Pearson regarded the airliner as a decoy. The 'monitoring' theme was touched on by Mann mainly to strengthen the spying assertion. For Pearson, it was the prime focus of attention. Johnson concentrated on the 'spying' and 'provoking' deceptions, seeing the United States as anxiously trying to destroy all hopes for peace, with the Soviet Union exercising restraint. Contradictions between the essays meant that if one was true, the other two must be false.

Being of Western origin, these articles naturally enjoyed a measure of credibility in the non-communist world. The media replayed them, often in a remarkably uncritical manner. After the Mann piece appeared, Thames Television quoted his article in a programme featuring such knights as Volkman, Keppel, Bernard and a German pilot, Rudolf Braunburg.[59] Jochem Schildt in *Stern*

(Hamburg) quoted Pearson, Mann (whom he described as one who formerly worked in a 'sensitive area' of the United States Air Force), Bernard and Keppel in a piece which reinforced the spying theme.[60] Even the sober British *Economist* treated Mann with respect and concluded that the author 'appears to have had access to high-level intelligence sources in the past, and he has raised some disturbing new questions'.[61] The most glowing endorsement appeared in a piece syndicated worldwide by the *New York Times.* Tom Wicker reviewed Pearson's article and concluded that it demonstrated America's guilt 'to a reasonable certainty'.[62] Perhaps this opinion encouraged Pearson's publisher, *The Nation*, to insert a full-page advertisement in the *New York Times* in October, 1984.

This unusual announcement advanced from Pearson's conspiracy theory as if it was a proven, cast iron case. ' . . . we have a national scandal. Why is Congress not probing this? And where is the American press?', it complained. 'No mainstream medium has published results of its own investitation. As far as we know, none has undertaken an investigation.'[63]

Perhaps what the publisher and editor of *The Nation* really meant was: 'Why has the West's mainstream media not joined the quest?'—the question asked by Mann. For it was not as though there had been any lack of unbiased research. In answer to *The Nation*'s query, the *New York Times* reported three days later that Leonard Downie Jr, managing editor of the *Washington Post*, had told them that the *Post*, like other major newspaper and TV networks, 'investigated the flight of KAL 007 within an inch of its life. We were very open-minded and sceptical about what happened. We tracked down an amazing number of tips . . . and they just didn't check out.' The *New York Times* assigned six reporters to look into the flight, according to Bill Kovach, the paper's Washington editor. Special attention had been given to the possibility that 007 was on a spying mission or, if not, that the United States Government knew that the plane was off course but failed to issue a warning. No evidence was found to support assertions that the United States authorities were in a position to warn the pilot as 007 entered Soviet airspace.[64]

In May, 1984, the *Sunday Times* in London produced a two-part report by Murray Sayle.[65] This concluded that 007's diversion had been accidental and explained in detail how it could have happened. Earlier, the British author Anthony Sampson had studied the whole affair in his book *Empires of the Sky*, and parts of the relevant chapter appeared in *Parade Magazine.*[66] Sampson examined all the evidence and discussed the various theories, but concluded that, in the absence of convincing evidence, the reason for 007's diversion remained a mystery. These pieces received no observable secondary publicity, indeed, Sayle's early research caused a British television company to drop plans for a documentary, because his findings exonerated the Americans.[67] Nor apparently did the negative findings of American journalists qualify as news, until prised out in response to *The Nation*'s challenge. Indeed, the *New York Times* evidently published Tom Wicker's endorsement of Pearson's theory even though its own experts on the subject knew it to be flawed.

Media reaction to Sampson's conclusions changed when he was published by the German magazine *Der Spiegel*. Sampson had agreed to share the authorship of a four-part series about 007 with the German writer Wilhelm Bittorf on the understanding that the two authors would produce alternating episodes. But *Der Spiegel*, in Sampson's words, 'ended up by mixing everything together and using my shared name to underwrite a good deal of nonsense'.[68] This 'nonsense' rehashed Mann, Pearson, and Keppel and introduced the charge that a telephone conference among Washington decision-makers at five o'clock in the afternoon of 31 August 1983, Washington time, resulted in a decision to put out the story that 007 was safe on Sakhalin, which had indeed been the initial press coverage. This was much more interesting to Western media than what had gone before. For instance, *LA Weekly* for 26 October–1 November 1984, referred to 'Anthony Sampson, one of the world's most respected journalists' producing findings which were 'stunning'.

If the *Spiegel* series strengthened the conspiracy lobby, events in London had meanwhile undermined Mann's credibility. The suggestions contained in his article over KAL's involvement in a spy mission and that the airline had deliberately put passengers and crew members at risk, were shown to be groundless.[69]

Defence Attaché's retraction appeared in their first 1985 edition, together with a 'balancing' article by James Oberg, a NASA space systems expert. He demolished P Q Mann's arguments, particularly those relating to the Ferret satellite and the Challenger space shuttle. Because of the large number of military satellites circling the earth at any one time, Oberg argued, 'the odds are at least ten to one that ONE of them will be near any arbitrary point on Earth in any arbitrary time interval'. As for the shuttle, the closest it ever came to 007 was 2,900 kilometres, and the maximum range of its line-of-sight radios was 2,280 kilometres, given the relative heights of Challenger and aircraft. While he was at it, Oberg pointed to frequent contradictions and lies in Soviet statements about the shootdown, and punctured some of Pearson's balloons, including his assertion that the 'Cobra Dane' radar 'would have been actively monitoring KAL 007 for practically the entire length of the flight'. Not only was Cobra Dane a line-of-sight radar, and thus out of range of 007's flight path, but its acquisition and tracking software, designed for missile tracking, was specially programmed to reject aircraft returns.[70] There was little or no replay of Oberg's article, except in the *Washington Times*.

Knights of the quest deserved critical responses to their obsessive and sometimes foolish theories. Anti-establishment and anti-American dedication is no substitute for serious investigation. This can be said without abandoning the agnosticism that inquiry into any incident of this sort demands. For even if evidence of some underlying conspiracy were to surface tomorrow, this would not affect the conclusion that some writers acted on compulsion, knowing in advance what they had to prove. They retreated to a medieval way of thinking. But none of the articles discussed above deserves the label of 'Soviet disinformation'. Various knights made use of Soviet source material, as this chapter has. But there is no evidence linking such authors to Soviet manipulation. Indeed, Soviet deception over 007 has been primarily overt. There has been little or nothing by way of forgeries or rehearsed witnesses. It would seem that the negative potential of resurrecting this painful subject, even for the purpose of strengthening the Soviet case, acted as a deterrent.

* * *

Professor Alexander Dallin's analysis of the shootdown, *Black Box*,[71] is a masterpiece of research. The rigour of his investigations caused him to discard the conspiracy theories of Johnson, Mann, Pearson, *et al.* as at best unproven and at worst 'gross misconceptions'.[72] Recognising that so much vital evidence is lost or unavailable and that definitive answers to the questions arising from the shootdown cannot at present be given, Dallin keeps an open mind. Additional evidence might one day surface which would point strongly to one theory or another. No one would want to disagree with this, nor is there any argument with his statement that 'the lingering doubts about the US role in what happened are bound to persist'.[73] As cases such as Rudolf Hess, presidential assassinations, and the Bermuda Triangle demonstrate, unsolved mysteries tend, if anything, to become increasingly fascinating with the passage of time.

For all his scholarship, however, Professor Dallin does seem impatient to know the answers. He congratulated the knighthood of the quest for their 'imaginative and determined' investigation, contrasting them favourably with 'at least a part of the American media [who] deserve a low grade for their failure to question government handouts and to challenge official interpretations'.[74] One wonders what grade Dallin might award Murray Sayle, and the other investigative journalists who probed the incident 'within an inch of its life', but found that the facts did not support any of the theories? 'A' for determination, one assumes, but failing grades for imagination.

Although Dallin's mind is open, it is as though his emotions are made up. After carefully evaluating the evidence as to why 007 flew off course in the way it did, Dallin concluded that, 'There is no "smoking gun".' Before summing up, as it were, for the jury, he temporarily stepped outside the constraints of academic judgement with the warning 'history is a fickle mistress, and logic may at times be a poor guide'.[75] After warning readers of the dangers of jumping to the conclusion that the incident arose out of a United States-sponsored intelligence mission of some sort, unless there was evidence to back it up, he then seemed to overlook his own advice. 'In fact', he announced, 'it must be acknowledged that with the passage of time this [United States-sponsorship] argument, unlike all others, looms stronger than before.'[76] Reviewing *Black Box* for the *New York Times Book Review*, Philip Taubman highlighted this statement, seeing it as the author's essential judgement.[77] In the search for higher truth, the British journalist Duncan Campbell ransacked Dallin's book, leaving the facts, the analysis, and the doubts untouched. He made off with just that one sentence and used it to support his theory in a *New Statesman* article published in the spring of 1985.[78]

* * *

Almost from the first day, the Soviet deception planners sought to strengthen the credibility of their simulations by replaying supporting material from the Western or Third World press. Initially, they relied upon communist sources—quoting party newspapers such as Austria's *Volksstimme*, Italy's *L'Unita*, Britain's *Morning Star*, Portugal's *Diaro*, France's *L'Humanité*, and Mexico's *El Dia*.[79] Then they borrowed from non-communist sources, distorting the meaning to suit their purposes.

On 14 September, the BBC Television show 'Newsnight' had broadcast a summary of an informal investigation by Britain's Civil Aviation Authority. The punch line of the show was the CAA's discovery of a plausible explanation of how 007 could accidentally have deviated off course, without the crew ever being aware of the error. CAA's theory was that the switch ordering the auto-pilot to accept instructions from the computer system (INS), which should have been changed to that mode at the handover point from local ground to Oceanic control, was never turned, or was only turned one notch, or was turned and didn't work. 007 might then have continued with its auto pilot accepting instructions from the original compass heading which was close—but not close enough—to the one programmed into the INS.[80] The hypothesis answered a question which the public had found baffling, and in doing so undercut some budding conspiracy theories. But this was not the way CAA's findings were reported by Marshal Kirsanov. Writing in *Pravda* on 20 September, the Marshal borrowed only from the introduction to the programme—the part where the hitherto unsolved mystery was summarised—and mendaciously claimed that the CAA's conclusions supported Moscow's. This showed TASS's trawl operation in action, providing Moscow with tidbits of potential use. The product of the quest was not missed by this trawl.

Johnson's article was summarised by *Trybuna Ludu* on 13 January 1984, and reproduced in full five days later by *Literaturnaya Gazeta*. Mann was quoted immediately by TASS and then by *Izvestia*.[81] Later his article was reprinted in Russian (with some changes and additions) in *Za rubezhom*.[82] On the anniversary of the shootdown, the *Washington Post* said that 'recent Soviet articles about the downing of the jumbo jet have relied heavily on Western accounts questioning the Reagan administration's version of the event, including a recent West German documentary, articles in *The Nation* and in the British publication *Defence Attaché*, in Japanese and Italian newspapers'.[83] The *New York Times* explained that 'in recent weeks *Za rubezhom*, a weekly of the Union of Writers, has carried entire articles from the British publication *Defence Attaché* and the American weekly magazine *The Nation* . . .'[84] The articles provided Soviet diplomats and spokesmen abroad with a convincing answer to criticism of the shootdown: refer to your own media, they could reply, and you will see the truth.

No doubt these Soviet replays were intended as much for domestic as international audiences. If deception planners feared that their simulations had been less than wholly convincing, their later confirmation by British and American 'scholars' would have helped to overcome doubts and to reinforce the credibility of official channels. In the non-communist world, the quest had already made some impact. Almost regardless of their highly speculative and in some cases mendacious character, the conspiracy theories had provided a higher form of truth for a constituency that was greatly in need of reassurance. Nowhere was this acceptance illustrated better than in Greece where in October,

1984, the Prime Minister, Andreas Papandreou, told the parliamentary group of his ruling PASOK party:

> In the case of the airliner which caused so much uproar, it is now a fact that it was on a spy mission for the American CIA, and that it actually violated Soviet airspace for the purpose of spying. What I am saying has been reported by authoritative newspapers in Britain, the United States and Canada.[85]

* * *

When Marshal Orgarkov and Soviet Deputy Foreign Minister Kornienko received Seymour Hersh in the Soviet defence ministry, they told the American Pulitzer Prize-winning investigative journalist that 'Your assignment is to find that it [007] was an intruder'.[86] Hersh found nothing of the sort, and his interviews in Moscow in the course of researching the shootdown convinced him that the Soviets had no evidence to support their conspiracy theories. Hersh was far too good a journalist to join the quest: he looked for evidence first; arrived at conclusions second.

Hersh was critical of the way the United States Administration had handled the public relations aspect of the event. However, his 1986 book, *The Target is Destroyed*, probably did more to dispel the myths concerning spying, provoking and failing to warn, than any other document. For here was a distinguished writer whose ability to ferret out facts, however disagreeable, was as much feared by conservatives as it was celebrated by liberals, and he demonstrated that the 'pilot error' explanation was most likely true. The book arrived too late to affect the Soviet deception campaign that followed the shootdown. It did, however, undercut the arguments of administration critics, and it will doubtless provide a valuable source for historians.

Another 1986 book came from the pen of R W Johnson, the Oxford political scientist who had written the very first of the knightly works, in the *Guardian*. In 1983 Johnson had written that the Soviet fighter pilot attacking 007 was 'hoping to minimise the loss of life for he refrains from pumping any more missiles into the plane'.[87] Johnson added that this was a major 'sacrifice', because the extra minutes thus allowed could be used 'to eject surveillance equipment . . .'[88] By 1986, Johnson was willing to concede that the two missiles fired were all that the fighter carried. There was no suggestion that the pilot, Major Kasmin, was concerned over loss of life, and the theory that the plane carried surveillance equipment which might have been ejected during its fatal descent was replaced by a horrifying account of the plight of the passengers.[89]

In 1983, Johnson had said that the Western media's theory that the Soviet pilots might have confused a Boeing 747 with a RC-135 was a 'gross libel'.[90] By 1986, 'it was fairly remarkable that the notion that the Russians had known they were shooting down an airliner should ever have gained wide currency',[91] and it was the Americans who were at fault for suggesting that Soviet pilots *had not* confused a Boeing 747 with a RC-135. Of course Johnson may claim that two-and-a-half years' research improved his understanding of the incident, in which case one may wonder why he wrote with such conviction in 1983, knowing absolutely nothing. One may also observe that his conclusions, while being a shade more tentative, were essentially the same in 1986 as in 1983, and that he continues to rely in his book on such doubtful sources as Ernest Volkman, of *Penthouse* magazine. In a 1987 paperback edition he takes Seymour Hersh to task for failing to behave like a knight proper.[92] In effect, he echoes Orgarkov and Kornienko: 'Your assignment is to find that it was an intruder.'

Not to be outdone by an internationally respected journalist and an Oxford don, David Pearson also turned his conspiracy theory into a book, published in 1987.[93] Reporting on its contents, a Reuters correspondent dismissed Pearson's arguments, noting that Hersh's book had 'concluded that the official United States version of events was essentially correct—the intrusion was accidental'.[94]

On 9 March 1988, Senators Edward Kennedy and John Kerry requested that the Senate Committee on Transporation investigate the KAL 007 tragedy, but this request was denied. At this stage knights of the entertainment world took over the quest. A 'made-for-TV' film, *Shootdown*, based on R W Johnson's book, was shown first on Canadian television on 26 November 1988 and on American screens two evenings later.[95]

The producers of this 'docu-drama' admit in their credits that 'Certain persons and events in this film are fictitious . . .', which may be one of the few honest statements in the piece. Their reconstruction of the actual shootdown of 007 adopts the most sensational propositions in Johnson's book—reports and allegations that he is careful to qualify as unproven—and presents them as 'fact'. In reality this film is 'faction', that combination of real life action (television clips showing the President of the United States and Secretary of State, and even Seymour Hersh) that persuade the audience that they are watching reality, overlaid with all the unproven, and in many cases refuted, stories from the quest. The film's lack of objectivity is overshadowed by its manipulative script, which twists a mother's grief over the loss of her son into suspicion and even hatred of the United States.

* * *

This analysis of the disinformation arising out of the KAL 007 shootdown suggests that the early Soviet deceptions had, at best, limited success outside the captive constituency served by the Soviet and East European official media and among the faithful elsewhere. The importance of these audiences should not be underestimated. For a ruling body that relies for its legitimacy on a monopoly of truth, it is essential to present 'a true picture of reality' that corresponds to and reinforces the Party's political imperatives. In this sense it was probably at least as important for Soviet propagandists to be able to trawl credible explanations from the Western media that could be replayed for Party and domestic consumption as it was to have such explanations accepted by Western audiences.

In any case, neither official Soviet statements nor blatant attempts to shift the blame to the United States were accepted at face value in the West. Nevertheless, the overt Soviet disinformation effort did manage to place in circulation three explanations of 007's flight that served Moscow's purposes. The nature of the incident raised enough unanswered questions to spur conspiracy buffs into action without assistance from the Soviet Union: witness the theories of fuel conservation and luring planes off course. However, it is a fact that the Soviet explanations did find their way into the writings of the questing knights. There is no evidence that these authors wrote under Soviet direction. Possibly the Soviet arguments acted as subliminal signposts in the minds of people seeking a particular truth. If this was the case, then the authors may remain unaware of the influence.

Considering the formidable covert resources at the disposal of the KGB, it is interesting that very little evidence points to secret deception. Perhaps the Soviets, seeing the progress of the quest, had the good sense to stand back. Keppel, Mann, Johnson, Pearson, *et al.* asked all the right questions, again and again, steadfastly refusing to accept United States explanations. By putting the United States Government on the defensive, these apparently independent Western critics did a better job than all the cohorts of the KGB's Service A.

Prime Minister Papandreou's statement stands as a monument to the success of the quest, and indirectly to the triumph of Soviet deception. For after the initial furore had died down, the Soviet Government was not subjected to the same degree of rigorous scrutiny. As a result, while the Soviet action may never be forgotten or forgiven, it is the United States that has been made to live under a cloud of suspicion.[96]

Endnotes

1. V Stepakov, 'The Leninist Ideological Legacy and Party Propaganda', *Kommunist* (Moscow), No. 16, 1965.
2. *Daily Mail, France-Soir* and *Die Welt* respectively.
3. *New York Times*, 2 September 1983.
4. *Baltimore Sun*, 2 September 1983.
5. Principal sources on the incident include: International Civil Aviation Organisation (ICAO), Montreal, Council (i) State letter LE4/19.4-83/130 dated 30 December 1983 and Attachment A—Resolution adopted by the Council on 13 December 1983, Attachment B—Report of Investigation; ICAO (ii), Air Navigation Commission, Document C-WP/7809: 1,818th Report to Council by the President of the Air Navigation Commission (dated 16 February 1984); Oliver Clubb, *KAL Flight 007: The Hidden Story* (Sag Harbour, NY: Permanent Press, 1985); Alexander Dallin, *Black Box: KAL 007 and the Superpowers* (Berkeley, Calif: University of California Press, 1985); Seymour M Hersh, *'The Target is Destroyed'* (New York: Random House, 1986); R W Johnson (i), '007; Licence to Kill?', *Guardian*, 17 December 1983; (ii), *Shootdown: The Verdict on KAL 007* (London: Chatto and Windus, 1986) and (iii), Updated Paperback Edition (London: Unwin Hyman, 1987); Peter Kenez, 'The Lesson of 007', in Joseph S Gordon, ed., *Psychological Operations: The Soviet Challenge* (Boulder & London: Westview, 1988); Thomas Maertens, 'Tragedy of Errors', *Foreign Service Journal* (Washington), September, 1985; 'P Q Mann', 'Reassessing the Sakhalin Incident', *Defence Attaché* (London), No. 3/1984; James Oberg (i), 'Sakhalin: sense and nonsense', *Defence Attaché* No. 1/1985; (ii) 'Sense and Nonsense: A Reader's Guide to the KE 007 Massacre', *The American Spectator*, October 1985; David Pearson (i), 'Kal 007: What the US Knew and When We Knew It', *The Nation*, 18–25 August 1984; (ii), *KAL 007: The Cover-Up* (New York: Simon and Schuster, 1987); Richard Rohmer, *Massacre* 747 (Markham, Ontario: Paperjacks, 1984); Anthony Sampson (i), *Empires in the Sky* (London: Hodder and Stoughton, 1984); (ii), 'What Happened to Flight 007?', *Parade Magazine*, 22 April 1984; Murray Sayle (i) 'Human Error, Or What Really Happened to the Korean 747', *Sunday Times* (London), 20 and 27 May 1984; (ii)'KE 007–A Conspiracy of Circumstance', *New York Review of Books*, 25 April 1985; Jeffrey St John, *Day of the Cobra: The True Story of KAL Flight 007* (Nashville: Thomas Nelson, 1984); 'Akio Takahashi', *The President's Crime: The Provocation with the South Korean Airliner Carried Out by Order of Reagan* (Moscow: Novosti Publishers, 1984).
6. See Chapter 10.
7. According to one source, of the 40 foreign radio stations broadcasting to the Soviet Union in the early 1980s, only 15 were partially jammed. See Dimitry Mikheyev, 'The New Soviet Man: Myth and Reality', paper to the Second International Congress of Professors World Peace Academy, 13–18 August 1985, p. 18.
8. Paul Lendvai, *The Bureaucacy of Truth: How Communist Governments Manage the News* (London: Burnett, 1981), p. 54; see also David K Shipler, *Russia: Broken Idols, Solemn Dreams* (Harmondsworth, England: Penguin, 1983), pp. 295–300.
9. Dallin, p. 92, agrees that the Politburo would have approved the guidelines.
10. Quoted Denis and Peggy Warner, *The Tide at Sunrise: A History of the Russo-Japanese War, 1904–1905* (New York: Charterhouse, 1974), pp. 402–13.
11. *Ibid*, p. 417.
12. Nathan Leites, *A Study of Bolshevism* (Glencoe, Illinois: Free Press, 1953), pp. 67–73.
13. *Ibid*, p. 71.
14. *Ibid*, pp. 67–73.
15. *Ibid*, pp. 321–3.
16. Quoted *Pravda*, 28 September 1983; *Globe and Mail* (Toronto), 29 September 1983. (Unless otherwise noted, quotations from Soviet media are from *Current Digest of Soviet Press* (Columbus, Ohio).
17. See US Congress, House, Subcommittee on Oversight of the Permanent Select Committee on Intelligence, *Soviet Covert Action (The Forgery Offensive)* (Washington, DC: US Government Printing Office, 1980); US Congress, House, Permanent Select Committee on Intelligence, *Soviet Active Measures* (Washington, DC: US Government Printing Office, 1982)—particularly evidence of Stanislav Levchenko, pp. 139–69; US Congress,

Senate, Subcommittee on European Affairs of the Committee on Foreign Relations, *Soviet Active Measures* (Washington, DC: US Government Printing Office, 1985); Ladislav Bittman (i), *The Deception Game* (Syracuse, NY: SURC, 1972); Bittman (ii), *The KGB and Soviet Disinformation, An Insider's View* (New York, London: Pergamon-Brassey's, 1985); John Barron (i), *KGB: The Secret Work of Soviet Secret Agents* (New York: Reader's Digest, 1974); Barron (ii), *KGB Today: The Hidden Hand*, (New York: Reader's Digest, 1983); Richard H Shultz and Roy Godson, *Dezinformatsia: Active Measures in Soviet Strategy* (New York, London: Pergamon-Brassey's, 1984); Baruch A Hazan, *Soviet Propaganda: a Case Study of the Middle East Conflict* (Jerusalem: Keter, 1976); Gayle Durham Hannah, *Soviet Information Networks* (Washington, DC; CSIS, 1977); Lendvai, cited.

18. Quoted *Valeurs Actuelles*, 9 July 1984; see also 'Chronicle: Greece's disinformation daily?' *Columbia Journalism Review* (November–December 1983); 'Pericles, Greece Needs You Back', *Wall Street Journal*, 19 June 1984; 'Greek Media and the Soviet KGB', *The Middle East Times*, 3–10 November 1984, pp. 1, 14.
19. Hannah, p. 14.
20. See John Clews, *Communist Propaganda Techniques* (New York: Praeger, 1964), p. 76.
21. *Pravda*, 4 September 1983, p. 5.
22. *Pravda*, 8 April 1984, celebrating 'Air Defence Day'.
23. Dallin, p. 92, footnote, quotes a Radio Liberty attitude survey on 274 Soviet travellers abroad, which suggested that most citizens who relied on Soviet news believed the deception while many who listened to Western broadcasts did not.
24. *New York Times*, 5 September 1983.
25. *Pravda*, p. 4; *Izvestia*, p. 5, 10 September 1983.
26. Reported *Pravda*, 20 September 1983.
27. The misquoted report was in *New York Times*, 3 September 1983, p. 6. There were many other deliberate misquotes from Western sources.
28. Dallin, p. 81.
29. See Werner Cohn, 'A Clear Provocation: Esoteric Elements in Communist Language', *Encounter*, May 1986, pp. 75–8.
30. ICAO (i), Attachment B, pp. 1, 55.
31. *Ibid*, p. 1.
32. ICAO (ii) reported that 'no unusual equipment or structural changes' on 007 were reported prior to the flight by airport staff at New York or Anchorage and that in the view of the plane's manufacturer, 'any modification to equip the airframe for intelligence gathering purposes would require substantial "outage" from service, and service records proved that this had not been the case'.
33. *Washington Post*, 25 September 1983.
34. *Washington Post*, 2 September 1983, p. 15; *New York Times*, 2 September 1983, p. 7.
35. Oberg (i), pp. 45–6.
36. *Globe and Mail* (Toronto), 10 September 1983, p 1.
37. *Defense Electronics*, March 1984, pp. 20–21.
38. These terms are described by Hazan, cited, pp. 19–28.
39. David MacFarlane, 'In the Shadow of the Cruise', *Saturday Night* (Toronto), December 1984, p. 20.
40. See, for instance, Noam Chomsky, 'The Cold War is a device by which the superpowers control their own domains. That is why it will continue', *Manchester Guardian Weekly*, 21 June 1981; E P Thompson, *Beyond the Cold War* (New York: Pantheon Books, 1982); E L Doctorow, 'It's a Cold War Out There, Class of '83', *The Nation*, 2 July 1983, pp. 6–7.
41. Stalin, quoted J D Atkinson, *The Edge of War* (Chicago: Regnery, 1060), p. 223.
42. Arthur Koestler, 'On Disbelieving Atrocities', *New York Times Magazine*, January 1944.
43. Judith Timson, 'A Death in the Family', *Saturday Night*, October 1984, p. 32.
44. St John, cited; Bruce Herbert, 'Terror Anniversary: KAL slaughter was planned', *New York Tribune*, 31 August 1984, p. 1B.
45. Rohmer, cited.
46. Reported *Globe and Mail*, 7 September 1983; the story is demolished in Oberg (ii), p. 39.
47. T Edward Eskelson and Tom Barnard, in *Denver Post*, 13 September 1983, distributed by Los Angeles Times—Washington Post News Service.

48. See Jim Motavalli, 'U2, KAL?' *New Haven Advocate*, 24 October 1984. On the U-2 deception, see Chapter 14.
49. *Washington Post*, 26 August 1984, p. 27; *New York Times*, 31 August 1984; *San Francisco Examiner*, 26 August 1984, quoted Dallin, p. 117.
50. Johnson (i), cited.
51. *Ibid.*
52. *Ibid.*
53. *Ibid.*
54. 'P Q Mann', cited.
55. Mann, p. 54.
56. See also *The Times* (London), 29 August 1984.
57. Pearson (i), cited.
58. Pearson (i), p. 124; the interview quoting Pearson was in *New Haven Advocate*, cited.
59. Thames Television, 'TV Eye-007—Licensed to Spy?' (TV Transcript: July, 1984), London, 1984.
60. *Stern*, 30 August 1984.
61. *Economist*, 16 June 1984.
62. Tom Wicker, 'Flight 007: A Damning Silence', *New York Times*, 7 September 1984, p. 27.
63. See *New York Times*, 25 October 1984, p. B 15.
64. *New York Times*, 28 October 1984, pp. 3, 6.
65. Sayle (i), cited.
66. Sampson (i) and (ii), cited.
67. See Oberg (ii), p. 40.
68. Anthony Sampson (iii), letter to the author dated 19 September 1985, quoted with permission.
69. *New York Times*, *The Times* (London), *Globe and Mail* (Toronto), all 20 November 1984; *Time*, 3 December 1984; *Washington Times*, 20 November 1984.
70. Oberg (i), cited, pp. 37–47.
71. Dallin, cited.
72. *Ibid*, p. 42.
73. *Ibid*, p. 97.
74. *Ibid*, p. 93.
75. *Ibid*, p. 55.
76. *Ibid*, p. 56.
77. Philip Taubman, 'Theories and Conspiracy Theories', *New York Times Book Review*, 21 April 1985, pp. 7, 9. (Taubman also reviews Clubb's book (cited), a knightly work which earns the comment 'His book is a polemic, and not a very good one.')
78. Duncan Campbell (i), 'What Really Happened to KE 007', *New Statesman* (London), 26 April 1985, p. 8. Campbell's earliest contribution was (ii) 'Spy in the Sky', *New Statesman*, 9 September 1983, pp. 8–9, which ended 'But it is clear that the airline passengers have been innocent victims of a long, secret, electronic cold war in the air.' Having committed himself so early in the game, he was inevitably pledged to the quest.
79. See *Pravda*, 5 September 1983, p. 5 and *Izvestia*, 6 September 1983, p. 4; *Izvestia*, 10 September 1983, p. 5.
80. Transcript, BBC2 TV 'Newsnight' Item, 14 September, 1983.
81. TASS quoted *Philadelphia Inquirer*, 19 June 1984; *Izvestia*, 30 August 1984.
82. *Za rubezhom* (Moscow), No 27, 1984, quoted Dallin, p. 116.
83. *Washington Post*, 1 September 1984, p. 24.
84. *New York Times*, 31 August 1984, p. 3; see also 'KAL One Year Later . . .', *Democratic Journalist* (Prague) 10/84, which quotes Eskelson, Bernard, Johnson, Mann, TV Eye, Volkman.
85. *Elevtherotipia*, 4 October 1984.
86. Hersh, cited, pp. 190–91.
87. Johnson (i), cited.
88. Ibid.
89. Johnson (iii), pp. 24–30.
90. Johnson (i).
91. Johnson (iii), p. 186.

92. *Ibid*, pp. 305–17.
93. Pearson (ii), cited.
94. Reuter report 'US knew jet astray, book says', *Globe and Mail* (Toronto), 10 September 1987.
95. Leonard Hill Films made-for-TV production, *Shootdown*, shown CTV, Toronto, 26 November 1988.
96. In August 1989 a US Court awarded damages against KAL on grounds of negligence, but Counsel for the victims did not allege any of the conspiracies discussed in this chapter. No doubt the decision will be appealed and legal actions will continue for many years.

CHAPTER 10

The Soviet 'Peace' Offensive

J A EMERSON VERMAAT

> *I do not believe that in all the world a man could be found who not only spoke of peace, but fought and struggled for it more than I did.*
>
> *Adolf Hitler*[1]

This chapter examines the Soviet use of 'peace' as a political weapon and the role of deception in its support, using the period 1977–1985 as the principal focus of attention. It is argued that deception enters into Soviet peace diplomacy in three ways: first, the Kremlin pretends to be offering peace as the United Nations understands the term—'to practise tolerance and live together in peace with one another as good neighbours'[2]—when in fact it is offering something altogether different. Second, the Soviets use deceptive and disguised sources for their peace propaganda. Third, Moscow manipulates Western fears of nuclear war while denying responsibility for those fears. These aspects will be examined in turn, beginning with the rival meanings of 'peace'.

* * *

The Russian word for peace is *mir*. This means both 'peace' and 'world'—the earth and those who live on it.[3] When socialism was established in Russia, *mir* was given an ideological meaning. It came to describe the conditions of Soviet society which were considered 'right and just'. In the Soviet view, real peace or *mir* was feasible only in the context of a socialist, that is, Soviet society. Marxist-Leninists contend that peace is historically determined by the economic formation of society. It is, therefore, an 'international principle of Socialism and Communism'.[4] Only the proletarian society is able to attain 'peace'.[5]

The Soviets' precondition for real peace is the victory of socialism, without which peace with capitalist states will only be a truce, an interlude.[6] As they see it, socialism eventually brings peace as the two are indissolubly linked.[7] *Mir* cannot be achieved unless capitalism has been uprooted and the whole world has been converted to Communism, since capitalism and the division of classes are seen as the source of all wars.[8]

Socialism does not mean 'pacifism'. A war advancing the cause of socialism—the Soviet world order—also serves the cause of 'peace' (*mir*). Such wars, then, are 'just' wars as opposed to 'imperialist' or 'unjust' wars waged by forces opposing socialism. Just wars are 'progressive' wars.[9] Lenin wanted pacifists to operate 'in the other, *i.e.* bourgeois camp'.[10] But pacifism within the socialist camp was to be vigorously suppressed. Lenin had no admiration whatsoever for 'peace resolutions of the social-pacifists, the would-be socialists who in reality are bourgeois phrase-mongers'.[11] Reformist pacifism and revolutionary Marxism are diametrically opposed.[12] Lenin even attacked the *disarmament* slogan of the left-wing pacifists, arguing in contradiction that to be opposed to war in general was a fallacy:

> Socialists cannot be opposed to all war in general without ceasing to be Socialists . . . Civil wars for the proletariat against the bourgeoisie for Socialism are inevitable . . .[13]

Proposals in favour of 'disarmament' made to the great powers by Karl Kautsky, were described by Lenin as 'the most vulgar opportunism', and 'bourgeois pacifism' which 'actually serves to distract the workers from the revolutionary struggle'.[14] Revolutionary socialists cannot demand disarmament in the face of 'a bourgeoisie armed against the proletariat'.

> That is tantamount to complete abandonment of the class-struggle point of view, the renunciation of all thought of revolution. Our slogan must be: arming of the proletariat to defeat, expropriate and disarm the bourgeoisie. These are the only tactics possible for a revolutionary class.[15]

Thus, the main party task is to use 'peace' to strengthen the proletariat militarily and the socialist states it controls, and to *disarm* the bourgeoisie and its states. But this endeavour is no longer acknowledged publicly in fora likely to be reported in the non-communist press. True believers, doctrinaire communist party members and the *apparat* they control, understand perfectly that, for the Communist Party of the Soviet Union (CPSU), 'peace' is virtually synonymous with communist victory. The mass of Soviet citizens probably do not share this extreme view, but they do accept the official argument that strong Soviet armed forces guarantee peace.[16] In the West and the non-aligned world, a widespread ignorance of Russian history and Marxist-Leninist theory deprives publics of an understanding of the Soviet view of 'peace'.

Thousands upon thousands of peace-loving Westerners were persuaded to demonstrate in the late 1970s against the adoption by NATO of 'enhanced radiation' nuclear warheads (the so-called 'neutron bombs'). Perhaps 10 times the number turned out in the early 1980s to protest against the deployment in Europe of Cruise and Pershing II missiles.[17] The peace they wanted was one of tolerance and good neighbourliness; with the exception of a handful of committed communists, some of whom organised major events, they had no idea that a large part of the peace campaign was being orchestrated by the Soviet Union, nor were they aware that Moscow's notion of peace was one of eventual world communist hegemony.

Since the operational purpose of Soviet peace diplomacy was to disarm the West while preserving Soviet power, the threat to peace, as understood by the West, had to be presented as entirely Western, especially America's nuclear forces. Boris Ponomarev, a former head of the International Department of the Communist Party of the Soviet Union, (ID-CPSU), which coordinates all Soviet foreign policy, non-ruling communist parties, front organisations, domestic and foreign communist media and the entire Soviet subversive effort,[18] wrote in 1981:

> The US President's decision on the full-scale production of neutron bombs, which are the most inhuman of weapons, is further evidence of where the threat of war comes from . . . An imperative of life today is to turn this (Western public) anxiety of the masses into a powerful barrier to the forces of aggression and confrontation.[19]

And, three years later, the Moscow journal *International Affairs* was explaining that 'the world anti-war movement is aimed, first and foremost, against the aggressive schemes and the arms race which the US ruling quarters try to impose upon other countries and peoples'.[20] Soviet deception on the meaning of the word 'peace' has therefore gone beyond mere dissimulation—the hiding of their own understanding of *mir*. It simulates a false picture in which the West's notion of peace is threatened only by the West's stubborn insistence on defending itself.

Naturally, so utilitarian a philosophy cannot tolerate domestic criticism of Soviet military power and policy, because, within the prevailing ideology this policy reinforces *mir*. Hence the suppression of unofficial peace activists within the Soviet Union and the self-serving justification:

> There can be no political or moral basis for an anti-war movement that is directed against the policy of the socialist governments because they consistently and steadfastly pursue the policy of peace meeting the vital interests of the peoples and therefore enjoying their unreserved approval.[21]

Thus, pacifism may only be used against 'imperialist' wars but never against 'revolutionary' wars. Pacifists within the Soviet Union commit the crime of 'anti-Sovietism'.[22] On the other hand, pacifists in the bourgeois societies of the West 'will help to demoralise the enemy'.[23]

This thesis has become the guiding principle of Soviet-oriented Communist parties, particularly in the nuclear age:

> With all the inconsistency of the pacifists, their campaign against nuclear war constitutes an important social factor which cannot be discounted.[24]

Whenever critics of the Western peace movement mention such issues as human rights in the Soviet bloc or the massive Soviet military build-up, they are denounced by Soviet propaganda as agents of Western intelligence, working against 'peace'.[25]

Soviet deception over the meaning of 'peace' extends to sister concepts—'peaceful coexistence', and 'neutrality'. The first, *mirnoe sosushestvovanie*, has the goal of avoiding a major war with the West (which the Soviet leadership wants no more than the West) while at the same time continuing the struggle by less dangerous means—psychological, ideological and political warfare, surrogate conflict in the Third World, intimidation through superior military power, and subversion. Peaceful coexistence does not mean a gradual integration or reconciliation of communist and capitalist concepts. On the contrary, this compromise is rejected as 'bourgeois liberalism'. There can be no ideological coexistence. The purpose of peaceful coexistence is not to avoid ideological confrontation but to establish *mir* on earth.[26] Peaceful coexistence constitutes a specific form of class struggle between socialism and capitalism.[27] It is the continuation of war by relatively peaceful means, although it does entail support of subversive and terrorist movements in non-socialist states. Peaceful coexistence is the means of achieving victory of socialism over capitalism without resorting to open warfare.[28]

The West was deceived in the 1970s, moreover, partly by its own wishful thinking but also by Soviet dissimulation,[29] into thinking that the era of détente would restrain Soviet activities in the Third World. Soviet and Cuban actions in Angola and the Horn of Africa, followed by the invasion of Afghanistan, brought Western illusions face-to-face with an unwelcome reality.[30]

The Soviet interest in 'neutrality' is purely tactical. Within the

Marxist-Leninist world view, which sees the gradual, inevitable and irreversible conversion of the capitalist countries to socialism, there can be no 'third world'. Non-alignment, neutralism, neutrality and all attempts to opt out of the historical process are, to the Soviets, simply nonsensical.[31] This has not prevented the Soviets from propagandising neutralist trends in the West or boosting its image in the non-aligned and Third World. Although they make no sense in the longer-term, the desire on the part of non-communist countries to find safety in such corners is of immense tactical value to the CPSU. India is a conspicuous example. For one thing, a neutral state cannot belong to any of the military pacts opposing Soviet world aims, such as NATO or ANZUS. Defection from such pacts weakens the West. For another, nominally non-aligned nations can more easily be coopted for propaganda, diplomatic and other crusades which are harmful to Western interests. They are likely to oppose the usually visible evidence of Western military power while ignoring the largely invisible Soviet strength.

So while 'neutrality' is a nonsensical concept in Moscow's strategic pattern, it can be used tactically both to improve the Soviet overall position and as a stepping stone between the capitalist and socialist camps.[32] Nuclear weapons-free zones are, of course, a manifestation of neutralist desires and are deceptively encouraged by the Soviets. For example, the Soviets have repeatedly propagated a nuclear-free zone in Northern Europe, although the only nuclear power posing a threat to Northern Europe is the Soviet Union itself. The Soviets, however, oppose any inclusion of their own territory in the concept of a nuclear-free zone. Their proposals appear to have propaganda objectives only, intending to mobilise public opinion against NATO. The Soviets commit considerable resources to the quiet and often indirect advocacy of neutralism in its varying forms, all of which are insincere and deceptive.[33] In Western Europe, they have focused particularly on Scandinavia, West Germany and the Netherlands.

* * *

The deceptive messages of peace are carried to non-communist publics by all manner of communication media, some overt, such as diplomacy, the press, radio and television; some covert or deceptive, such as the 'fronts'. The latter are mass organisations which the general publics in the West are not supposed to associate with Moscow. They pass themselves off as neutral, but the positions they take usually coincide with official Soviet positions or tactical aims. They comprise the second element in Soviet 'peace' deception.

The Communist International, or Comintern, was for 24 years

the principal Soviet front organisation. From its foundation in 1919 until its disbandment in 1943, the Comintern repeatedly voiced Russia's desire for peace. But this 'peace' was always defined according to Marxist-Leninist ideology. For instance, during the period of the Nazi-Soviet alliance, August/September 1939 to June 1941, the Comintern energetically opposed the war being fought by Poland, Britain, France and other allies against Germany. This was an 'imperialist' war. Communist parties in France, Britain, the United States and elsewhere were ordered to mobilise public opinion against the war with Germany. Even 'strikes for peace' were organised, particularly in the United States, with a view to keeping that country out of the war with Germany.[34] But the moment Hitler broke the pact with Stalin and invaded the Soviet Union, the nature of the war changed abruptly. It was now a 'just' or 'progressive' war: communists worldwide were required to mobilise unconditional support.[35]

The demise of the Comintern in 1943 was not the end of Soviet propaganda and deception efforts. The organisational gap was quickly filled the same year when the task of handling foreign communist parties was transferred to the newly-created 'International Department of the Central Committee of the CPSU' (ID-CPSU) led by Dmitri Manuilski and, after him, by Georgi Dimitrov—who had earlier been the Comintern's general secretary. Boris N Ponomarev was deputy. Ponomarev, who had been a member of the Comintern's Executive Committee between 1936 and 1943, later became functional head of the ID-CPSU. During his long reign his department demonstrated great ingenuity in the deceptive arts.[36]

Between 1945 and 1958 a whole new network of Soviet fronts was created in the West. However, before this process was anywhere near complete, relations between East and West deteriorated rapidly in the 1947-1949 period. The Soviets improvised with existing assets, particularly such Western communists as France's Maurice Thorez and Italy's Palmiro Togliatti, and managed launch a massive propaganda campaign in the West resulting in major demonstrations for peace and culminating in the famous 'Stockholm Appeal' of March 1950, which was allegedly signed by some 500 million people.[37] This was an appeal against the arms race in general and the atomic bomb in particular.

Whilst such topics, set within the West's notion of peace, would be above reproach, the 'peace offensive', like those which were to follow, advanced the Soviet concept of *mir* while appealing deceptively to Western hopes. Although basically an event orchestrated by the Communists, many non-communists signed the Appeal and participated in major peace conferences and Soviet-controlled peace organisations. This, of course, was in the direct interests of the Soviets. The 'unity of action' concept had earlier been developed by Comintern General Secretary Georgi Dimitrov and the Soviets now emphasised it in the following way:

> Particular attention should be devoted to drawing into this movement trade unions, women's, youth, co-operative, sports, cultural, educational, religious and other organisations, and also scientists, writers, journalists, cultural workers, parliamentary and other political and public leaders who act in defence of peace and against war.[38]

This peace offensive was apparently intended to awaken pacifist and neutralist sentiments in West Europe.

The most important of the new Soviet fronts created in the post-war period was the World Peace Council (WPC). This emerged in November, 1950 from the 'World Committee of Peace Partisans' (WCPP) which was founded at the 'First World Peace Congress' in Paris and Prague, April, 1949.

The WPC had its first headquarters in Paris but was expelled by the French Government in 1951 for 'subversive' and 'Fifth Column' activities. The WPC subsequently moved to Prague and, in 1954, to Vienna. In 1957 the Austrian Interior Minister instructed the WPC to leave Vienna on the ground that it had 'directed its activities against the interests of the Austrian State' and had 'interfered in the internal affairs of countries with which Austria has good and friendly relations'. To circumvent expulsion, the WPC quickly created a legal front to enable it to continue its work under the name of 'International Institute for Peace' (IIP), still in Vienna.[39] In 1968 the WPC's secretariat moved to Helsinki.

The WPC consistently followed the Soviet line. Its President, the Indian Communist Romesh Chandra, indicated in July, 1975 that 'the Soviet Union invariably supports the peace movement. The WPC in its turn positively reacts to all Soviet initiatives in international affairs'.[40] In June, 1981 Soviet President Leonid Brezhnev personally presented Mr Chandra with the Order of Lenin.[41] A statement made by Chandra that same year illustrated this spokesman's creative ability in attempting to present the Soviet view of *mir* in a form acceptable to non-communists:

> In the hands of Socialism force has for the first time in human history become an instrument for safeguarding peace and social progress. If it does exist, it is, figuratively speaking, not a threat *to* peace but a threat of peace in the name of peace.[42]

On another occasion, Chandra defended an act which can hardly be conceived as a contribution to world peace, that is, the invasion of Afghanistan:

> The progressive world public, and the fraternal Soviet Union which extended its help to Afghanistan in the latter's hour of need, are on their side. Exercising its sovereign right, the legitimate Afghan Government asked the Soviet Union for military assistance. The Soviet troops are in Afghanistan to help its people to uphold its independence.[43]

Throughout the 1977-1985 peace offensive, the WPC and other Soviet fronts were controlled through the 'International Social Organisations Section' of the ID-CPSU led by deputy chief Vitaly Shaposhnikov, who was also on the WPC's Presidential Committee.[44] Communist parties carefully selected pro-Soviet individuals (communists and fellow-travellers) for positions throughout the WPC structure. The CPSU's Central Committee has a strong say in the staffing of the WPC's secretariat. Within this secretariat, an ID-CPSU representative—in 1985, Tair Tairov—had ultimate control and was recognised by the organisation as the final authority, whose powers included the right of veto.[45] Another ID-CPSU functionary, Oleg Khakhardin, acted as link-man between the WPC and the Soviet Peace Committee—the Soviet affiliate. The Soviet Committee was inevitably a CPSU organisation, without any international or independent pretensions.[46] Yet, in spite of so obvious a character, deception was nonetheless attempted. Yuri Zhukov, then head of the Soviet Peace Committee, told Reuters, 'We're an independent popular organisation; we're not involved in politics'.[47]

Funding was another area of deception. Mr James Lamond, a British member of parliament who is also a WPC vice-chairman, recited the official Soviet line that the WPC was financed by individual subscribers. In response, Ms Ruth Tosek, who had worked as 'a former senior interpreter at several of the Moscow-controlled organisations', [48] assured him that 'all the funds of these organisations, in local and in hard currency, were provided above all by the Soviet Union but also by other East European satellite countries on the basis of set contribution rates, paid by the governments of these countries through various channels'.[49]

In February, 1981, as part of a constant effort to penetrate the United Nations system, particularly its Non-Governmental Organisations, the WPC sought to upgrade its status with the Economic and Social Council (ECOSOC). Its application for 'consultative status' had to be withdrawn, however, after the WPC contravened paragraph 8 of ECOSOC's Resolution 1,296 (XLIV) requiring associated organisations to make a full and accurate declaration of their financial budget, including a proper account of funding. The United Kingdom delegate pointed out that 'the representative of the organisation (WPC) carefully avoided answering specific questions put to him by members of the Committee on that point. It is clear, however, that the WPC has received large-scale financial support from government sources, and has gone to great lengths to conceal that fact from the Committee'.[50]

As an international organisation, active in some 140 countries, the World Peace Council had vast experience and penetrating power. Moreover, new, subsidiary fronts enabled it to deceive different audiences and to re-establish credibility with the media and professional or social groups.

It was Brezhnev who proposed the establishment of an 'International Committee of Scientists for Peace'.[51] In 1983 Soviet scientists addressed an 'Appeal to the Scientists of the World'. This reiterated a similar 'appeal' by Soviet scientists made in 1981.[52] Yuri Chazov, the late Brezhnev's personal physician, was a signatory to both appeals. In 1981, he became the President of the newly founded 'Soviet Committee of Physicians for the Prevention of Nuclear War', now affiliated with the 'International Physicians for the Prevention of Nuclear War' (IPPNW). The idea of organising 'Physicians for Peace' was first launched by the Soviets at the end of 1980. Among those directly involved in the initiative and its subsequent 'internationalisation' were the ubiquitous Boris Ponomarev and Georgi Arbatov—neither of whom were medical doctors—and a retired general of Soviet military intelligence—the GRU—Mikhail Milstein, who worked with Arbatov in the Institute for the United States and Canada, one of the ID's many organisations.

One of Moscow's most interesting moves was the organising of a group of retired senior generals and admirals from NATO countries, members of the very alliance that Soviet peace offensives were designed to weaken. The group manifested itself as 'Generals for Peace and Disarmament' and was soon successful in presenting Soviet positions without being identified as part of the Soviet propaganda offensive against the West.

The idea of organising 'Generals for Peace' originated at the World Peace Council's 'World Parliament of the Peoples for Peace' held in Sofia in 1980. Three former NATO military experts submitted a statement protesting at the United States Presidential Directive 59. These experts were Marshal Francisco da Costa Gomez, former President of Portugal, retired General Nino Pasti of Italy, former senior NATO officer, President of the Superior Council of the Italian Armed Forces, and Senator, and retired Admiral Antoine Sanguinetti, former Deputy Chief of Staff of the French Navy. All three had close ties to the WPC. Da Costa Gomez was Chairman of the Portuguese Peace Committee and a Vice-President of the WPC; Pasti, who was elected to the Italian Senate as an 'independent' running on the Communist Party ticket,[53] was President of *La Lotta per la Pace*, member of the Italian National Coordinating Committee for Peace, and an active WPC member; Sanguinetti was a member both of the French Peace Council and WPC.

As the organisation grew, its new members seemed divided into two categories: overt Moscow supporters such as Greece's generals Koumanakos and Tombopoulos, respectively WPC presidential committee member and Vice-President of the Greek Committee for the Struggle against the American Bases, and 'innocents' whose motivations were probably as varied as they were obtuse. In 1985, the group numbered 13, with members from West Germany, Norway, Portugal, Britain, Greece, Netherlands, Italy, France, and Canada. Admiral Lee, of the United States, acted in harmony with the group in 1981, but there was no American member. Instead, the group seems content to co-operate informally with retired Admiral Gene LaRocque's Washington-based Center for Defense Information.[54]

From the outset, 'Generals for Peace' was directed to counter NATO's cruise and Pershing II deployment. They made their first public appearance as a group at a Hague press conference on 25 November 1981. Individual 'Generals for Peace' were frequently invited to East European peace events organised by various 'peace councils' or other pro-Soviet groups. Press coverage in Eastern Europe was usually substantial and positive.[55] The WPC's outlets also gave them publicity.

In 1981 the book *Generäle für den Frieden* (Generals for Peace) was published by Pahl Rugenstein Verlag in Cologne, a German Communist Party (DKP) outlet. Receiving an annual subsidy of 60,000,000 German marks from East-Berlin, the DKP pursues a rigid pro-Soviet line.[56] The book's editor, Professor Gerhard Kade, was a Vice-President of the Vienna-based International Institute for Peace (IIP) (which between 1957 and 1968 had fronted for the WPC) and was also actively involved in the Prague-based Christian Peace Conference (CPC), another Soviet front established to mobilise 'religious circles' for Soviet causes. The book was hailed in the Soviet Bloc press and in the *World Marxist Review*.[57] Other booklets by Nino Pasti and Germany's General Gert Bastian were published by the WPC's Information Centre, Helsinki.[58]

During the group's first three years Kade and the IIP managed and co-ordinated major meetings and initiatives of the group in Vienna. At the 1982 United Nations Special Session on Disarmament the 'Generals for Peace and Disarmament', as they were now called, presented a memorandum that substantially echoed the Soviet position.[59]

In January 1983 the group held one of its meetings in Vienna where possibilities of a meeting with 'retired generals' from Eastern Europe were discussed. During 1982/3 some members of the group had made extensive trips to Eastern Europe where they had met GRU General Milstein, Georgi Arbatov and others. Further preparations for the envisaged meeting were made in June 1983 at the WPC-sponsored 'World Assembly for Peace and Life, Against Nuclear War' in Prague where the retired generals prepared their own 'issue group' report.

In February 1984, the meeting between retired NATO and Warsaw Pact generals finally took place in (as always) Vienna. The findings of that meeting were loudly hailed in the Soviet-controlled propaganda outlets. For example, *Moscow News* reported that 'the generals who served in the opposite camps were unanimous in all their assessments. They recognised the facts that Washington's course towards achieving military superiority was the main issue threatening peace.'[60] It is, of course, a routine Soviet theme that any Western attempt to recover parity and prevent the further erosion of the East-West balance in Moscow's favour should be defamed on the deceptive grounds of being an attempt to 'achieve military superiority'. After the conference, Vladislav Kornilov, Secretary of the Soviet Peace Committee, who had been present, said that 'the aims of the (Western) group . . . are identical to those of the Soviet Peace Movement'.[61]

Other subsidiary fronts set up by the WPC have included International Commissions of Enquiry into the 'Crimes of the Chilean Junta' and 'Israeli Crimes against the Lebanese and Palestinian Peoples', Commissions on Mass Media and Information, Scientific Research for 'Peace', European Security and Cooperation, Non-Alignment, Parliamentarians, Trade Unions, and so on. WPC co-operates with parallel front organisations, particularly the Afro-Asian Peoples' Solidarity Organisation—which has taken over many Third World revolutionary duties previously inspired by WPC, and the World Federation of Trade Unions. There is a shared interest in the 'International Trade Union Committee for Peace and Disarmament'.

The 'Committees for European Security and Cooperation' were designed to pull down 'the barriers artificially erected during the Cold War years'.[62] This theme denies that the West has anything to fear from the Soviet Union, arguing instead that the danger comes from 'nuclear war', which an aggressive America may unleash. West Europeans, in this scenario, would find true security under Soviet protection—that is, out of NATO and disarmed.

* * *

It goes without saying that the urge for peace, the real living power behind the Western peace movements, is entirely indigenous. The Soviet role has been to redirect and manipulate these democratic and peaceful forces, and this formed the third component of Soviet deception.

Soviet planning deceptively to take over and indirectly to control Western peace activities between 1978 and 1985 took place in the early 1970s. The new movement was to be patterned after the highly successful anti-Vietnam War movement which deeply impressed the Soviets.[63] The impact of public opinion had made itself felt and, as the Soviets saw it, forced the Americans to withdraw from Vietnam. As early as 1965, the Vietnamese Communists had indicated that public opinion rather than a military victory would compel American withdrawal.[64] Plans to repeat this tactic and to mobilise public opinion in the West through a massive peace movement were worked out at the WPC sponsored 'World Conference of Peace Forces' in Moscow in 1973.

The 1973 Congress resulted in much better co-ordination and co-operation between various 'anti-war movements' aiming at 'mobilising public support for disarmament demands'.[65] Communist Party leaders and pro-Soviet activists received instructions to create 'a pluralistic front of peace forces' and co-ordinate all their efforts. This was revealed in 1975 by a Belgian peace activist who participated in the Moscow peace congress.[66] In all this the WPC operated through its 'Continuing Liaison Committee' chaired by Romesh Chandra, which was intended to internationalise the 1973 Peace Congress. An attempt was made at the 'Conference of Communist and Workers' Parties' in East-Berlin, June 1976, to unify Communist Parties behind the Moscow line, particularly those parties which subscribed to the 'Eurocommunist' view. The conference resulted in a 'new impetus for broad interaction of Communists'.[67] The previously dissenting Dutch Communist Party (CPN) made a pro-Soviet turn and, in close co-operation with high ranking ID-functionaries, key members of the Party began to prepare a major peace campaign in the Netherlands. Shortly after the Berlin Conference, the CPN made a strong plea for the unification of Communist parties in the struggle for peace.[68] Then, in July, 1977, President Carter announced that the United States planned to produce the new enhanced radiation warhead—the 'neutron bomb'. The new alliance of 'peace' forces had an issue, and the issue would be turned into a crusade.

The campaign against the neutron bomb has been well documented.[69] It demonstrated how Western concerns for real peace could be manipulated by Soviet threats that the deployment of these warheads 'would gravely endanger detente'.[70] It also showed how diplomacy and overt propaganda worked closely with covert front activity, and how easily the Soviets turned Western concerns for peace into support for *mir*.

From the start, Radio Moscow presented the weapon as a horrific new means of mass destruction. During two weeks in August 1977, the theme consumed about 13 per cent of Soviet broadcast time. The inhumanity of a weapon which 'killed people and left property intact' was emphasised. *Pravda* carried a regular column on its foreign page entitled 'No to the Neutron Bomb'.[71]

The WPC immediately set the stage for a massive peace propaganda campaign. An 'Action Week' was organised for 6–13 August 1977. Most West European Communist Parties began campaigning, too. On 8 August 1977, 28 Communist Parties from Europe, the United States, and Canada published a joint declaration in which they condemned the new weapon. In Amsterdam, the CPN took the lead by launching the 'Initiative for a Broad Movement' to 'Stop the Neutron Bomb', a fact that *Pravda* reported with obvious approval. Nico Schouten, a prominent member of the Dutch Communist Party, became secretary of the new movement which attracted many non-communists, clergymen and members of parliament, among them Social-Democrats.[72] With a view to internationalising the initiative, Romesh Chandra consulted the main initiators in Amsterdam in December, 1977.[73] The following year he travelled to the United States to attend the inauguration of the WPC's US affiliate—the United States Peace Council.

Both the anti-neutron bomb campaign and its successor, the campaign to prevent NATO deploying cruise and Pershing II missiles, were carefully prepared and co-ordinated. There is little doubt that the second and larger offensive owed much in its organisational techniques to the lessons of the neutron bomb campaign experience. Soviet Embassy staff and intelligence officers in The Hague, Holland, and various European capitals played an important role in stimulating the peace movement in the 1977-1983 period. In 1982, an important meeting of Soviet diplomats from various countries was held in the Soviet Embassy in The Hague.

The purpose of that meeting was to co-ordinate activities to stimulate peace and anti-nuclear movements in NATO countries of Western Europe.[74]

Summarising Dutch intelligence findings, the Dutch Minister of the Interior, Mr Jaap Rietkerk, told parliament in December, 1982:

> Before and after each important peace manifestation, a number of people frequently consulted important party officials from the Soviet Union and other East European countries. The frequency and timing of these consultations was such that one could not avoid the conclusion that these were preparatory as well as reviewing consultations. The whole pattern of consultations would justify the interpretation of these talks as being regular deliberations with a view to assessing the development of the peace movement insofar as it was under control.
>
> In this period important Party officials and Party delegations from the Soviet Union and other Eastern Bloc countries paid frequent visits to the Netherlands and acted in the manner as indicated above. There were also activities on the part of international Communist front organisations, East European intelligence services, and East European diplomats.[75]

The Soviet Party leader, Leonid Brezhnev, was no doubt making the same point when he reviewed CPSU Party activities over the preceding period at the 26th CPSU Congress in 1981:

> Our Party and its Central Committee have worked actively for the further expansion and deepening of all-round co-operation with the fraternal Parties. During the period under review, members of the political Bureau and secretaries of the Central Committee alone have received several hundred delegations from other Parties. In their turn, representatives of the CPSU participated in the work of Communist Party congresses and other Party functions abroad. *We have regularly briefed fraternal Parties on our internal developments and our actions in the field of foreign policy.*[76]

Very little Soviet propaganda any longer relates to communist theory or the supposed advantages of living under socialism. It relates instead to contradictions within the target society, and to the self-interest of individuals or groups within that society. Since survival is a very powerful interest, and since nuclear war threatens people's survival, *fear* of such a conflict and the uncertainty that accompanies it can cause groups or whole societies to seek escape—at virtually any cost. The Soviets' appreciation of Pavlov's experiments on conditioned reflexes has coloured their psychological operations.[77] Stimulus and response, mushroom cloud or dangling skeleton and fear, underlie their psychological strategy of 'peace'. But, as Michel Tatu has said, since the missile threat aimed at Britain by the Soviet Union in 1956, direct nuclear threats have rarely been issued by the Soviets. Instead, they have usually given veiled warnings about 'the danger of War' (supposedly accidental) which might result from Western defensive deployments.[78]

Soon after achieving leadership of the CPSU, Mikhail Gorbachev renewed this campaign with an appeal which seemed to be addressed largely to Western Europe:

> Today we call upon all states of Europe and other continents to rise above the differences and become partners in fighting a new danger that threatens the whole of mankind—the danger of nuclear extinction.[79]

The power to coerce in this manner depends, as the Soviets apparently see it, upon their possessing overwhelmingly superior armed forces across the whole spectrum of violence. For the Soviets, a favourable 'correlation of forces'—a concept which embraces relative political, ideological, economic, social as well as military strengths—entitles them to important political and psychological dividends. Writing at the end of 1984, V Matveev, Vice-President of the Soviet Peace Committee and *Izvestiya* political propagandist, sought to exploit these advantages in these stark terms:

> In the nuclear age, there is a second force that can be added to the force of reason—that of self-preservation. If some politicians in the West lack the force of reason, the feeling of self-preservation acts in the same way as the force of reason.[80]

By catering to European fears and desires for peace in a terrifying nuclear age, Soviet propaganda made it seem that NATO and the United States, and more particularly the doctrine of nuclear deterrence, represented the principal dangers. At the same time, Soviet leaders constantly advanced peace proposals. The intended effect was to convince West Europeans that the Soviet Union was the best partner for peace and that Europe should find security 'with' and not 'against' the Soviet bloc. This, presumably, is what Matveev was aiming at in his reference to 'feeling of self-preservation'.

* * *

How well did Soviet deceptions in the name of 'peace' work in the 1977-85 period? So far as Western peace movements were concerned, they appear to have worked quite well. Few activists questioned the Soviet meaning of 'peace', accepting it as identical to their own understanding of the word. Soviet links to organisations such as the World Peace Council were recognised to greater or lesser extent but, although it was, their sincerity was seldom questioned. They were given the benefit of the doubt. Soviet success in manipulating Western movements, particularly in spreading fear, is best measured by comparing the objectives set out by the WPC in its *Programme of Action* for 1981,[81] with the subsequent actions of Western movements.

The 1981 programme instructed peace activists to 'co-operate' with all relevant United Nations committees and agencies, particularly the Disarmament Commission and Committee and the United Nations Centre for Disarmament, the non-governmental Special Committee on Disarmament, UNESCO and its NGOs in Paris. Also listed for this attention were peace institutes throughout the non-communist world, international, regional and national religious organisations, especially the World Council of Churches, the media, international trade unions, youth and student organisations. Activists in politics were also encouraged to form peace groups in their parliaments.[82]

According to the programme, 'broad front' tactics were to be employed so that the peace movement would have the appearance of a 'common united mass action' leading to 'mass solidarity actions on the widest national and international scale'. Local concerns were to be highlighted: neighbourhoods close to the future sites of Cruise or Pershing II missiles were to be mobilized. Implicitly, the disarmament issue was to be concentrated exclusively against Western (imperialist) arms: no criticism was to be levelled at Soviet military strength. The peace offensive was to be linked with 'liberation', the issue of Third World 'development', the 'new economic order' and the 'New International Information Order'.[83]

The themes to be propagated, according to this WPC directive, were *fear*, to create uncertainty in the West ('never before has there been so great a danger of a world nuclear holocaust'), the danger posed by *Western* weapons—especially neutron warheads, Cruise and Pershing II missiles, the deliberate misinterpretation of Presidential Directive 59, to demonstrate an American desire to wage a nuclear war confined to Europe, the importance of 'nuclear-free zones', the price of armaments and its impact on 'development', high taxation and 'galloping inflation', and the value of a 'no-first-use' treaty.[84] The theme of a nuclear 'freeze' appeared for the first time in the 1983 WPC directive.[85]

Readers may judge from the events of 1982–5 how well the WPC succeeded in transmitting these instructions into indigenous Western peace movements, having them assimilated and then reiterated as though they were the fruits of original domestic initiatives. The themes and activities were adopted with most thoroughness in West Europe. But in the United States, the United States Peace Council, established as recently as 1978, nevertheless claimed prime responsibility for organising the 1982 peace rally in New York City. This event attracted more than a million protesters—the greatest demonstration in the country's history. While the Peace Council's claim may contain an element of bravado, and while many non-communist peace groups undoubtedly played important roles, the fact that Michael Myerson, Executive Director of the United States Peace Council, should have made such a claim is significant.[86]

Not every Western movement adopted all these WPC issues. Particularly in Western Europe, a number of peace movements also voiced criticisms of the Soviet arms build-up. Although married to the New Left notion of two morally symmetrical superpowers locked in a mad arms race, E P Thompson's European Nuclear Disarmament (END) movement at least insisted that the East as well as the West was responsible for the war danger, and should be brought under pressure by world and Eastern public opinion to disarm.[87] For this heresy, END became the target of increasingly hostile Soviet attacks. In his asssessment of the Western peace movement at the end of 1984, V Matveev wrote:

> . . . some are trying to batten on the anti-war drives and use them for aims that have nothing in common with defence of the world against a nuclear catastrophe. I am speaking of attempts by some people like Thompson in England to bring Cold War motives and tendencies into the anti-war movement . . .[88]

The idea of END encouraging the Cold War was utterly absurd: Matveev's motive, presumably, was to discredit a group which had dared to step out of Moscow's line. Similarly the Greens in Germany and the Interchurch Peace Council (IKV) in The Netherlands more than once criticised the human rights situation and 'militarisation' in East Bloc countries. On balance, however, since peace movements primarily targeted the United States and not the Soviet Bloc, most of their activities still favoured the Soviet position.

* * *

The deception of Western peace movements was a means to an end, not an end in itself. How far did the 1977–85 offensive carry the Soviet Union towards its political objective of weakening NATO *vis-á-vis* the Warsaw Pact?

At one level, it is arguable that reactions against the peace offensive in, for instance, West Germany, Britain and the United States, may have contributed to the return of conservative governments. In France, under a socialist government, Soviet deception seems not to have succeeded, even against the peace movement, and French leaders refused to be intimidated. If the Soviets really believed that their peace offensive could prevent the deployment in Europe of Cruise and Pershing II missiles, or convert American policy towards a nuclear 'freeze', they must have been disappointed.

Supporters of the NATO policy can claim with some justification that, by rejecting pressures from their peace movements, Western governments brought the Soviets back to the negotiating table. Arms control succeeded, these might argue, because the deployment of Cruise and Pershing IIs had gone ahead on time, demonstrating to Moscow that serious bargaining was more productive than coercive public diplomacy.

In the short term, then, the 1977–85 peace offensive did not advance Soviet political goals, even though peace movements had been deceived. But the Soviets seem to plan for the long term, and it is in the longer view that they may reap benefits from some of the secondary effects of their offensive.

In NATO countries, it was at least possible that some of the conclusions reached by peace lobbies had been endorsed by much wider segments of the populations. It seemed possible that faith in military strength and defence had been eroded. Specifically, NATO's policy of deterrence no longer went unchallenged. For some, it was the nuclear weapons that posed the threat, not the Soviet Union. And in Western Europe there was ambivalence in apportioning blame for international insecurity between the Americans and the Russians.

* * *

After Mikhail Gorbachev became Secretary General of the CPSU in 1985, Soviet 'peace' diplomacy underwent important changes. 'New Thinking' became the password for international affairs and the internal reform programme improved the image of the USSR. New faces appeared in all the key departments: Anatolii Dobrynin, for 24 years Soviet Ambassador to the United States, replaced Boris Ponomarev as head of the International Department; Alexander Yakovlev, after 10 years as Ambassador to Canada, took control of propaganda. These two men possessed an intimate understanding of Western audiences and under their influence Soviet peace propaganda became more flexible and potentially more effectvie. To non-communists, appeals were devoid of ideological overtones: the West was addressed in its own idiom.

In the Soviet Peace Committee, the abrasive Yuri Zhukov was replaced by the smooth Genrikh Borovik and in the WPC, Romesh Chandra was pushed sideways to make room for a new Secretary

General, Johannes Pakaslahti, a Finnish professor of communication with decidedly pro-Soviet views. The Committee and the Council became instruments for courting a broader range of Western opinion as the CPSU worked hard to dissolve Western notions of a 'Soviet threat'.

A change for the better in Soviet foreign policy, or the start of a new deception campaign based on 'peace'? Time will tell.

Endnotes

1. Quoted Max Domarus, *Hitler: Reden und Proklamationen* (Weisbaden: R Löwitt, 1973), Vol. 2, pp. 615–16.
2. *Charter of the United Nations*, San Francisco, 26 June, 1945, preamble, lines 13–14.
3. S I Ozhigov, *Slovar Ruskogo Yazyka* (Dictionary of the Russian Language) (Moscow: Izd. Ruskiy Yazyk, 1981), p. 312; and H Bax, 'De Russische vrede' (The Soviet Peace), *Ons Leger* (Our Army), June, 1982, p. 40 ff.
4. Georg Klaus and Manfred Buhr, *Marxistisch-Leninistisches Wörterbuch der Philosophie*, Vol. I (Reinbek bei Hamburg: Rowohlt Taschenbuch Verlag, 1972), p. 429.
5. *Bolshaya Sovyetskaya Entsyclopediya*, Vol. 16, 3rd ed., (Moscow: 'Sovyetskaya Entsyclopedia', 1974–81), p. 308.
6. V I Lenin (i), *Collected Works*, (Moscow: Progress Publishers, third printing, 1977), Vol. 26, p. 386.
7. G V Shredin, 'Marxist-Leninist Doctrine on War and the Army', *Soviet Military Review*, No. 1, January, 1978, p. 19.
8. Marx-Engels Werke (MEW), Vol. 4, (East Berlin: Dietz Verlag, 1980), p. 479; Gottfried Kiessling, *Krieg und Frieden in unserer Zeit*, (East Berlin: Militärverlag der DDR, 1977), p. 128.
9. *Cf.* N Kulikov, 'On Just and Unjust Wars', *Soviet Military Review*, No. 3, March 1977, p. 8 ff.
10. Lenin (ii), *Collected Works*, (Moscow, Progress Publishers, second printing, 1976), Vol. 45, p. 507.
11. Lenin (i), Vol. 23, p. 237.
12. Lenin (i), Vol. 23, p. 194.
13. Lenin (i), Vol. 23, p. 95.
14. Lenin (i), Vol. 23, p. 96.
15. Lenin (i), Vol. 23, p. 81; lest it be argued that these early pronouncements no longer govern Soviet policy, see *Kommunist*, No. 12, 1983, pp. 19–29. On p. 26 the editorial writer says: 'It is no secret that at first certain comrades were unable to grasp the nature of the anti-war movements . . . were not always able to overcome their prejudices against pacifists and ecological organisations with their inconsistent and contradictory arguments, and did not perceive the members of these movements as their objective allies in the struggle for peace.'
16. See, for instance, Hedrick Smith, *The Russians* (New York: Ballantine, 1977), pp. 402–34.
17. See J A Emerson Vermaat, 'Moscow and Europe's Peace Movements', *Problems of Communism*, November–December 1982, pp. 43–56.
18. For a study of the International Department, CPSU, see Robert W Kitrinos, 'CPSU ID', *Problems of Communism*, September–October 1984, pp. 47–75.
19. Boris Ponomarev, *The War Danger: Its Source and How to Stop It* (Moscow: Novosti, 1981), pp. 3–4.
20. A Sovetov, 'Peace Built on Strength—A Doctrine of International Terrorism' *International Affairs* (Moscow), No. 2, 1984, p. 60.
21. Sh Sanakoyev, 'Public Forces and Disarmament', *International Affairs*, No. 10, 1982, p. 20.
22. *Cf.* W van den Bercken, 'De Sovjethouding in het Ontwapeningsvraagstuk' (The Soviet

Attitude towards Disarmament), *Internationale Spectator* (The Hague), April, 1981, p. 198.
23. Lenin (i) Vol. 45, p. 507.
24. A S Milovidow and V G Koslov, eds., *The Philosophical Heritage of V I Lenin and Problems of Contemporary War* (Soviet Military Thought Series, No. 5), (Washington,: US Government Printing Office, undated but originally published in Moscow by Voyenizdat in 1972), p. 37.
25. *Cf. Pravda*, 6 September 1982; A Lebedev, 'The Public in World Politics', *International Affairs*, No. 9, September, 1982, p. 72; *News from the USSR* (The Hague), Press Bulletin of Soviet Embassy, 29–30 March 1982, p. 4.
26. *Bolsh. Sov. Ents.*, Vol. 16, p. 314.
27. Robert H McNeal, ed., *Resolutions and decisions of the CPSU*, Vol. 4 (Toronto and Buffalo, University of Toronto Press, 1974), p. 207.
28. Klaus and Buhr, Vol. 1,·p. 435.
29. See Joseph D Douglass Jr. (i), 'Soviet Strategic Deception,' *Defense Science 2002+*, August 1984, pp. 87–99.
30. Adam B Ulam, *Dangerous Relations: The Soviet Union in World Politics 1970–1982* (New York, Oxford: Oxford University Press, 1983), pp. 263–6; Richard Pipes, *US—Soviet Relations in the Era of Détente* (Boulder, Co.: Westview Press, 1981), pp. 88–96.
31. See P H Vigor, *The Soviet View of War, Peace and Neutrality* (London, Routledge and Kegan Paul, 1975), pp. 178–94.
32. Vigor, pp. 189–90; Joris J C Voorhoeve, 'Pacifist–Neutralism in Western Europe,' *Conflict Quarterly*, Vol. II, No. 4, 1982, pp. 14-30.
33. See, for example, *Disarmament: The Peoples and the United Nations* (Helsinki, WPC, 1982), pp. 15–17; L I Brezhnev, *Interview to 'Der Spiegel' Magazine* (reprint) (Moscow: Novosti, 1981), pp. 28–30; Andrei Gromyko, *Safeguarding Peace is the Main Task* (Moscow: Novosti, 1981), pp. 22, 28; Soviet Committee for European Security and Cooperation and Scientific Research Council on Peace and Disarmament, *The Threat to Europe* (Moscow: Progress, 1981), pp. 47, 48, 70, 71.
34. Anthony Cave Brown and Charles MacDonald, *On a Field of Red: The Comintern and the Coming of the Second World War* (New York: G P Putnam's Sons, 1981), p. 524 ff.
35. *Cf.* A Rossi, *Le Pacte Germano–Soviétique: L'Histoire et le Mythe* (Paris: Collection de la Revue 'Preuves', 1954), p. 83 ff.
36. Kitrinos, pp. 50, 57, 64.
37. Thorez and Togliatti jointly announced in 1949 that they would welcome the Soviet Army when it entered their countries 'in its battle against the aggressor'. *For a Lasting Peace, For a People's Democracy* (Cominform newspaper) Bucharest, 1 and 15 March, 1949, quoted Herbert Romerstein, *The World Peace Council and Soviet Active Measures* (Hale Foundation, 1983), p. 8; the claim of 500 million signatures is made in *World Peace Movement, Resolutions and Documents*, published by the Secretariat of the WPC (no place, no date) circa 1954, p. 47, quoted Romerstein, p. 8.
38. Quoted from *Cominform Journal*, 29 November 1949 by Iain Phelps-Fetherston, *Soviet International Front Organizations: A Concise Handbook* (New York, London: Frederick A Praeger, 1965), p. 17.
39. Phelps-Fetherston, pp. 9, 10, 32; Witold S Sworakowski, ed., *World Communism: A Handbook 1918–1965* (Stanford, California: Hoover Institution Press, 1973), p. 488.
40. Romesh Chandra in *New Times* (Moscow), July 1975, quoted *World Peace Council: Instrument of Soviet Foreign Policy: Foreign Affairs Note*, (Washington, DC: Department of State, April 1982), p. 5.
41. *XX Century and Peace* (Moscow), August 1981.
42. *World Marxist Review*, No. 1, January 1981.
43. *The Truth About Afghanistan. Documents, Facts, Eyewitness Reports*, (Moscow: Novosti, 1981) p. 196, 197.
44. Kitrinos, pp. 51, 57, 58, 67, 68.
45. Declassified CIA study, House of Representatives, Ninety-Fifth Congress, First and Second Sessions, 'International Communist Front Organizations' in *The CIA and the Media: Hearings before the Subcommittee on Oversight of the Permanent Select Committee on Intelligence* (Washington, DC: US Government Printing Office, 1978), p. 577.

46. Clive Rose, *Campaigns Against Western Defence: NATO's Adversaries and Critics* (London: Macmillan, 1985), pp. 56, 62, 91, 140, 141, 150, 152, 166, 222, 231, 232.
47. Syndicated Reuter report 'Soviet Peace Movement Claims it's Independent', *Daily Gleaner* (Fredericton), 10 December 1984.
48. Ruth Tosek, letter to the Editor 'WPC finances', *New Statesman* (London), 17 October 1980, p. 22.
49. *Ibid.* Soviet involvement in supporting and funding peace movements was confirmed in a May 1988 article in *Literaturnaya Gazeta* by Professor Vyacheslav Dashichev of the USSR's Institute for Socialist Economics. See *Manchester Guardian Weekly*, 29 May 1988.
50. UN/ECOSOC, Doc. E/1981/29, Annex I (16 March 1981), p. 12.
51. *Neues Deutschland* (East Berlin), 23 March 1982.
52. *New Times* (Moscow, No. 19, 1981, p. 5; and No. 16, 1983, p. 10.
53. See *Daily Telegraph* (London), 8 February, 1982; Rose, p. 293.
54. The Washington-based Center for Defense Information (CDI) was formed in 1973 as a spin-off project from the Institute for Policy Studies (IPS). Former military officers, intelligence officers and academics form the staff. The Center's Director, Gene LaRocque, wrote a Foreword for an official Generals for Peace publication—*The Arms Race to Armageddon: Generals Challenge US/NATO Strategy* (Leamington Spa, England: Berg Publishers, 1984).
55. See for example, *Pravda*, 7 August 1981; *Komsomolskaya Pravda*, 3 November 1979 and 18 July 1981; *Rude Pravo* (Prague), 25 June 1981.
56. *Verfassungsschutzbericht 1983* (Bonn: Ministry of the Interior, 1984), p. 37.
57. *World Marxist Review*, No. 9, 1981 (Kade was identified as Vice-President, IIP).
58. Nino Pasti, *Euro-Missiles and the General Balance of NATO and Warsaw Pact Forces* (Helsinki: WPC, no date) Circa 1980; Nino Pasti, *Euro-Missiles and the Balance of Forces: Propaganda and Reality* (Helsinki: WPC, 1983); Gert Bastian, *Nuclear War in Europe?* (Helsinki: WPC, no date) Circa 1981.
59. Generals for Peace and Disarmament, *Memorandum submitted to the Delegations of the Second Special Session on Disarmament of the United Nations General Assembly* (New York, 1982—mimeographed).
60. *Moscow News*, 27 May 1984.
61. *New Times*, No. 24, June 1984, quoted Rose, p. 76.
62. Y Konstantinov 'For Peace and Security in Europe', *International Affairs* (Moscow), No. 5, 1981, p. 56.
63. *Cf.* Georgiy Shakhanazarov, 'Policy of Peace and Our Time', in *Soviet Policy of Peace* (Moscow: USSR Academy of Sciences, 1979), p. 64; Vitaliy Korinonov, in *The Policy of Peaceful Coexistence in Action* (Moscow: Progress, 1975), pp. 52, 53; see also Chapter 7.
64. Statement by Dang Quay Minh, representative of the Vietnamese National Liberation Front, in Moscow, reported *Washington Post*, 28 May 1965.
65. Alexander Kalyadin, 'The Struggle Against the Threat of War and the Arms Race and For Disarmament', in *Public Opinion in World Politics* (Moscow, Soviet Peace Committee/ 'Social Sciences Today', Soviet Academy of Sciences, 1976), p. 65.
66. *Gazet van Antwerpen* (Antwerp), 25 November 1982.
67. Vadim V Zagladin, *Europe and the Communists* (Moscow: Progress, 1977), p. 84 ff.
68. *DeWarrheid* (Amsterdam), (Dutch Communist Party Newspaper), 13 September 1976.
69. See Sherri L Wasserman, *The Neutron Bomb Controversy* (New York: Praeger, 1983); Rose, pp. 108–10; Richard H Shultz and Roy Godson, *Dezinformatsia: Active Measures in Soviet Strategy* (New York: Pergamon-Brassey's, 1983), pp. 127–30; Gerhard Wettig, 'The Peace Movement in Western Europe: Manipulation of Popular Perceptions', in *Contemporary Soviet Propaganda and Disinformation: A Conference Report* (Washington, DC: Department of State Publication 9536, 1987), pp. 161–88.
70. Rose, p. 110.
71. Rose, p. 109.
72. See J A E Vermaat, 'De Oosteuropese Relaties van de Nederlandse Vredesbeweging' (The East European Relations of the Dutch Peace Movement), in *Ons Leger* (Dutch Armed Forces Journal), October-November 1981.
73. *Neues Deutschland* (East Berlin), 24–25 December, 1977.
74. Information provided to the author by a source within NATO.

75. *Handelingen Tweede Kamer der Staten-Generaal* (Dutch Parliamentary Record, Zitting (Session) 1982–3, No. 12 (16 December 1982), p. 1279.
76. *Information Bulletin* (Prague), No. 7, 1981, pp. 21, 22 (emphasis added).
77. See Stefan T Possony, 'Communist Psychological Warfare', W F Hann and J C Neff, eds., *American Strategy for the Nuclear Age* (New York: Doubleday, 1960).
78. See Michael Tatu, 'Public Opinion as Both Means and Ends for Policy Makers', *Adelphi Paper No. 191* (London: IISS, 1984), pp. 26–32; also Chapter 13.
79. Mikhail Gorbachev, addressing Warsaw Pact leaders in Warsaw, April 1985, quoted Serge Schmenann, New York Times Service, *Globe and Mail* (Toronto), 29 April, 1985.
80. V. Matveev, 'Disarmament: An Ideal, A Practical Orientation', *The Democratic Journalist* (Prague, No. 12, 1984, p. 4.
81. World Peace Council, *Programme of Action, 1981* (Helsinki: WPC, not dated), issued 1981.
82. *Programme of Action 1981*, pp. 11, 13, 14, 15, 31, 32, 33, 36, 37, 38.
83. *Ibid.*, pp. 6, 7, 12, 15, 17, 33.
84. *Ibid.*, pp. 3, 4, 12, 13, 14.
85. World Peace Council, *Programme of Action, 1983* (Helsinki: WPC, no date) ('adopted', November, 1982), p. 4, 5, 8.
86. Michael Myerson, interviewed by Les Harris, in Stornoway Productions, *Agents of Deception*, broadcast by CTV, 4 May 1986.
87. See E P Thompson and Dan Smith, eds., *Protest and Survive* (Harmondsworth: Penguin, 1980).
88. Matveev, *op. cit.*, p. 4.

Part Two

Studies in Western Deception Operations

Western Approaches

. . . a prince who deceives always finds men who let themselves be deceived.

Niccolò Machiavelli[1]

There is no more right to deceive than there is a right to swindle, to cheat or to pick pockets . . .

Walter Lippmann[2]

These two quotations illuminate an enduring tension in Western approaches to deception. On the one hand, there are strong ethical strictures against deception, rooted in the West's religious teachings and moral philosophy. The Ninth Commandment, for example, states clearly, 'Thou shalt not bear false witness . . .'[3] Yet it is equally clear that deception has a long history in the West, dating from classical times.[4] The writings of Francis Bacon, Edmund Burke, Alexis de Tocqueville, and Lewis Carroll attest—in their criticism of deception—to its longevity and persistent presence in political life.[5] In this century, its use has spread from party politics to business and advertising'[6] and to war and diplomacy. Western democracies have resorted to deception readily in time of war, and with only slightly less enthusiasm in the 'Cold War'. In short, insofar as Western experience is concerned, ideals are one thing, practical affairs are somewhat different.

As the forthcoming case studies illustrate, this dichotomy between ethics and practice can have serious consequences when deception fails, or when it is discovered. This is particularly so when, as these cases reveal, the domestic audience becomes one of the primary targets of a deception. While the immediate embarrassment of the perpetrator may be short-lived, the long-term consequences may be much more serious: the erosion of the public confidence and trust in elected officials which is essential to a healthy democracy.

Machiavelli, the Florentine diplomat whose writing describes the nature of political life on the eve of the Reformation, is credited with articulating an 'operating code' for the use of deception in politics and diplomacy.[7] 'The one who knows best how to play the fox comes out best', he wrote, 'but he must be a great simulator and dissimulator.'[8] While many politicians since have been accused of following his advice to the point of being 'Machiavellian', it is only in the 20th Century that deception has been 'institutionalised' within government. In this regard, the First World War was something of a watershed. The British used deception both on the battlefield, to facilitate military operations, and on the home front, to mobilize and sustain public support for the war. General Sir Edmund Allenby demonstrated the potential of military operational deception. In October 1917, just before the Third Battle of Gaza, he arranged for a set of false plans to fall into German-Turkish hands. Together with other information which played upon preconceived ideas, the documents persuaded the Turks to alter their defensive preparations. This allowed Allenby to achieve surprise and victory.[9]

Perhaps the most successful deception practised against the British domestic audience (and many overseas) was the report of the committee of historians and lawyers under the chairmanship of Lord Bryce. It charged the German Army with unspeakable atrocities during the 'rape of Belgium'. Published in 1915 and subsequently translated into 30 languages, it was influential in branding the Germans as the aggressors, and thus served its purposes well. It was not until 1922 that a Belgian commission of inquiry concluded that none of the report's charges could be substantiated.[10]

The exposure of the Bryce report, together with the post-war mythology which tended to exaggerate the wartime accomplishments of the British Ministry of Information, may have helped to blur the distinction between propaganda and deception.[11] It is important to emphasise, however, that, unlike propaganda, there was no single, central organization for deception planning and execution in the First World War. The two examples cited above, however successful, were *ad hoc* efforts.

During the Second World War, deception became, in Michael Handel's words, 'the focus of formally organised staff work'.[12] That this occurred may be attributed in large measure to the British Prime minister, Sir Winston Churchill, who brought to the task a zealous, if often misguided, enthusiasm for clandestine, unorthodox action.[13] Under his direction an array of deception organisations flourished; they can be mentioned but briefly here. London

Controlling Section, part of the British Joint Planning Staff, was responsible for creating and co-ordinating strategic deception. As such it was not an executive body; its deception plans were carried out by the Allied military forces, the British intelligence community and other special, secret services.[14] Two of the latter were the XX-Committee and the Political Warfare Executive (PWE). The first of these supervised the use of captured German agents as conduits to pass disinformation to German intelligence.[15] The PWE's main task was to use various means of communication to conduct subversion in Germany and the occupied countries, particularly among the enemy forces. Deception played a role here in the form of 'black' propaganda; for example, radio broadcasts whose form and content suggested a German rather than a British origin.[16]

While some of the deceptions were very simple, involving no more than the re-playing of a rumour, others—designed to assist major operations—were complex and elaborate, involving many people in the military, politics, diplomacy and the secret services. In 1943, for example, the British carried out Operation MINCEMEAT, the object of which was to persuade the Germans that the Allies planned to invade Greece and Sardinia, when in fact the real target was Sicily. This deception operation, often referred to as 'The Man Who Never Was', involved planting false plans on a corpse supplied with the plausible 'legend' of a British officer. The body was allowed to wash ashore in Spain where, as hoped, the plans fell into German hands. The deception complicated German planning, although they never ruled out Sicily as a likely target. The following year, the Allies conducted the FORTITUDE SOUTH deceptions, which were intended to convince the Germans that the main invasion of Europe would come at the Pas de Calais, the Normandy landings being no more than a diversion. To this end the deception planners created in South-East England a fictitious army group, whose existence and alleged purpose was suggested by notional physical evidence (camps, vehicle parks, dummy landing craft), controlled leaks of radio communications, and air bombardment of the supposed invasion target. But it was the planted reports by German agents under British control that clearly persuaded the German High Command to believe the deception plan.[17] It is this degree of co-ordination and sophistication that sets this period of deception activity apart from those that preceded it. The other feature which stands out at this time is the central role of intelligence. In many cases, it was the Allies' ability to 'read the enemy's mail', and thus to discern his weaknesses, that made successful deception possible.[18]

The end of the war saw a rundown and reorganisation of British intelligence and other special services. Officially, the London Controlling Section was disbanded along with the XX-Committee and PWE.[19] But the rapid emergence of the 'Cold War' forestalled the complete eradication of deception capabilities. That much is clear. The problem for historians of this contemporary period is that information on clandestine organisation and activities is incomplete and often unreliable.

What seems to emerge from the literature is the decentralisation of British deception capabilities. Missions and resources were 'parcelled out' to the Foreign Office, the Secret Intelligence Service (SIS), or the Armed Forces, as the need or opportunity arose. This would be in keeping with a return to peacetime, the 'Cold War' notwithstanding, and with an approach to government management which, by tradition, has not been remarkable for tidiness and efficiency. If there was some office or committee responsible for central planning and direction of deception, it is not discernible from the available evidence.

As Simon Ollivant points out in Chapter 11, the PWE was the first wartime organisation to be emulated, albeit in somewhat truncated form. The Foreign Office's Information Research Department (IRD), which operated from 1948 to 1977, specialised in 'grey' propaganda. It arranged for publication, in British and foreign media of selected unattributable stories, usually based on intelligence. To the extent that the actual source of the stories was concealed and that IRD was able to facilitate the re-playing of these stories in the foreign media, the IRD certainly had the organisational potential for deception.[20] But there is no information available to researchers that demonstrates that IRD engaged in 'black' propaganda or outright deception. That was left to 'operational' bodies, those elements of the intelligence community and the Armed Forces actually engaged in the front lines of the 'Cold War'. In 1946 the SIS absorbed the covert action responsibilities of the wartime Special Operations Executive, and 'formed a new Special Operations Branch and Political Action Group'.[21]

One of its most effective political action assets was Sharq-al-Adna, a Cyprus-based radio station which broadcast to the Arab world. Managed by an SIS 'front', the Near East Arab Broadcasting Company, and run by a director and staff genuinely sympathetic to their audience, the station transmitted supposedly 'untainted' news, entertainment, and straight commercial advertising. Apparently immensely popular, it is said to have had a larger listening audience than Cairo Radio in the 1950s. True to its 'cover', it even generated revenue from the advertising it carried. If this latter benefit created a minor bookkeeping problem for the SIS accountants, this was undoubtedly outweighed by the opportunities for deceptive propaganda that the station offered. Nonetheless, Sharq-al-Adna was 'blown' by the Suez Crisis, during which the British military took it over as a psychological warfare asset. Insensitive to its real value in this regard as a supposedly independent station, they changed its name to 'Voice of Britain' and its mission to anti-Nasser propaganda. Its credibility compromised irreparably by these shortsighted actions, Sharq-al-Adna was closed permanently at the end of the Suez War.[22] Deception in that operation is the subject of Michael Handel's contribution in Chapter 13.

* * *

Prior to 1941, the American approach to deception was similar to that of pre-war Britain; there was no single agency responsible for or routinely engaged in deception. That is not to say that deception had never been used. As noted earlier, it had come to play a part in politics and commerce. But it was war that brought out America's talent for deception, although until the Second World War this usually could be attributed to the initiative of particular individuals rather than to policy and organisation.

During the Revolutionary War, George Washington made effective use of military tactical deception on the battlefield, and supplemented this by planting disinformation—in the form of agent reports and false documents—on the British commanders.[23] Timothy Webster, a Pinkerton Detective Service agent, worked as an *agent provocateur* for the Union during the Civil War. Posing as a rabid Confederate sympathiser, he penetrated and compromised secret rebel groups in Baltimore. His daring 'escapes' from Union forces not only failed to arouse suspicion; they added to the credibility of his 'legend'.[24]

The Second World War gave the United States the *entrée* into large-scale deception operations. The Americans were, of course, major partners with the British in the FORTITUDE strategic deceptions. They also created their own counterpart to the London Controlling Section: Joint Security Control, in the bureaucratic structure of the Joint Chiefs of Staff. The Joint Security Control never gained authority or access comparable to the London Controlling Section, but did educate some American military commanders in the art of deception, and ran a number of operations with a degree of success. The first of these, Plan WEDLOCK, diverted Japanese attention from the American assault from the South Pacific on the Marianas Islands. It simulated the build-up of a joint American-Canadian force in the Aleutians.

Employing bogus radio traffic, air and naval attacks, information planted on Japanese agents under control, leaks to the media, and misleading physical evidence, WEDLOCK played on Japanese anxieties about a possible thrust from the North Pacific. This persuaded them to retain and even increase forces deployed on the Kurile Islands at a time when forces were needed to meet the American thrust from the South.[25]

In the Office of Strategic Services (OSS), the United States had concentrated all those functions—intelligence collection, special operations, and psychological warfare—which the British had assigned to three distinct organisations: the SIS, the Special Operations Executive and PWE. The OSS's Morale Operations subsection worked in collaboration with PWE in the field of 'black' propaganda and, like its British counterpart, it used various means to attack and undermine enemy morale. These means included clandestine radio stations which purported to broadcast from within Axis-held territory, the distribution of forged documents, and the initiation of rumour campaigns. The record does not indicate whether these operations produced the desired effect on the enemy. It does suggest that these operations developed a life and momentum of their own, which sometimes was at variance with official Allied policy. Moreover, these activities may have complicated the intelligence analysis process; on more than one occasion OSS-planted disinformation rumours were picked up by Allied intelligence, which was unaware of the original source of the information.[26] This kind of 'blow-back' is a risk inherent in deception operations where co-ordination is poor, as was sometimes the case during the war.

If, during the Second World War, the Americans were junior partners to the British in the field of deceptions, that position changed rapidly after the war. Like Britain, the United States demobilised rapidly at war's end, and this included the OSS. It was de-activated in October 1945, its functions either terminated or transferred to other government departments.[27] From about 1947, the Cold War brought a rapid re-mobilisation of American clandestine warfare capabilities. The United States brought to this endeavour a greater degree of centralised co-ordination than is apparent in British efforts at this time. The two most significant developments were the creation in 1947 of the National Security Council (NSC) and the Central Intelligence Agency (CIA). The NSC consisted of the President, Vice-President, the Secretaries of State and Defense, a small staff, and other advisers as needed. Its role was (and is) to advise the President on the integration of domestic, foreign and defence policies in order to facilitate effective co-operation between those agencies which play a role in national security affairs.[28]

In the *interregnum*, between the demise of the OSS and the creation of the CIA, United States foreign policy-makers had perceived a need to respond covertly to Soviet subversion in Europe and elsewhere. Both the State Department and the military eschewed involvement in covert action. But a vaguely-worded 'catch-all' clause in the National Security Act gave the NSC the authority to direct the CIA 'to perform such other functions and other duties related to intelligence affecting the national security as the National Security Council may from time to time direct'.[29] So, by law and by default, the CIA became the 'lead agency' for covert action. Responsibility devolved upon the Clandestine Service, known officially from 1952 as the Directorate of Plans, and later as the Directorate of Operations. Initially, the principal activity was covert psychological warfare, but the range of activities gradually expanded to include paramilitary operations. By the mid-1950s, covert action came to dominate the priorities of the Agency's Clandestine Service and to consume a majority of the Agency's budget.[30] Inevitably, many of these operations involved deception to a greater or lesser degree.[31] In Chapter 12 Trevor Barnes describes operations in Europe in the 1950s: in Chapter 15 David Charters discusses deception in support of paramilitary operations.

Under the NSC's direction, the CIA co-ordinated the activities of national level intelligence services (military, State Department, etc.) and provided a central intelligence evaluating body and common services that could best be provided by a single national agency.[32] American foreign policy objectives, events overseas, and the internal dynamics of the organisation itself, fostered the growth of the CIA. The requirement for national strategic estimates particularly on the Soviet Union, and the CIA's mandate as the over-arching national intelligence service, allowed the Agency to move into fields that other intelligence services declined to cover. Technical collection was one of these, and the U-2 high-altitude reconnaissance programme marked the beginning of the CIA's ascendancy in this field.[33] The deception plan covering the U-2 programme, and attempts to deceive after Francis Gary Powers' U-2 was shot down in 1960, form the subjects of Chapter 14.

The armed forces of the Western powers, particularly those of Britain, France and the United States that became involved in military operations related to the Cold War, did not overlook the importance of deception. In both the French and the (later) American campaigns in Indo-China, and the British counter-insurgency in Malaya, armies or secret services provided special 'psy-ops' teams to attempt to influence or deceive hostile forces or their supporters. For the most part, these operations lie outside the scope of this book, being military in character and lacking an 'extra-diplomatic' dimension. Sometimes, however, low-intensity conflicts did give rise to deceptions far removed from the battlefield. One of these, the so-called 'uncounted enemy' in Vietnam, is reassessed by Michael Hennessy in the final case study of this volume.

* * *

On the wall of the main lobby at CIA headquarters is the Biblical inscription: 'And ye shall know the truth and the truth shall make you free' (John 8:32). It is an apt motto for the intelligence service of a democratic society. But once such a service proceeds to conduct its missions, especially when these involve deception, contraditions emerge. The chapters which follow illuminate those contradictions clearly, demonstrating the difficulties of 'squaring the circle' in a way that reconciles ethical ideals and operational practice.

Endnotes

1. John Plamenatz, ed., *Machiavelli: The Prince, Selections from the Discourses and Other Writings* (London: Fontana/Collins, 1972), p. 107.
2. Walter Lippman, *Essays in the Public Philosophy* (Boston: Little, Brown, 1955), p. 128.
3. Exodus, Chapter 20, line 16. In Chapter 23, line 1, Moses also handed down to the Hebrews the admonition not to propagate false reports or cooperate with evil men by giving information known to be false.
4. Michael I Handel. 'Introduction: Strategic and Operational Deception in Historical Perspective', *Intelligence and National Security*, Vol. 2, No. 3 (July 1987), pp. 2–5.
5. Francis Bacon, *The Essayes or Counsels Civill and Morall* (London: Everyman's Edition, 1983), p. 17; Edmund Burke, *Reflections on the Revolution in France* [1790] (Garden City, New York: Dolphin Books, 1961), p. 181; see the quote from de Tocqueville's *Oeuvres Complètes*, Tome I, Vol. I, p. 413 (a 1951 edition of his collected works), in Melvin Richter, 'Tocqueville's Contributions to the Theory of Revolution,' in Carl J Friedrich, ed., *Revolution* (New York: Atherton Press, 1969), p. 88; Lewis Carroll, *Through the Looking Glass, and What Alice Found There* (London: Macmillan, 1955), p. 114. Alice's conversation with Humpty Dumpty was an allegorical, but pointed attack on the use of deceptive language in British politics.
6. Sissela Bok, *Lying: Moral Choice in Public and Private Life* (New York: Pantheon, 1978), p. 85; Jeffrey Schrank, *Deception Detection: An Educator's Guide to the Art of Insight* (Boston: Beacon Press, 1975), pp. 1–35 *passim*; the essays on televised political debates, and 'Advertising: the Merchandising of Illusion', in Reo M Christenson and Robert O McWilliams, eds., *Voice of the People: Readings in Public Opinion and Propaganda*, 2nd ed. (New York: McGraw-Hill, 1967), pp. 359–73, 501—28, *passim*; see also Robert Spero, *The Duping of the American Voter: Dishonesty and Deception in Presidential Television Advertising* (New York: Lippincott and Crowell, 1980).
7. Bok, p. 136.
8. Plamenatz, p. 107.
9. Handel, pp. 6–11.
10. Phillip Knightley, *The First Casualty* (New York: Harcourt, Brace, Jovanovich, 1975),

pp. 82–4. The 1,200 depositions upon which the report was based have since disappeared.

11. *Ibid.*, pp. 82, 84–5, 104–7. Knightley notes that some of the most conspicuous deceptions originated with war correspondents, rather than with any officially-sanctioned body. See also Michael Balfour, *Propaganda in War 1939–1945: Organisations, Policies and Publics in Britain and Germany* (London: Routledge and Kegan Paul, 1979), pp. 3–10.
12. Handel, p. 20.
13. Christopher Andrew, *Secret Service: the Making of the British Intelligence Community* (London: Heinemann, 1985), p. 448; Anthony Cave-Brown, *Bodyguard of Lies* (New York: Harper and Row, 1975), pp. 4–5, 45, 47.
14. Cave-Brown, pp. 4, 45–58, 268–77.
15. John C Masterman, *The Double Cross System in the War of 1939–1945* (New Haven: Yale University Press, 1972).
16. Balfour, pp. 89–102.
17. On FORTITUDE, see Cave-Brown *passim*; 'An Eyewitness Report on the *FORTITUDE* Deception', in Donald C Daniel and Katherine L Herbig, eds., *Strategic Military Deception* (New York: Pergamon, 1981), pp. 224–42. T L Cubbage, 'The Success of Operation Fortitude: Hesketh's History of Strategic Deception', *Intelligence and National Security*, Vol. 2, No. 3 (July 1987), pp. 327–46. On MINCEMEAT, see the 'sanitised', authorized account: Ewen Montagu, *The Man Who Never Was* (Philadelphia: Lippincott, 1954). Scholarly analysis and critique may be found in Handel, pp. 66–73; Klaus-Jürgen Muller, 'A German Perspective on Allied Deception Operations in the Second World War', *Intelligence and National Security*, Vol. 2, No. 3 (July 1987), pp. 301–16; and a response by David Hunt in Vol. 3, No. 1 (January 1988), pp. 190–94. See also F H Hinsley, *et al.*, *British Intelligence in the Second World War* (London: HMSO, 1984), Vol. 3, Part 1, pp. 77–80, 120n.
18. This point is made repeatedly by many scholars and analysts. See, for eg., Handel, p. 22, who asserts that the 'Ultra' intelligence (German signal intelligence intercepted and decrypted by the British) 'was the single most important means of facilitating deception available to the Allies'. See also, Balfour, p. 100, and Andrew, pp. 487–8.
19. Andrew, pp. 488–9; Cave-Brown, p. 805. There is some evidence to suggest that LCS, or an agency using the same name, still existed in 1950.
20. Lyn Smith, 'Covert British Propaganda: the Information Research Department 1947–77', *Millenium*, Vol. 9, No. 1 (Spring 1981); see also Anthony Verrier, *Through the Looking Glass: British Foreign Policy in the Age of Illusions* (London: Jonathan Cape, 1983), p. 52; see Jonathan Bloch and Patrick Fitzgerald, *British Intelligence and Covert Action: Africa, Middle East and Europe Since 1945* (Brandon, Ireland: Brandon Book Publishers, 1983), pp. 90–95 for a more sinister interpretation of IRD's mission and accomplishments, based on some sources of doubtful reliability.
21. Andrew, p. 488. According to M R D Foot, *SOE: an Outline History of the Special Operations Executive 1940–46* (London: British Broadcasting Corporation, 1984), pp. 245–6, the SOE was formally closed down on 15 January, 1946 and SIS absorbed some of its personnel.
22. Verrier, pp. 149–50, 153–4, 157. Bernard Fergusson, *The Trumpet in the Hall 1930–1958* (London: Collins, 1970), pp. 258–75. Bloch and Fitzgerald, pp. 95–6, 121–2 also identify the Arab News Agency (Cairo) Ltd. as an SIS front during this period.
23. William R Corson, *The Armies of Ignorance: the Rise of the American Intelligence Empire* (New York: Dial Press, 1977), pp. 509–11.
24. *Ibid.*, pp. 525–8.
25. Katherine L Herbig, 'American Strategic Deception in the Pacific: 1942–44', *Intelligence and National Security*, Vol. 2, No. 3 (July 1987), pp. 260–30.
26. Bradley F Smith, *The Shadow Warriors: O S S and the Origins of the C I A.* (New York: Basic Books, 1983), pp. 174, 205–6, 233–4, 276–7; R Harris Smith, *OSS: the Secret History of America's First Central Intelligence Agency* (Berkeley: University of California Press, 1972), pp. 12–13.
27. Anne Karalekas, *History of the Central Intelligence Agency* (Laguana Hills, California: Aegean Park Press, 1977), p. 6.
28. This is a paraphrase of the NSC's role as outlined in the National Security Act of 1947.

See Scott D Breckinridge, *The CIA and the US Intelligence System* (Boulder, Colorado: Westview Press, 1986), pp. 8, 14–19.

29. Karalekas, pp. 26–8; The National Security Act (1947), Public Law 253, 26 July 1947, Section 102(d) (5).
30. Karalekas, pp. 28–38, 45–6; Breckinridge, pp. 32–33.
31. Breckinridge, pp. 222–3, 244; see also Gregory F Treverton, *Covert Action: the Limits of Intervention in the Postwar World* (New York: Basic Books, 1987), pp. 15–17, 26–7, 29; John M Orman, *Presidential Secrecy and Deception: Beyond the Power to Persuade* (Westport, Connecticut: Greenwood Press, 1980). Its title notwithstanding, Darrel Garwood, *Under Cover: Thirty-Five Years of CIA Deception* (New York: Grove Press, 1985) is an exposé about covert action, and does not examine the deception element in any systematic way.
32. The National Security Act of 1947, Public Law 253 26 July 1947, Section 102(d) (4).
33. Karalekas, pp. 15–23, 55–9.

CHAPTER 11

Protocol 'M'

SIMON OLLIVANT

The essence of lying is in deception, not in words.

John Ruskin[1]

The election of a Labour Party government in Britain in July, 1945 ensured that British foreign policy in the immediate post-war period would be influenced by the strong emotion and ideological sympathies towards the Soviet Union that had characterised the Labour Party since the 1920s. In place of Churchill's outspoken suspicion of Soviet intentions, there was a commitment to trust. The goal was a lasting and peaceful settlement in Europe between the war-time Allies.

The process of disillusion took two years. A series of failed conferences, Soviet obstruction and mounting anti-Western propaganda from Moscow finally convinced British policy-makers of the fundamental hostility of Stalin's regime, and of the need for positive action to resist what threatened to become an irresistible extension of Communism westward across Europe. This need shaped British policy in both the overt and covert spheres from late 1947 onwards. It was given public expression by Foreign Secretary Ernest Bevin in the House of Commons in January 1948, but at the same time a more rigorous policy of covert anti-Communism was being developed, particularly in Germany, where the political situation was most fragile and the confrontation between the former allies most intense. It was a development that many in the Foreign Office had eagerly anticipated, and it took its first hesitant steps in the curious episode, now largely overlooked, of Protocol M.

*   *   *

Protocol M was the name given to a document of uncertain provenance that was first made public in January, 1948. It appeared to be the blueprint for a campaign of communist industrial subversion and sabotage in the Ruhr, the industrial heart of Germany and now within the British zone of occupation. The primary targets were the transport system of the region, including that involved in food distribution, and the iron and steel industry. Its stated objective was the disruption of United States Secretary of State George Marshall's European Recovery Programme (the Marshall Plan), then awaiting ratification by the United States Senate.

The Protocol was a detailed document which gave the names of agents involved and set out a three-stage timetable. The first stage, lasting until the end of December, 1947, was set aside for co-operation with the Social Democratic Party (SPD)[2] in agitating for a plebiscite on the future of the Ruhr in order to play on German fears of Allied plans for de-industrialising Germany. The second stage, for January and February, 1948, was the organisation of strike cadres in key plants and regions. The third stage was the organisation of a general strike, scheduled for early March. This programme was to be backed up by a campaign of propaganda aimed at denouncing the Marshall Plan as 'monopoly capitalist enslavement' and at exploiting strikes already taking place in France and Italy.[3]

There were a number of interesting features in the document. It appeared to confirm the message of the recently founded Cominform[4] when it identified the Soviet Union as supporting the fight against monopoly capitalism 'with all and every means', and it identified the Cominform itself as the co-ordinating body behind the campaign. Although the key targets were the transport workers union at Dortmund and the iron and steel workers union, no opportunity to infiltrate other unions was to be missed. The SPD was also singled out for special attention: 'It is quite agreeable to the central executive committee that the Social Democrats should, at first, hold the important positions in the joint action committees'; but the weaknesses of the party were to be identified and 'ruthlessly exploited'.

The text of Protocol M was printed in full in the Berlin evening paper *Der Kurier* on 14 January 1948. In the London *Times* of 15 January it was reported that the British authorities had known of the document for 'some time', and that 'while it is emphasised that no proof exists of the authenticity of the document it is considered in British circles here [Berlin] to be genuine'.

The *New York Times* took up the story in greater detail on 16 January. Its report made three points. First, that 'it is understood that the British have been watching for some time a suspicious courier service plying between the communists in Berlin and the

Ruhr'. Second, that the Foreign Office in London had confirmed the authenticity of the document, although it would not produce evidence to support this for fear of compromising its sources. And third, that the State Department in Washington had said that it had been informed of the Protocol 'a month or two ago' and that it came from 'people we consider reliable'; it was not, however, prepared to comment on the authenticity of the document.

On 17 January the *Times* returned to the theme of corroborative evidence when it reported a Foreign Office spokesman as stating that 'further information had reached the British Government confirming their belief in the authenticity of the document'. The spokesman also revealed that the Foreign Office had made copies of the Protocol available to other Western governments before it had been published in *Der Kurier*.

On the following day, a Sunday, the *New York Times* devoted several pages to the new concepts of the "Cold War" and 'East-West confrontation'. It reported that publication of the Protocol had further widened the split between communist workers in favour of economic reconstruction and their more orthodox colleagues. But the British press, despite official assurances, remained cautious. On 19 January the *Times* referred to the 'alleged' plan of the Cominform and noted that Soviet newspapers were variously attributing its authorship to the British Secret Service and to the SPD. It also told its readers that although there was indeed industrial unrest in the Ruhr 'those who know the area intimately are convinced that the unrest is not politically inspired'. The popular British papers had given the Protocol prominent treatment: 'Plot to Wreck Marshall Plan' was the headline in the Labour *Daily Herald* on 15 January but it was still careful to note that German leaders 'considered the plan with some suspicion'.

By this time Parliament had reassembled after the Christmas recess. On 21 January, the day before an important and much-anticipated foreign policy debate, one of the House's two communist MPs challenged the Foreign secretary to make a statement about the document. Bevin's deputy, Hector McNeil, replied that the text of the Protocol had been made available to the press and that he had nothing further to say except that the Government believed it to be genuine. Challenged again as to whether or not he himself had seen the original of the Protocol, McNeil replied delphicly: 'I naturally would not make any assertions to the House of Commons without making such examination as lies within my power.'[5]

Protocol M then largely disappeared from the news until early April. On Sunday 11 April, the *New York Times* printed a front-page article under the headline 'British Declare Protocol M a Fake; Red Plot in Ruhr Ruse by German'. Quoting a 'completely reliable source'. the paper reported that the British authorities in Germany, 'after careful detective work', had discovered that the Protocol was the work of an anti-communist German. It described the 'bonanza of fake documents' currently available in Germany, and quoted British sources to the effect that the document had been 'dropped like a hot potato when investigations indicated that some mistake might have been made'.

On 19 April, in response to four parliamentary questions, Hector McNeil again addressed the House on behalf of the Foreign Secretary. He made five points. The first was that the British authorities had held a copy of the Protocol in their hands for 'several days' before it was published, and had been making enquiries as to its authenticity. The second was that these enquiries had found no reason to suspect that the document might not be genuine. Thirdly, he said that the Foreign Secretary had, nevertheless, ordered 'careful and exhaustive investigation' into the document's antecedents. Next, that these investigations had eventually led to a German who claimed to be the author of the Protocol, although his testimony was not entirely reliable. And finally, that although the document itself was now 'in doubt' the substance of the plan was corroborated by current developments in Germany and by other information 'already in our possession'. The exchange ended with angry references to the Zinoviev letter.[6] One week later a further parliamentary question brought the admission by Christopher Mayhew, a junior Foreign Office minister, that the identity of the supposed author of the Protocol was known, but would not be made public.

The British press, despite some angry comments in provincial newspapers, made surprisingly little of Protocol M's exposure. But with the imminent implementation of the Marshall Plan and growing evidence of the open hostility of the Soviet Union, public opinion was perhaps not in the mood for news that would bring comfort to the East. It was generally accepted that the Foreign Office had been deceived by a hoax document, and with public interest being absorbed by the tightening Russian blockade of Berlin, the incident was soon forgotten. With the advantage of hindsight, however, the story appears both more interesting and more significant.

* * *

Much of this significance lies in the timing of Protocol M's appearance and the circumstances which surrounded it, for it is now clear that the years 1947 and 1948 were of profound importance both in the development of East-West relations and in the shape and direction of British foreign policy. A brief sketch of the background to the story of Protocol M will help to make this connection more clear.

Britain was fortunate at this time to have as its Foreign Secretary a man able to play a role, and exert an influence, in international affairs beyond that which the nation's depleted resources could, by themselves, have commanded. That Ernest Bevin, a tough former trade union leader, could have exerted this influence, and in doing so win the respect and loyalty of the patrician Foreign Office, is all the more remarkable given the extremely limited options open to him at the beginning of 1947.

Although Bevin was, from his trade union experience, a convinced anti-communist, he nevertheless felt that he had a duty to do everything possible to preserve the wartime alliance. He knew also that if a breach with the Soviet Union did occur, Western Europe could rely on no firm commitment to its survival from the United States. He was shackled by Britain's desperate economic plight, which denied him the resources to fulfil all the nation's international commitments. Furthermore, he knew that the situation in Occupied Germany, particularly in the industrial regions of the British Zone, was critically unstable, and that the financial drain of the occupation was already more than the British Treasury could stand: and always at his back, there was the Left Wing of the Labour Party, traditionally friendly towards the Soviet Union and inclined to regard him as a class traitor and a puppet of a conservative Foreign Office.

During the course of 1947, Bevin's situation saw significant improvement on a number of fronts. The Marshall Plan, of which Bevin was an enthusiastic proponent, offered the first real hope for European recovery and with it an end to the conditions favourable to the spread of Communism. The Foreign Secretary also won broad support for his policies in the Parliamentary Labour Party, while his critics, although always vocal, were diminished in number. But with the Soviet Union, Bevin could make no progress. 1947 was, for many, the year of disillusion. The breakdown of the foreign ministers' conference in Moscow in April, and the Russian walk-out from the Marshall Plan talks in June, were followed by the formation at the beginning of October of the Cominform, a Soviet-controlled organisation with the stated intention of co-ordinating European resistance to Marshall Aid, and which, for the first time, gave public Soviet endorsement to a Europe divided into two hostile camps. But even then Bevin felt obliged to give the wartime alliance one last chance to reach an agreed settlement in Europe. He approached the London Foreign Ministers' Conference of November and December 1947, resolved that, if agreement did prove impossible, the failure should not be laid to the account of the Western powers.[7]

Meanwhile, in Germany, where the realities of East-West confrontation were a daily experience, there were serious doubts that the Government understood the problems facing the occupying powers. The majority of German people, particularly those in urban areas, were living on the brink of starvation; refugees from the East had added almost 20 per cent to the population of the British Zone; strikes and demonstrations over food shortages were widespread; and many aspects of allied policy, such as the dismantling of industrial plants in pursuit of reparations, were bitterly resented. Of all these problems, and of the fears of the coming winter, the communist press made gleeful propaganda.

The London conference foundered, like its predecessors, on the rock of Soviet obstruction. It was against this background, with the Chiefs-of-Staff, the Foreign Office, members of the British administration in Germany and even German political parties, all calling for positive steps to challenge the flood of Soviet propaganda,[8] and with Bevin at last feeling free to go over to the offensive in his foreign policy, that Protocol M made its dramatic appearance in Berlin.

At this time, January, 1948, the administration of the British Zone (which, with the British sector of Berlin, was known by the initials CCG(BE)—the British Element of the Control Commission for Germany) was something of a hybrid; part civilian, part military, part British, part German. The Zone was divided into four *Länder* (Lower Saxony, North Rhine-Westphalia, Hamburg and Schleswig-Holstein) which corresponded approximately to the division of administrative responsibility at corps level following the German surrender in 1945. From 1946, the administration of the *Länder* was nominally a civilian structure, but the shortage of suitable personnel meant that in practice many officials were either transferred or seconded direct from military service.

Overall responsibility in the CCG(BE) remained vested in the person of the Commander-in-Chief and Military Governor. By 1948 this post had been entrusted to a soldier chosen for his administrative abilities. General Sir Brian Robertson, who was widely experienced, both in military administration and in the higher echelons of industry, had served his apprenticeship as Deputy Military Governor, in charge of the day-to-day running of the Zone, and was Bevin's personal choice.

Within this structure, responsible German government was steadily developing. In April 1947, elections in the British Zone had returned administrations in each *Land* headed by a minister-president, and with clearly defined, mostly local, areas of responsibility. The organisation of party politics was also well advanced: the three main parties were the Christian Democrats, the SPD and the Communist Party (KPD), with strong independent, centrist and liberal representation in some areas.

But the real ruler of British Germany was Ernest Bevin. In London, the process of demilitarising the administration of the occupied territory had proceeded more rapidly than in Germany itself: control had passed at an early stage from the War Office to a semi-autonomous body called the Control Office for Germany and Austria, which was headed by a junior minister with the post of Chancellor of the Duchy of Lancaster, but which, after a a year of prickly co-existence, was absorbed into the Foreign Office in April 1947. The German Section, thus swollen into 11 departments, was effectively a government within the Foreign Office and Bevin, never one for committee decisions, preferred to inform, rather than to consult with, his Cabinet colleagues on German affairs. Bevin and the German section dealt directly with General Robertson at CCG(BE).

In Germany the British establishment was equally substantial, reaching almost 26,000 in 1946.[9] Of the divisions responsible for information and intelligence functions, the most important was the Intelligence Division. Alone accounting for almost 4,000 personnel, this division was based at Herford under the command of a major general and maintained regional intelligence officers in each *Land* and an advanced headquarters at Berlin. The Political Division, based in Berlin, the Public Relations and Information Services Division and the divisions responsible for industry and economics were also important sources of information. Together they generated an immense tide of intelligence material. Special dispatches, correspondence between London and Berlin and liaison with the Chiefs-of-Staff supplemented the regular intelligence, political intelligence and economic intelligence summaries and analyses that reached London week after week.

Of the 11 departments in the German Section of the Foreign Office the most important with regard to both intelligence and policy-making was the German Political Department, but much of the information coming in from Germany was first digested by the Foreign Office Research Department (FORD) which had been created in 1943 by merging the old Political Intelligence Department with the Foreign Research and Press Service. FORD's weekly intelligence summaries provided an important briefing on intelligence developments in Germany but their product, rated 'Confidential' rather than 'Secret' or 'Top Secret', did not detail sensitive items of intelligence. Much more secret, even to this day, were the activities of the Information Research Department (IRD), set up to research and conduct a positive propaganda campaign against selected targets, increasingly the Soviet Union. In January, 1948, the IRD was not yet fully functional but the individuals involved in its creation, and the policy needs it served, have considerable relevance to the story of Protocol M.

Faced with so much specialist information, Bevin inevitably came to look to a small number of individuals whose opinion he especially valued, and in doing so he put together a team of particularly high calibre. In the Foreign Office these were Sir William Strang, who was in overall charge of the German section; Gladwyn Jebb, an economic specialist and so of particular interest to a trade unionist and former Minister of Labour; Ivone Kirkpatrick, supervising Under-Secretary in the German Political Department; and the Head of the Department, Patrick Dean. In Germany there were Christopher Steel, political adviser to the Military Governor, and Cecil Weir, another specialist in economics. Above all, there was the Military Governor himself, Brian Robertson. Robertson was almost certainly Bevin's most trusted adviser when it came to German affairs; he was a man of exceptional ability and accepted in the Foreign Office as the only man who could, occasionally, deter Bevin from a course of action on which he had set his mind.

It was by this curious and complex government of occupation that the affairs of the most populous part of Germany were managed in the first, critical years of peace.

* * *

The story of Protocol M is not easy to follow through the Foreign Office Records. Despite the enormous quantity of information reaching London from Germany, there seems to have been no systematic preservation of documents. Intelligence summaries survive only in the most haphazard way and in no consistent series. Furthermore, many items, either singly or in batches, remain closed to public access over and above those classes and groups subject to *en bloc* restrictions, and the inconsistency of document preservation makes it extremely difficult to detect any purpose or pattern in these additional closures.

Nevertheless, there are some indications of more than usual censorship of papers dating from the end of 1947 and the beginning of 1948. Most notable are 15 items from the German Political Department files of early 1948 which have either been retained in the Foreign Office or are closed until at least 1990.[10] Two of these items are specifically identified in the Foreign Office's own indexes as concerning Protocol M.[11] Another potentially interesting group consists of the first seven papers (covering January to March) of Bevin's own Press and Propaganda file of 1948,[12] and there are many other files with papers relating to this period which are subject to extended closure. This is standard practice in the case of papers relating to the IRD, so some of the missing items may well relate to the process by which that department was set up, probably over the period November, 1947 to January 1948.[13] This may well be the significance of the gap in the Bevin Papers. It is interesting to note, at this stage, that some papers specifically relating to Protocol M appear to have been of similar sensitivity.

Accessible Foreign Office papers relating to Protocol M are therefore thin on the ground. There is no record of any intelligence evaluation of the document, and no specific mention of its existence prior to 15 January. Reports reaching the Foreign Office after that date were concerned principally with assessing reactions to the document. In their January summary the Berlin intelligence staff reported the reaction of the headquarters of the Communist-controlled Socialist Unity Party (SED):[14] 'The publication of "Protocol M" released a flood of vituperation which amply confirmed reports of the uproar it caused in the Lothringerstrasse (Ulbricht [the Party leader] is said to have lost his temper completely).'[15] But in Hamburg the Protocol had made a different impression. The authenticity of the document was the subject of 'the utmost scepticism' in the city, and 'even the bitterest opponents of the KPD have not seen fit to make any use of it for political purposes . . . On the whole, Protocol M has fallen very flat in Hamburg.'[16]

This note of disappointment was echoed in the Foreign Office itself. On 27 January the British Ambassador in Moscow reported that the Protocol had been denounced as a forgery in *Pravda*, and that much was being made there of the apparent scepticism of even the British press. Grace Rollestone, a member of the German Political Department, minuted: 'Unfortunately, all too true about scepticism of the press'.[17]

But senior Foreign Office officials seemed reluctant to share the press's caution. Tough new anti-communist regulations were being prepared for the CCG(BE) and in cabling instructions to Robertson in Berlin Kirkpatrick added 'the news of the communist sabotage plot in the Ruhr reinforces the need for early action'. This was written on 15 January. Until at least mid-February, the Foreign Office was distributing copies of the Protocol to anyone who asked for them.[18]

News that Protocol M was about to be exposed publicly reached the Foreign Office by 10 April. On that day William Strang cabled the Military Governor with an outline of the forthcoming *New York Times* article. The Foreign Secretary, he said, had given instructions that absolutely no comment should be made in answer to questions.[19] Three days later Bevin himself cabled Robertson:

> I have to answer questions in the House of Commons on 19 April about the so-called Protocol M and you will realise that the situation is very awkward. The full circumstances of the decision to publish this document and of the investigations made in Germany as to its authenticity are not clear to me from the records available here. The divisions concerned appear to be the Political and Intelligence Divisions. Before making reply in parliament, I must be fully apprised of all the aspects of the case. I shall want you therefore when I see you in Paris to place me in possession of the whole story so far as it is known at your end. You will no doubt bring with you such responsible members of the Political and Intelligence Divisions as will be able to testify at first hand and such documentations as will enable them to answer questions in detail.[20]

The meeting took place in Paris on 14 April. Robertson brought with him his political adviser, Christopher Steel; Bevin was accompanied by Ivone Kirkpatrick, and the meeting was reported by Frank Roberts, Bevin's private secretary.

Robertson began by saying that the British authorities had obtained their copy of the Protocol from a 'reliable informant' in the SPD called Kielgast, from whom the Berlin press had also got their copies. He insisted that there had been no British statement actually vouching for the document's authenticity, but that no attempt had been made to prevent publication of it for three reasons. First, that there was no reason to believe that the Protocol was not genuine. Second, that the SPD (who were the party officially favoured by the British government) were 'running' the story and the authorities did not wish to embarrass them. And third, that the Protocol was 'very good anti-Soviet propaganda, more particularly at the time it appeared, last January'. It was only later, Robertson reported, that the British had caught up with a man called Hahn with whom Kielgast had had some dealings and who had formerly acted as an agent for the Americans, at one time producing quality information. After first denying it, Hahn eventually admitted to the authorship of the Protocol and the British, although still unsure if he was telling the truth, passed on the information to the American commander General Lucius Clay, from whose office, it was assumed, it had been leaked to the *New York Times*. Robertson went on to suggest a line to take in the House of Commons and made the points which formed the basis of Hector McNeil's statement on 19 April.[21] With this report, and the subsequent parliamentary exchanges, Protocol M drops from the Foreign Office records, but it was clearly not the whole story since Robertson's account contains a number of inconsistencies.

Robertson advised Bevin to insist that the government 'had never asserted that they were absolutely convinced that the document was genuine', only that it 'seemed' genuine.[22] So far as official statements in either London or Berlin were concerned, this was probably strictly true. Nevertheless, both the *Times* and the *New York Times* had been given to understand by the Foreign Office that new evidence had shown the Protocol to be genuine.[23] But if such new evidence had been forthcoming, why did Bevin order, as McNeil claimed he had, 'exhaustive investigations' into the document's antecedents?[24] And if the Secretary of State had taken such a personal, and contradictory, interest in Protocol M in January, it is strange that he should appear to have known so little about it in April. Given the close relationship that existed between Bevin and Robertson, it seems unlikely that either set out to deceive the other, and so Bevin's confusion and Robertson's report are likely to represent accurately the information available to each of them at that time. There is no record of the contributions to the Paris meeting of either Steel or Kirkpatrick, and yet it is probable that their knowledge of the episode went considerably further than that of either of their principals. The full story of Protocol M would seem to lie within the Foreign Office.

* * *

What is immediately striking about the handling of Protocol M is the speed with which it was given Foreign Office endorsement, at a time when, given the great number of forgeries known to be in circulation in Berlin, common sense would have suggested extreme caution. There seem to be two possible explanations for this. The first is that the document was believed to be authentic; the second that, although questionable, it served some immediate policy purpose. These two possibilities will be considered in turn.

With the advantage of hindsight it seems quite clear that Protocol M was not what it claimed to be. Definitive proof of its authorship is lacking, but it is not unreasonable to accept that it was indeed the work of Hahn. He had been selling information to the Americans but, as Robertson reported, 'had rather dried up'.[25] The times were propitious to forgers. As the *New York Times* reported on 11 April 1948: 'both sides are eager to pay money for all kinds of falsified information concerning each other on the mere assumption that it would be a bad gamble to let a possibly good thing go'. The selling price of such a piece of information would have made a considerable difference to a German in the Winter of 1947–8.

Nor does there seem any reason to doubt the connection with the SPD man Kielgast. Robertson spoke of the SPD 'running' the story, and at least two days before *Der Kurier* published the Protocol, the SPD leader Kurt Schumacher referred in a speech to politically motivated strikes that were expected to break out in March.[26] The SPD had the most to fear from the Communists. The party in the Soviet Zone had been split by communist infiltration and by the subsequent formation of the SED, and Protocol M spelt out in obliging detail the communist attitude to the SDP in the Western Zones.

Nevertheless, without the advantage of hindsight, the Foreign Office may well have been satisfied that the document was genuine. They certainly appeared to behave as though they did. Copies of the Protocol were distributed to friendly governments before it was made public on 15 January; on that day Kirkpatrick quoted it as an additional reason for cracking down on the Communists in the British Zone; a copy was placed in the House of Commons library, and as late as mid-February, English translations were being made available to anyone who requested them.[27]

But, on closer inspection, this confidence was not deeply rooted. In the first place, investigations were continuing in Germany as to the Protocol's authenticity, and this would have been known in the Foreign Office. Secondly, Bevin made no reference to the document in the House of Commons on 22 January. This is a surprising omission. In January, 1948 Bevin's chief preoccupation was to win first cabinet, and then parliamentary, approval for a new foreign policy initiative which would for the first time publicly acknowledge the Soviet threat to Europe and call for a 'spiritual union' of the West to resist the spread of Communism.[28] Both in Cabinet, on 8 January, and in the House on 22 January, he took pains to spell out the reality of Soviet anti-Western propaganda and subversion. He specifically referred to the Cominform's campaign against the Marshall Plan and to politically motivated strikes in France, but made no mention of the Protocol.[29] While it is possible that the Foreign Office did not know of the Protocol as early as 8 January, the same cannot be said of the debate in the House two weeks later, only the day after McNeil's endorsement. It would seem, therefore, that the Foreign Office had advised the Secretary of State against publicly associating himself and, by extension, his policy, with Protocol M.

There is a third indication that government confidence in the document went neither very far nor very deep. At the beginning of February, a secret report was prepared by the joint intelligence chiefs of the Services for a meeting of the Chiefs of Staff on 9 February. The subject was 'Germany: the possibility of disaffection and disorder'.[30] Although the contents remain classified, it has been established that Protocol M was discounted because its authenticity was held to be in doubt.[31]

From this it seems fair to draw two conclusions: that the document taken up by the Foreign Office was known at an early stage to be, if not absolutely a forgery, at least seriously questionable; and that consequently the decision to promote it was not so much a simple error of judgement as a deliberate act of policy. This raises in turn the question of what policy purpose this decision was designed to serve.

All the Foreign Office statements on Protocol M, both those officially recorded and those reported in the newspapers, have one feature in common—the theme of corroborative evidence. This was neatly summarised in McNeil's statement to the House on 19 April:

> There have been developments in Germany which correspond to statements included in the document, and there are strong indications that even if the document is not itself authentic, it has been compiled from authoritative Communist sources, and this is corroborated by information already in our possession.[32]

This corroborative evidence was never produced in public but it is quite clear that the threat of subversion and unrest in the British Zone was taken seriously. There was good reason for this. It was clear to almost everyone in the British Government that the only hope for the economic recovery, and the consequent restoration of political stability, of Europe lay in the Marshall Plan. Consequently, the instructions given by the Cominform at the time of its inauguration in October, 1947 to Western Communist Parties, particularly those of France and Italy, to use every means at their disposal, including strikes and local violence, to defeat the Plan was as uncompromising a declaration of communist hostility to Western hopes for Europe as could under the circumstances be imagined. In Germany this anxiety was reinforced by the desperate food shortages in urban areas, shortages which had already caused widespread strikes. In the middle of January, 1948 FORD reported strikes under way or 'incipient' in Cologne, Duisburg, Gensungen, Oberhausen, Essen, Solingen and Bremerhaven. And a damaging strike in Hamburg had only just come to an end.[33]

Opinion was divided about the political nature of these strikes.

Most intelligence officers rejected the idea of a co-ordinated political campaign of strikes, and put the main blame on food shortages. But communist involvement was sufficiently high profile to be disturbing. A report on the Hamburg dock strike revealed that the leaders of all six workers councils involved were members of the KPD, and that the technique of dominating such councils was increasingly being used as a means of controlling trade unions.[34] And the KPD were never slow to exploit a strike, whatever the original grievance. On 6 January Rudi Wascher, Deputy Chairman of the North Rhine-Westphalia KPD executive, told a conference of KPD officials that 'unless all the signs were misleading, strike reports indicated that a great struggle was beginning. The trade unions should take the lead in this struggle.'[35] One piece of evidence, which may be important, survives. In keeping with the random nature of the Foreign Office files, it lurks in a monthly intelligence summary from North Rhine-Westphalia dated 3 January 1948.[36] At the end of a report on the communist infiltration of trade unions, there is the following annex:

> A Russian directed propaganda and sabotage organisation (FAST BOWLER) has come into existence and a report on this subject has already been circulated. This report is the most sensational we have received for many months, but our investigations have so far failed to reveal any trace of the perons mentioned as cell leaders in Dortmund or Wuppertal.

This may well be the first appearance of Protocol M, although in the absence of any subsequent, or previous, record of 'Fast Bowler' the identification cannot be certain. The Protocol certainly names leaders in Dortmund but, although it gives names of agents in the iron and steel workers union—and Wuppertal was an important centre of those industries—it does not mention that city by name. If this was not Protocol M, it may well have been part of the corroborating evidence that so closely resembled the Protocol as to appear to authenticate it.

Whatever the justification, the threat of communist subversion in the British Zone was taken seriously in Whitehall. We have already seen that there was enough evidence, even excluding Protocol M, to disturb the Chiefs of Staff at the beginning of February.[37] At the same time, Bevin was writing to the Minister of Defence to ask what the effect on British military strength would be of a withdrawal of civilian labour.[38] At the beginning of March, the Foreign Office received the first of a series of weekly reports on the possibility of unrest in the British Zone, prepared by the Ministry of Defence. It noted that, in February, some communist workers had confidently been forecasting a general strike, but that 'the publicity given by Protocol M to the choice of March for a general strike may have discouraged the KPD from pursuing such a timetable'.[39]

It now seems likely that British fears were exaggerated. Popular support for Communism was minimal and fear of the Russians intense. But Communism has seldom spread by public acclamation and circumstances in the British Zone in January 1948 resembled the textbook requirements for revolution: political uncertainty, weak institutions, near starvation and industrial unrest. Communists occupied powerful positions among the workforce and had a clear mandate from the Cominform to obstruct economic recovery. It was a threat that had to be taken seriously. It is not unreasonable to suggest, therefore, that Protocol M was exploited to expose and even head off a seriously perceived threat of communist subversion in the CCG(BE) zone in the spring of 1948.

There were, of course, other advantages to be gained from publishing and promoting Protocol M at this time. As late as early January 1948, Bevin's cabinet colleagues were pressing him not to be too anti-Soviet in his foreign policy speech of 22 January. They were afraid of alienating not only the backbenches of the Labour Party but also fellow socialists in Europe.[40] Protocol M was an opportune reminder of the reality of the Soviet threat and of the communist attitude to Socialist parties. Publicity of this sort also made it easier for a British social-democratic government to implement some decidedly undemocratic measures against the communists in the CCG(BE). Furthermore, the Foreign Office may also have had one eye on the United States. The vital Marshall Plan was still making slow progress through Congress (it did not finally receive the presidential signature until early April), and a timely reminder of communist opposition to the economic recovery of Europe can only have helped Europe's friends in Washington. The Foreign Office saw to it that the American press was well briefed about Protocol M.

* * *

The use of such a dubious document as Protocol M reflects a significant change of attitude in the Foreign Office, both as regards relations with the Soviet Union and the means of achieving policy objectives.

At the end of the war in Europe there was a sharp distinction between the military establishment in Britain, who were prepared to recognise the Soviet Union as the new adversary, and the Foreign Office, who were determined to extend the wartime alliance into the peace. There were always exceptions to this latter attitude, including Frank Roberts whose reports from Moscow (when British Minister there, before he assumed his appointment as Bevin's Private Secretary) left no room for doubt as to the reality of Stalin's feelings towards his former allies, and Christopher Warner, then head of the Northern Department (which dealt with the Soviet Union) who had written an influential memorandum in April 1946, calling for a 'defensive offensive' against the Soviet Union.[41] However, it was not until 1947 that this view gained general support. Even then, the Foreign Office had not ruled out the possibility of reaching an understanding with the Soviet Union. In June, 1947, the British Ambassador in Moscow suggested that the time had come to implement Warner's 1946 proposals for political warfare against Communism. Warner replied that the time was not right; an open breach should be avoided until at least after the London Conference in November and December.[42]

The failure of that conference brought to an end a two-and-a-half year search for a European settlement in co-operation with the Soviet Union—a search which had finally outlived even the hopes of its most idealistic supporters. Bevin was under pressure from the Foreign Office for a more aggressive stand against Communism and the Chiefs of Staff were pressing on him 'the desirability of reviving political warfare machinery and undertaking certain "black" operations'.[43] Although Bevin did not need convincing, he had no intention of setting up a British CIA. Outlining his ideas for a propaganda counter-offensive to the Cabinet on 8 January 1948, he told his colleagues that 'the only new machinery required will be a small section in the Foreign Office to collect information concerning communist policy, tactics and propaganda and to provide material for anti-communist publicity . . . .'[44]

This small section was the IRD. The initiative in setting up such a group had come from a junior Foreign Office minister, Christopher Mayhew, who had discussed it during the Autumn of 1947 with Warner, Sir Orme Sargent (then head of the Foreign Office) and Ivone Kirkpatrick. Warner was to be the first supervising Under-Secretary, and Kirkpatrick was given the job of establishing and staffing the new department.[45]

At this time Kirkpatrick was also deeply involved in drafting the new official policy towards the Soviet Union. His mood, and that of the German Political Department, was by now decidedly hawkish. When General Robertson suggested in a cable that the most effective anti-communist propaganda in Germany would be a speedy restoration of prosperity, the department's reaction was impatient. That was a long-term solution; what was needed now was a present answer to the tide of anti-Western publicity and a more aggressive promotion of the Western point of view. 'With the present dearth of information in Germany', noted a senior member of the department, F J Leishman, 'I am sure we can do a great deal to help the German press by making suitable material available to them.' Patrick Dean agreed that such material ought to be supplied but he also wanted more repressive measures against the KPD: 'We should not hesitate to hit back hard and use fairly unscrupulous methods.' Kirkpatrick added: 'Yes, I think that Berlin underestimates the extent to which we could really embarrass the Communists by well-conceived pressure.'[46]

Given the mood in the Foreign Office and among the Chiefs of Staff, at the end of 1947, could the whole document have been a British fabrication? This seems unlikely for two reasons. The first is that the German provenance of the Protocol, and its involvement in the first instance with the SPD, seems reasonably well established. The second is that this would have been at odds with the direction in which the Foreign Office, and Kirkpatrick in particular, were moving. The new IRD was not to be involved with fabrications, but with propaganda based on carefully selected factual material.[47] And Protocol M did closely match known and stated communist objectives as set out by the Cominform. Certainly the Protocol was not the work of the IRD—the department was not yet operational in early January 1948—but it is most probable that it was handled by those who were in the process of setting it up.

A decision was taken in the Foreign Office to make use of the document, but there is no surviving record of who it was who took that decision. The presence of Christopher Steel at the Paris meeting in April 1948 suggests that it was he, as would be expected with such a potentially important document, who passed the Protocol on to the Foreign Office department with which he was in closest and most regular contact—the German Political Department; and the cautious attitude of Berlin throughout the whole episode also suggests that his advice was that it should be treated with caution. In London, the man with overall responsibility for the German Political Department was the supervising under-secretary, then also the midwife to the infant IRD, and the fourth participant at the Paris *post mortem* on Protocol M—Ivone Kirkpatrick. With his close involvement in the development of both the covert and overt policies towards the Communists in Germany, Kirkpatrick was well placed to appreciate what was subsequently agreed in Paris to have been 'very good anti-Soviet propaganda'. With Cabinet backing for an anti-communist offensive, and the

IRD not yet in operation, the temptation to run Protocol M may have proved irresistible.

* * *

As an exercise in deception, the Protocol M episode was apparently neither very spectacular nor very successful. The document was widely regarded with suspicion and was exposed as a fake within three months. In few histories of the "Cold War" does it receive more than a footnote. But the story is not without significance in the development of post-war British policy towards the east, both in terms of particular policy needs and of the general application of propaganda techniques.

Although the Foreign Office was clearly disappointed by the scepticism with which Protocol M was received in some quarters, and embarrassed by the circumstances of its exposure three months later, it is unlikely that they had ever looked for a long-term acceptance of the document. In the first place, the purposes the document was expected to serve were immediate and short-term. Whatever the true state of affairs in the CCG(BE), there was sufficient industrial unrest and discontent for any political involvement aimed at delaying or preventing economic recovery to be an unacceptable threat. In addition, the Foreign Office did have reason to believe that the publicity given to the Protocol had some effect on discouraging those political zealots who might have been preparing to exploit that discontent to their own ends.[48]

At the same time, the major political obstacle to the new policy towards the Soviet Union, on which the Foreign Secretary was now determined, lay in the traditional loyalties and sensibilities of the Labour Party. In terms of timing, the publication of Protocol M on the eve of the announcement of a policy that effectively declared a state of political warfare with the Soviet Union and the Communists in the CCG(BE) cannot be faulted. It is impossible to quantify the precise effect of this one incident on a process which had been underway for some time, but the importance of the shift within the Labour Party should not be underestimated. At the end of January 1948, with the disillusion of the Prague coup and the Berlin blockade still to come, the American Chargé d'Affaires in London reported to Washington that 'virtually the whole British Labour movement has, step by step, abandoned its sentimental attitude towards Soviet Union and ranged itself behind Bevin. . . . This remarkable change in its way almost as significant as the change in US from isolationism to internationalism.'[49] Protocol M can certainly lay claim to having been a step.

In tactical terms, the handling of the Protocol reflects a more astute understanding of the nature of propaganda on the part of the Foreign Office than had previously been the case. Until 1947 it had been the usual practice of the British, and the Americans, to counter Soviet propaganda claims with carefully reasoned refutations. To these the Soviets paid little attention; by the time one charge had been shown to be false it had already been succeeded by several others. This lesson was not lost on the Western Allies. At the beginning of April 1948, the Foreign Office issued a series of basic directives to define the way in which the British authorities in Germany should deal with the Communists and the Soviet Union. These directives had been under discussion, principally between Kirkpatrick and Robertson, since the middle of January. They contain a revealing passage on the state of Foreign Office thinking on propaganda:

> Although you must disseminate clear and cogent answers to misrepresentations about Britain, you must not make the fundamental mistake of being drawn into concentrating too much energy in dealing with subjects selected by the Communists. Counter-propaganda is bad propaganda. In general it should be taken in the stride of positive, offensive action.[50]

The value of a document like Protocol M lay in its immediate effect, a headline such as 'Plot to Wreck Marshall Plan' in the Labour *Daily Herald* of 15 January counted for very much more than a minister's statement three months later that the document might not, after all, be genuine. If the Foreign Office calculated that the incident would soon be forgotten, they were right. By mid-April, Marshall Aid was guaranteed, the Czechoslovak Government had been overthrown in a Soviet-backed coup, and Anglo-Soviet relations had been emotionally soured by the loss of a British transport plane in Germany, with many people on board, following a collision with the Soviet fighter that had been harassing it. The despatch of Protocol M scarcely registered in the newspapers.

Within the analytical framework of this volume, deception in this instance can be detected at two levels: that of the instigator, Hahn, and of the subsequent exploiter, the Foreign Office. Hahn used dissimulation to hide the true authorship of his Protocol and simulation to make it appear to be the work of the Cominform. Although the overall message of the forgery—that the Cominform and German communists conspired to damage the German economy and oppose the Marshall Plan—was essentially true, simulation carried this a stage further, with detailed plans of action and the naming of conspirators. While these inventions or exaggerations may have provided the element of sensation that made the document saleable (first to Hahn's direct paymaster and then to the SPD, the Foreign Office and the news media), it seems likely that they also undermined the Protocol's credibility. Nowhere was acceptance wholehearted, and in some places, such as Hamburg, the deception failed completely.

There is no evidence that anyone in the Foreign Office designed and implemented a deception operation, as such. It does seem likely that the Protocol, though regarded with suspicion, was nevertheless accepted as manna from heaven as a means of 'surfacing' a story, believed to be substantially true, which had previously lacked the spark of sensationalism to catch the public imagination. The Foreign Office could have withheld comment, allowing the story to run on its own momentum. Instead, the department apparently engaged in what might be termed 'deception once removed' by giving the Protocol a stamp of approval and permitting official channels to be used as a transmission belt for doubtful material. The Foreign Office's keen interest in audience reactions—their monitoring of the operation—was the characteristic behaviour of deceivers, whether professional magicians or aspiring propagandists. These and other features of the story indicate that the Foreign Office's handling of the Protocol was more skilful than first impressions might suggest. The publicity which the Protocol gave to the whole question of communist opposition to the European Recovery Programme made it easier for the authorities to proceed with specifically anti-communist measures in the CCG(BE). By the time Protocol M was discredited, communist organisations in British Germany had been seriously weakened.

The Foreign Office was also careful to distance the Foreign Secretary from the whole episode. There is no record of it having been discussed in Cabinet, and no mention was made of it in Bevin's public speeches. Backing for the document was given in the Commons by Hector McNeil, and in April the pieces were picked up by McNeil and Christopher Mayhew. So successful was this policy that even when the deception was about to be exposed, Bevin had difficulty piecing the story together from the records available to him.[51]

* * *

The Foreign Office was probably wise not to involve Bevin with Protocol M; it was not the sort of operation that would have appealed to him. At the end of the Paris meeting on 14 April, the Secretary of State 'laid down that it was a mistake to publish or to appear to vouch for documents of this kind,'[52] and these instructions appear to have been followed. Protocol M did not set a precedent: the IRD, which continued to operate until 1977, concentrated firmly on factual material. But it did mark a new phase in the Foreign Office's dealings with the East; not only in the more self-confident and more aggressive pursuit of its policies, but also in the growing awareness that it would soon have to devote to the war of words the skill and energy it had devoted to the war of arms that had so recently come to an end in Europe.

Endnotes

1. John Ruskin, *Modern Painters*, Vol. V, part IX, Chapter Seven.
2. The SPD, despite its name, had strong ideological links with Marxism. It was, nevertheless, the party most favoured by the Labour Government in London.
3. For an English language translation of Protocol M, see Appendix II to Otto Heilbrunn, *The Soviet Secret Services* (London: Allen & Unwin, 1956).
4. The Communist Information Bureau. Founded on 5 October 1947, the Cominform was comprised of communist parties in full power (excluding Albania) together with those of France, Italy and Czechoslovakia, and had the declared aims of consolidating communist control in Eastern Europe and of opposing the Marshall Plan.
5. Hansard, Parliamentary Debates, Vol. 446, 21 January 1948, col. 187.
6. Ibid., Vol. 449, 19 April 1948, cols. 1434–5. The Zinoviev Letter, of 1924, which was purported to be a directive from the head of the Comintern (Zinoviev) to the British Communist Party, urged the party to pressure the Labour Party to ratify the Anglo-Soviet treaty, to intensify subversion in the Armed Forces, and to prepare for the expected British revolution. Intercepted by British intelligence, the letter was leaked to the press, and brought down the Labour government. The dramatic circumstances arising from exposure of the letter inevitably raised questions as to its authenticity. On this, see Christopher Andrew, *Secret Service: the Making of the British Intelligence Community* (London: Heinemann, 1985), pp. 298–313.
7. A Bullock, *Ernest Bevin: Foreign Secretary 1945–1951* (London: Heinemann, 1983), p. 490.
8. The main parties joined together in an appeal to the British Government for a campaign to counter the 'irresistible flood' of Soviet propaganda. FO371/64278 (C16120/76/18).
9. M Balfour and J Mair, *Four-Power Control in Germany and Austria 1945–6* (London: Royal Institute of International Affairs, Survey of International Affairs, 1956), p. 38.
10. FO371/70477 (C /412/413/463/529/530/546/564/571/626/627/741/931/1057/1201/1929/1/18). (In the case of documents at the Public Record Office, the PRO file number is followed by the Document's original departmental classification number.)
11. C571/1929/1/18.
12. FO800/498 (Bevin Papers, PRS/48/1–7).
13. L Smith, 'Covert British Propanganda: The Information Research Department 1947–77', *Millennium* ix, 1 (Spring 1981), pp. 67–70.
14. Formed from the KPD and a rump of the SPD. Designed to overcome the unpopularity of the KPD in the Eastern Zone, the merger split the SPD in Russian-occupied Germany.
15. FO 371/70613A (C1065/108/18).
16. FO371/70641B (C1225/333/18).
17. FO371/70477 (C661/1/18).
18. FO371/70477 (C165/1/18 and C1040/1/18).
19. FO800/467 (Bevin Papers, GER/48/14).
20. FO800/467 (Bevin Papers, GER/48/15).
21. FO371/70478 (C3135/1/18/G).
22. *Ibid.*
23. See above, notes 3–5.
24. See above, note 5.
25. FO371/70478 (C3135/1/18/G).
26. FO371/70616 (FORD weekly background notes on Germany, 125, 22 January 1948, C125/121/18).

27. See above, note 12.
28. This was not a new idea in itself, only in terms of official British policy.
29. Hansard, Vol. 446, 22 January 1948, col. 393.
30. DEFE4/10 (JIC(48)8(O)F).
31. Private information.
32. Hansard, Vol. 449, 19 April 1948, col. 1435.
33. FO371/70616 (FORD weekly background notes on Germany, 124, 15 January 1948, C124/121/18).
34. FO1014/144 (SECT/130/13/S).
35. FO371/70616 (FORD, C124/121/18).
36. FO371/70613A (C234/108/18/G). The file contains two December summaries, one from Hamburg and one from Schleswig-Holstein, three items from North Rhine–Westphalia (including the annex), two monthly reports from Berlin covering December 1947 and January 1948, and six weekly reports, also from Berlin, covering 4 June to 11 July 1948.
37. See above, note 16.
38. FO800/467 (Bevin Papers, GER/48/7).
39. FO371/70478 (C2023/1/18/G).
40. Bullock, p. 517.
41. V Rothwell, *Britain and the Cold War 1941–1947* (London: Jonathan Cape, 1982), p. 258.
42. *Ibid.*, p. 280.
43. FO371/71648 (N34/31/G).
44. CAB 129/23 (CP(48)8).
45. Smith, pp. 68–9. Kirkpatrick, who had wartime experience in propaganda, had already created an Information Policy Department in the post-war Foreign Office. Staffed by personnel drawn from the disbanded Ministry of Information, the IPD supervised and provided 'guidance' to information officers at diplomatic posts. It also provided the Foreign Office with specialised expertise to alert staff and policy-makers to the 'propaganda dimension' of foreign policy. On this and other efforts by the British Government to establish a peacetime propaganda capability, see, M Ogilvie-Webb, *The Government Explains: a Study of the Information Services* (London: Allen and Unwin, 1965), pp. 67, 81–82, 86; and J B Black, *Organising the Propaganda Instrument: the British Experience* (The Hague: Martinus Nijhoff, 1975), pp. 14–22, 30.
46. FO371/70477 (C163/1/18).
47. Smith, p. 69.
48. See above, note 18; and in Heilbrunn, cited, pp. 83–4, the author in 1956, while accepting that the Protocol's authenticity was 'doubtful', nevertheless quoted from it as though its contents were reliable.
49. *Foreign Relations of the United States 1948*, iii, pp. 1069–77 (quoted in Bullock, p. 549).
50. FO371/70478 (C2182/1/18/G).
51. See above, note 20.
52. FO800/467 (Bevin Papers, GER/48/15).

* Documents in the British Public Record Office are cited with the permission of Her Majesty's Stationery Office.

CHAPTER 12

Democratic Deception: American Covert Operations in Post-War Europe

TREVOR BARNES

> *I would say that since the war, our methods—ours and those of the opposition—have become much the same. I mean you can't be less ruthless than the opposition simply because your government's policy is benevolent, can you now?*
>
> 'Control' in *The Spy Who Came in from the Cold*[1]

On a Spring day in 1954, a spruce and clean-shaven American walked into the lobby of the Travellers' Club in Paris. He was Robert Amory Jr—tall and elegant, with the cosmopolitan self-confidence of a man familiar, both as a student and professor, with every straggly blade of grass in Harvard Yard. Amory, however, had left Harvard two years before to join the Central Intelligence Agency as its Deputy Director of Intelligence. He had come to the French capital from Washington to discuss an ambitious and top secret plan to bribe the French Chamber of Deputies with largesse of $700,000. The American Government was apprehensive that the French, who had been procrastinating for three years about whether or not to join a proposed European Defence Community, would finally make up their minds in 1954 to reject the whole scheme.

The issue had come to a head because, at the Big Four Conference in Berlin in January, the Soviet Union had proposed a pan-European Agreement to replace the European Defence Community, and the Americans were very concerned that their hopes of a coherent block of European nations united against the Russian threat would be dashed. In a private guest room at the Travellers' Club, Amory had a long discussion with the CIA's station chief in Paris and one other man about whether or not to implement the plan. The CIA had already built up a network of politicians and other agents of influence in France who could be relied upon to help. Bribery had worked on occasions in the past so why not again? But what about the uncertain opposition to the Defence Community: the unideological alliance of Gaullists and Communists? Could they be split? The three men debated for some time. They finally concluded that the risks were too great to justify the high costs. In the absence of this particular CIA operation, the French Chamber of Deputies finally voted against the Defence Community in August 1954.[2]

The plan to bribe the Chamber of Deputies is now an historical curiosity, an example for some of the CIA's hubris at the height of the "Cold War", of the Agency's willingness to consider almost any grandiose stratagem to further the interests of the United States. However, the plan was atypical only in its scale. Just after the Second World War, the CIA instituted a campaign in Europe of what became known as 'covert action'—secret operations stopping short of war carried out by one government to further its interest against those of another. The CIA plan to bribe the French Chamber of Deputies was just one of a myriad in a general programme of covert action in post-war Europe.

Covert action itself has received little serious attention, being the stuff of sensational journalistic exposés or self-serving memoirs by former members of intelligence services—either disillusioned with, or providing an apologia for, their erstwhile profession.[3] The golden thread running through almost all covert action is deception, but because covert action has, for the most part, not been studied seriously and because deception in peacetime has not received much attention, the important links between the two have been largely ignored. 'Covert action' is frequently a form of peacetime deception by another name, and in the noses of journalists smells much sweeter.[4]

A deception requires both the concealment of the real and the presentation of the false. Consequently it is technically possible for an operation to be 'covert', that is to say concealed, without being a deception. When, however, absolute concealment is impossible or undesirable, as was the case in each of the operations discussed in this chapter, the simulation of false 'cover' is essential—and such operations are therefore deceptive.

Europe was where America initiated post-war deception operations of this type and the lessons learned there were applied later around the rest of the globe. Viewing covert action from this perspective provides more than a realisation that America was practising deception in peacetime without realising it.[5] These Western operations were viewed at the time as a riposte to activities by the Soviets which were intended to deceive the West. If the KGB was the past master of deception through operations like 'the Trust' back in the 1920s, (see pages 17–18), which convinced the Western Powers that their help to the anti-Bolshevik rebels was bolstering rather than weakening Lenin's regime, then the Central Intelligence Agency (and Britain's MI6 and Foreign Office) were determined to give as good as they got in the aftermath of the Second World War.

The aim of all the deception operations practised in "Cold War" Europe was to undermine the forces of Communism, seen as having strong links with the Kremlin and antipathetic to the interests of America and its allies. If this was the overall objective, in keeping with United States or allied foreign policy in the area, each deception operation had, of course, its own specific aim, whether it was opposition to a particular Soviet 'front' group, defeating a communist party in a given election or imbuing a propaganda radio station with credibility.

The Americans had four principal targets to deceive. First of all there was the Kremlin. By giving covert aid to the people and groups opposing Communism, the CIA obviously hoped the Russian authorities would believe that the forces of democracy were stronger than they actually were in countries like France, Italy and Greece. The Russians might ponder the wisdom of engineering a seizure of power by communist parties if the opposition appeared more powerful than originally envisaged. The reverse side of the deception was that if the Soviet Union was not successfully deceived they would be able to exploit covert American support as a key theme in their international propaganda. Secondly, the Soviet Bloc population at large was to be deceived for a different purpose by American radio propaganda. Radio Free Europe and Radio Liberty were ostensibly independent of the United States Government but were actually funded in secret by successive American administrations. Here the deception was designed to try to ensure the credibility of Western propaganda. The third audience was in Western Europe, in whichever country the Americans organised the operation: the public at large, local bureaucracies, politicians, intelligence services, cultural elites, trades unions—any element to which some message was addressed by means which intruded upon domestic affairs. Since such intrusion, if known, would be resented, the first task of deception was always to conceal it, by covering links between certain domestic organisations and the United States. The messages themselves were not always deceptive, but they could become so when, for instance, they overstated threats from the Soviet Bloc or indigenous communists. The last audience to be deceived was the domestic one in

America. The duping of American public opinion about certain activities of the CIA abroad was to become a major theme of CIA exposés in the 1970s.

This chapter examines political deception in Italy between 1948 and 1958, the work of the 'Anti-Cominform', and the establishment of American broadcasting stations designed to reach audiences beyond the 'Iron Curtain'.

* * *

In 1945 Europe lay shattered and despondent, exhausted by the paroxysms of war. Millions of homes and factories had been destroyed by bombing and what remained of the European economies was out of joint for peacetime, having been geared to production for war. Hunger and discontent stalked the Old Continent. With Stalin's divisions looming on the eastern horizon and no settlement of the German question in prospect, political uncertainty was rife. Across the Atlantic the political health of Europe was viewed with grave apprehension, and there were genuine fears in Washington that unless action was taken and the possibility of economic collapse averted, a vacuum would be created which would suck communist parties into power. The crisis in Greece in early 1947 (above all the prospect of communist insurgents seizing power) had prodded President Truman to announce the Truman Doctrine on 12 March: 'I believe that it must be the policy of the United States to support free peoples who are resisting attempted subjugation by armed minorities or by outside pressures,'[6] It was an unofficial declaration of war against Communism everywhere, containing an open-ended promise to supply aid to any group seen to be opposing the Kremlin. On 5 June 1947 General Marshall, the American Secretary of State, called on the European nations to draw up a scheme for economic recovery to be funded by the United States. It was called the Marshall Plan and was aimed to contain Communism by alleviating the worst of Europe's economic woes. The Truman Doctrine provided the military impetus for opposing Stalin, the Marshall Plan the economic. While Congress procrastinated over the Plan, however, Europe's condition deteriorated. One factor contributing to this decline—as viewed in Washington—was Soviet-funded covert action in Italy, France and Greece. Stalin's minimum aim was to sow seeds of discontent against American policies like the Marshall Plan by propaganda, whilst his optimal objective was for communist parties to take power in these countries. At first, an atmosphere of frustrated impotence prevailed in Washington towards Soviet covert action because the American Government possessed no agency to counter it.

The United States' wartime foreign intelligence agency, the Office of Strategic Services, was dismantled in 1945. In January 1946 Truman created a feeble replacement, called the Central Intelligence Group. This was renamed the Central Intelligence Agency and placed under the control of the National Security Council just over a year later. The fledgling CIA had enough difficulties establishing a spy network and recruiting staff to produce meaningful intelligence estimates, without being burdened with any other responsibilities. But alarm inexorably increased about the crisis in Europe. In August 1947 the head of the CIA, Rear Admiral Roscoe Hillenkoetter, sent Truman details of top secret documents obtained clandestinely from the safe of a Soviet satellite's chief of mission and later microfilmed. The chief of mission proposed strike movements in Italy, France and Belgium; a general strike in Italy to overthrow De Gasperi, the Italian Prime Minister; and a victorious drive of communist partisans towards Salonika. The organising meeting of the Communist Information Bureau (*Cominform*) was held a month later. *Cominform*'s role was to co-ordinate action and propaganda against the Marshall Plan and, more generally, subversion against the West.[7]

The major headache in Washington, however, as the cold Winter of 1947-48 stole over Europe, was Italy, and concern was centered on the parliamentary elections scheduled for the following spring. At the joint initiative of James Forrestal, the strongly anti-communist Secretary of Defense, and General Marshall, and with support from George Kennan, the influential head of the State Department's Policy Planning Staff and originator of the so-called theory of 'containment', President Truman approved a remarkable document at a NSC meeting on 19 December 1947. It was NSC 4. The top secret annex, NSC 4–A, instructed the head of the CIA to undertake covert psychological activities to counter Communism. A Special Procedures Group was set up three days later to co-ordinate the work, and so began America's first peacetime deception operation abroad to influence the outcome of another country's election.[8] It was not described as deception, of course: deception was simply inherent in the whole project as part of the mechanics. The CIA would be supplied with funds to conduct secret psychological warfare against the Soviet-supported Italian Communist Party and front groups, and to assist the campaigns of the Christian Democratic candidates in the election. 'Our methods' were beginning to resemble 'those of the opposition'.

One of the key figures involved in the planning and execution of the Italian operation was a veteran of the wartime Office of Strategic Services who had worked in Italy during the War, James

Jesus Angleton. Tall, gaunt and prematurely grey, Angleton had edited a poetry magazine while at Yale and had known both Ezra Pound and T S Eliot. After the war he worked for the CIA in the Italian capital. Angleton and his colleagues cabled Washington with an estimate that 10 million dollars would be needed for the operation to be a success. The figure was apparently accepted without question. A major problem then arose: how to transfer the money from the United States to the grateful recipients in Italy without even a whiff of American Government involvement. To avoid curious questions from the Bureau of the Budget, the Economic Stabilization Fund was selected as the source of the money. The Fund had been created after the war to minimise swings in the value of the dollar and other currencies. Later, covert action would be financed by funds buried deep in the CIA's overall budget. The Secretary of the Treasury, John W Snyder, who controlled the Economic Stabilization Fund, was approached and agreed to co-operate.[9]

Ten million dollars were withdrawn in cash from the Fund and then laundered through individual bank accounts, whose owners in turn transferred the money—thus disguised as a donation—to a motley group of front organisations. These either passed the money directly on to the CIA and its front groups in Italy or else bought Italian lira for use in anti-communist campaigns. Angleton and the Special Procedures Group provided funds covertly to Italian centrist parties, the money distributed according to CIA assessments of the parliamentary seats most vulnerable to communist pressure. CIA cash helped pay for local election campaigns, including meetings, campaign literature and canvassing. There were also some bribes to voting officials disguised as 'bonuses'. In addition, Angleton and his colleagues organised a propaganda campaign which covered the whole monochromatic range from white, through varying shades of grey, to black. Posters and pamphlets were distributed in the streets and stories planted in the newspapers. Anonymous publications painted a brutal portrait of Russian troops raping and pillaging in Germany—a fate, trumpeted the propaganda, soon to be shared by Italy should it choose a communist government. Other material spread compromising rumours about the personal and sex lives of Italian Communist Party (PCI) candidates. The covert campaign of support was mirrored by overt action. Here, of course, no deception was involved. Wheat and other food was supplied to reduce shortages; Truman made dire threats about reductions in aid if the Communists took power; special radio broadcasts were directed towards Italy: Cardinal Francis Spellman and the American Sympathizers for a

Free Italy organised lobbying by Americans of Italian descent of their relations who were about to go to the polls.

In the election of 18 April 1948 the Christian Democrats won an overall majority of 40 seats. The CIA and leading American policy-makers considered the operation a success in terms of preventing the Communists gaining power. It was also successful as a deception, justifying the elaborate precautions taken to launder the secret funds. The evidence suggests that the Russians quickly discovered from their agents in Italy about the surge of funds to the centrist parties, but the espionage reports were distrusted because the spies were thought to be too close to the Italian Communist Party, and their reports a clever ploy to wheedle more money from the Kremlin. A top intelligence official was despatched from Moscow to investigate the reports but he concluded that the new funds came not from the Americans but from the Vatican, and that, besides, it was by then too late to reverse the electoral trend. The deception against the Italian people also worked to the extent that there were no leaks at the time or for many years afterwards about CIA involvement. The number of people deceived by those parts of the propaganda that were deceptive, such as extremes of anti-communism painted by the CIA-inspired posters and pamphlets, is debatable. Some, however, were certainly influenced.[10]

In Washington pressure quickly built up for more clandestine action, and thus deception, to thwart creation of a separate covert action department and the result—after prolonged haggling between the protagonists as to who should control it—was the innocuously titled Office of Policy Coordination (or OPC). The document, NSC 10/2, which founded the OPC referred explicitly to the 'vicious covert activities of the USSR, its satellite countries and communist groups to discredit the aims and activities of the United States and other Western powers'.[11] Deception was recognised to be integral to the OPC's work for its operations were to be 'so planned and conducted that any US government responsibility for them is not evident to unauthorised persons and that if uncovered the US government can plausibly disclaim any responsibility for them'.[12] The man chosen to head the OPC was Frank Wisner, who had served with the Office of Strategic Services in Eastern Europe during the Second World War. Wisner was a rich, clever Southerner. He suffered occasional bouts of verbal diarrhoea, however, recalled uncharitably in the memoirs of Kim Philby from the time when this celebrated Soviet agent was stationed in Washington. Philby quotes Wisner on the covert CIA funding of front groups: 'it is essential to secure the overt co-operation of people with conspicuous access to wealth in their own right'—that is, noted a man from Britain's Foreign Office, rich people. Philby described Wisner acerbically as 'balding and running self-importantly to fat'.[13] But whatever his conversational and cosmetic defects, Wisner was an extremely diligent worker and proud of the organisation he was to build up. The OPC began life in 1949 with 302 staff. Three years later it had 2,812.[14]

In Italy, the deception operation to manipulate the 1948 election was only the prelude to further involvement in the political arena. Funds continued to be siphoned to centrist politicians but they seemed, by 1953, to have had little effect on the long-term communist vote. The new ambassador to Rome, Clare Booth-Luce, who was rabidly anti-communist, was very alarmed by the result of the 1953 poll in which the Communists took 37 per cent of the vote, as against 40.1 per cent for the Christian Democrats. Sensing support in Washington from President Eisenhower's special assistant, C D Jackson, who had been involved during the Second World War in American psychological warfare against the Nazis, Booth-Luce prepared a top secret estimate of the situation. Copies were sent to Eisenhower, John Foster Dulles, the Secretary of State, and his brother, Allen, who was head of the CIA. Booth-Luce concluded that 'if vigorous political action is not taken . . . within two years Italy will be the first Western democratic nation, by legal democratic procedures, to get a communist government'. She pressed for a programme of urgent, covert action to allow parties of the centre to catch their breath, followed by their complete reorganisation and reconstruction.[15]

The response from Washington was enthusiastic and so began what was—in the words of the man who masterminded the operation, William Colby—the CIA's 'largest covert political action programme undertaken until then or, indeed, since'.[16] The aim was to prevent the Communists winning the 1958 election. According to CIA estimates, the Russians were tossing 50 million dollars a year into the coffers of the Italian Communist Party (PCI) and its front organisations. The CIA gave Colby substantial sums, perhaps as much as 20 to 30 million dollars a year. Funds were laundered through various organisations to the Social Democrats and Christian Democrats; anti-communist propaganda was printed and distributed; the services of the Roman Catholic hierarchy and the Church's social organisations were also sought, and certain eminent figures—among them Cardinal Giovanni Montani—received CIA money. Booth-Luce was closely involved, meeting at least once a week with Colby, and cables were often sent daily to Washington, where the operation was supported strongly. On the whole, there was agreement about the general approach although discord surfaced in the CIA about whether or not to fund the Italian Socialist Party (the so-called 'opening to the left') and about the advisability of buying up anti-American newspapers. There was another clash over how best to orchestrate the 'de-Stalinisation' speech of Khrushchev given at the 20th Soviet Party Congress in February 1956. The CIA had obtained a copy and Ray Cline, a CIA deputy director, argued strenuously that the text should be published complete, while certain others, like James Angleton and Frank Wisner, believed the material should be released piecemeal in Italy to have the maximum impact on the communist vote. Cline won. As it was, the 1956 revelations sent a tremor through the whole communist movement in Italy leading many voters to spurn the PCI—especially after events in Poland and Hungary.[17]

Another strand in the deception was secret advice and information given by the CIA to the new General-Secretary of the Christian Democrats, the dynamic and dictatorial Amintore Fanfani, on how to modernise and reorganise his party. Fanfani implemented many of the suggested reforms in areas where it suited his interests.

The electoral result in May 1958 showed no clear-cut defeat for the Communists, who lost only three seats. The Christian Democrats gained twelve, while victory at the ballot was muddied even further by more triumphs at the polls than expected for the socialist parties of Nenni and Saragat. What loss of public support there was for the PCI was probably as much attributable to the events of 1956 in Eastern Europe as to the deception operations of the American government. Colby, however, remained convinced that the CIA's activities in Italy between 1953 and 1958 demonstrated that a 'long term strategy of covert political help to democratic forces can work' and in retrospect he believed the politicians, like Fanfani, who co-operated with the CIA, were not corrupted by the association.[18] There is no evidence available to indicate how much the Soviets discovered about the CIA's role. Since the programme covered a long period, and because the funds supplied were so large, the KGB almost certainly knew something of the CIA's operation. However, if that was the case, they did not choose to parade the CIA's involvement as a major theme of communist propaganda against the centrist parties. On the other hand, the Italian public was clearly misled at the time about the nature and extent of American involvement in their domestic politics. This deception continued until the early 1970s when the delicate web which the CIA had woven around portions of the Italian body politic was partially unravelled.[19]

* * *

When Stalin set up the Communist Information Bureau (Cominform) in 1947, the new organisation was regarded by the West as simply a reincarnation of the Communist International or Comintern, which had been abolished during the Second World War. The Russian dictator wished to use the Cominform, like its predecessor, as a whip to lash unity of purpose into Europe's communist parties. That purpose was quite cynical: to service the interests of the Russian Motherland—in the East by encouraging the development of submissive satellite states, and in the West by undermining the former Allies' resolve to unify Western Europe against a perceived Soviet threat. Much of the Cominform's work in Western Europe was propaganda, spouting forth from front organisations and a myriad of subsidised publications. To counter this propaganda apparatus—and it was by no means centred on the Cominform alone—the West spent millions of dollars on developing an unofficial 'Anti-Cominform'. Britain was in advance of the United States in this field. There was grave apprehension in London about the quantity and quality of Stalinist propaganda flooding Western Europe just after the War. The British response was to establish the Information Research Department, or IRD as it became known, by a Top Secret Cabinet directive in 1948. In London the research staff of the IRD collected information which could be exploited as anti-communist propaganda. Various sympathetic trade unions and other groups were then used to place articles around the world under the by-lines of well-known writers and journalists. The IRD also attempted to expose Stalinist front groups, producing a weighty tome called *Facts about International Communist Front Organizations* and a monthly bulletin called *International Organizations*.[20]

The CIA, because it was far more generously funded than its British peers, built up a much larger propaganda network to match the Russians. Within the Agency it was dubbed irreverently 'The Mighty Wurlitzer'—one musician, Washington, performing on a colossal instrument of propaganda. The idea behind 'The Wurlitzer' was that Washington should be able to touch the 'key' for, say, Bombay, plant a story in a non-American newspaper or with a certain wire service, and then touch the other 'keys' so that outlets around the world would pick up the story and publicise it.[21] The success of 'The Wurlitzer' depended on deception as to the source of the story: few editors of credible newpapers or wire services would have cared to run an item that they knew had been deliberately placed with them by a foreign government. The CIA attempted to overcome this resistance by infiltrating agents into all the major news agencies. There were several in Reuters for example.[22]

Other roles of the unofficial 'Anti-Cominform' were to subsidise magazines and newspapers directly and to fund groups which were influential in the making of opinion and were non-communist. Some were set up with the specific task of countering the influence of a Soviet front in a particular area, for example student politics. Alternatively, these groups were encouraged to propagate splits in communist-dominated organisations such as trades unions. Almost all activities of the 'Anti-Cominform' were predicated on deception: deception of the American public as to what exactly their government was doing; deception of many recipients of CIA funds; deception of the wider European public about the strength of democratic forces in their countries (so, the argument ran, boosting their morale to resist Communism); attempted deception of the Soviets too, not only to persuade them that opposition to their efforts to undermine Western resolve was stronger than they thought, but also to prevent the Russians being able to exploit CIA involvement as anti-American propaganda.

So few documents have been released by Western governments relating to the 'Anti-Cominform' that it is difficult to trace an exact history of who began a particular operation at a particular time. One key event, however, took place in 1951 when the CIA's International Organizations Division was created. Its progenitor was Thomas Braden, who had been appointed the year before as special assistant to the then assistant head of the CIA, Allen Dulles. In 1951, all the CIA's operations to assist non-communist trades unions or other front organisations were divided amongst the various CIA national desks. There was no central co-ordination in Washington. Brash and confident, the young Braden proposed to Dulles that an International Organizations Division be created. His motives, as he admits openly, were mixed. He was empire-building as well as intent on reducing duplication and inefficiency. At a meeting of the CIA division chiefs, Braden's proposal was unanimously rejected but Dulles managed to overturn the decision and the then head of the Agency, General Walter Bedell Smith, approved. Braden, full of hardheaded liberal sentiment, presided over the next three years' rapid growth of the International Organizations Division, assisted and then succeeded by a forceful young American of slightly more pragmatic temperament, Cord Meyer.[23]

Subsidies to the non-communist press had begun in France, Italy and Germany almost as soon as the War was over. In France, for example, funds started to flow after a hero of the Resistance (probably Henri Frenay, although his identity has still not been confirmed) approached the Americans and explained his fears about the overweening influence of Stalinist propaganda in France. Two newspapers known to have received secret CIA subsidies thereafter were the independent socialist *Franc Tireur* and the Parisian *Le Combat*. Usually the editors did not know of CIA financing, which as a matter of principle went to papers only of the centre and centre-left. Braden's Division of the CIA was especially enthusiastic about 'openings to the left'—whether in politics, union affairs or publications. Communism could not be defeated through support for the forces of the centre and right alone. Braden explained the reasoning in a television documentary.

> You have an ideal like equality which is possessing a whole generation of people who are poor and struggling and coming out of a war and are told . . . that by joining such and such a party or such and such a group, they can achieve equality. The only way to combat it [i.e. the idea] unless you want to go to war . . . is to hit it with another good idea. The non-communist left idea is just as good as the communist idea. In my opinion it's a lot better. So we supported, the CIA supported, the non-communist left in Europe and everywhere else.[24]

Things were not so clear-cut in practice. Some in the CIA were suspicious of the Agency having intimate links with socialists of any hue and in Italy, for example, support for particular groups and individuals clashed directly with the strategy of aid to the Christian Democrats alone.

If there were no existing publications supporting 'the democratic point of view', the strategy of the International Organizations Division was simply to create them. The most famous example is perhaps *Encounter*. This magazine, in turn, issued from another CIA-financed organisation, the Congress for Cultural Freedom. Washington was particularly apprehensive after the War about Soviet front groups that rallied the Old World's intellectuals behind the banners of Communism and anti-Americanism. Europe's intellectual elite had to be won over to the non-communist, democratic camp and the Congress for Cultural Freedom was designed to play a key role in this battle.

The guiding spirit behind the Congress was Melvin J Lasky, a Trotskyist from New York's City College who had renounced his communist faith and worked for the magazine *New Leader* from 1941. In 1948 he went to Germany to work for the United States High Commission. With the encouragement of General Lucius Clay, then in charge of the United States Zone and credited with organising the Berlin Airlift, Lasky founded a literary magazine called *Der Monat*. Its success enthused Lasky. Using his burgeoning contacts amongst German intellectuals and with the American Government, Lasky organised the launch of the Congress for Cultural Freedom in June 1950 at the Titania Palace Theatre in the United States Zone of Berlin before an audience of 4,000 people. The aim of the new group was a modest one: to 'defend freedom and democracy against the new tyranny sweeping the world'.[25] The Congress immediately started political seminars and student exchanges across Europe. It published literature in various guises in support of the democratic, CIA-funded youth organisations. The Congress also launched the French magazine *Preuves* and in 1953 the intellectual review *Encounter* (referred to above) under the editorship of Irving Kristol. *Encounter* grew into one of the most influential journals of liberal opinion in the West. What the European public did not know—nor indeed many of the recipients—was that most of the funds for the Congress and its offshoots derived from the CIA. Melvin Lasky has consistently denied any knowledge of CIA financing but on the other hand Thomas Braden says 'the man in charge' was a witting agent. CIA money for the Congress was largely funnelled through the International Director of the American Federation of Labor (AFL), Jay Lovestone—like Lasky an ideological apostate, who had at one stage led the United States Communist Party before starting clandestine work for the American Government. Funds for other magazines and publications (and relatively few names have surfaced), were passed on through various American foundations which considered participation a matter of patriotic duty. It is certain that the Congress and *Encounter* could not have exercised the immense influence they did, for example, in sustaining the right wing of the British Labour Party in the 1950s, if it had been publicly known that they were financed by American intelligence.[26]

In the field of union affairs, the CIA tended to be sucked in on the wake left by American labour unions. In London in 1945 the AFL representative in the World Federation of Trades Unions told Paul Nitze, later to be director of the United States Policy Planning Staff, that it was only a matter of time before Europe fell to the Communists. The World Federation of Trades Unions (WFTU) had been founded in Paris earlier that year as a result of an initiative by the Soviets. Both the British Trades Union Congress and the United States Congress of Industrial Organizations became members. The WFTU, however, soon became a front for the Russians and, as "Cold War" tension sharpened in 1948, the British and American participants withdrew because of the blatant pro-Soviet line of the WFTU. The American and British Governments were further concerned by the monolithic power wielded by the communist-dominated unions in France and Italy, particularly in Sicily. Stalin manipulated these unions skilfully to further Soviet interests, for example, by fomenting strikes to disrupt Marshall Plan aid or leading campaigns against the rearmament of Western Europe.

Attempts to counter this Soviet activity were at first restricted to American trades unions. Immediately after the War, the International Ladies' Garment Workers Union tried to foster non-communist unions in France but their leadership was hampered by lack of funds. It is difficult to discover exactly when the CIA entered the lists: the Agency used many of the same personnel and conduits as cover. Two key personalities were Jay Lovestone, referred to already in connection with the Congress for Cultural Freedom, and Irving Brown, the AFL–CIO European representative. Lovestone and Brown almost certainly took part in late 1947 in a secret assessment of the strength of the Communists in Europe. Three veterans of the OSS and three officials of the AFL travelled to Paris, Marseille, Rome and Palermo. Action, the group concluded, was urgent and imperative. The first move was to supply secret funds to Léon Jouhaux, the French union leader, so that his moderate *Force Ouvrière* group could split away from the communist-dominated *Confédération Générale du Travail.* The CIA provided some finance for the initial break on 19 December 1947 through David Dubinsky of the American garment union. Afterwards, when funds began to run low, the CIA furnished subsidies to the *Force Ouvrière*, amounting to about one million dollars a year in the early 1950s. Simultaneously, bribes and strong-arm tactics were employed to smash the communist grip over the dockers' unions at Marseille where a series of crippling strikes took place in 1947 and 1948. Similarly, former OSS men used the Mafia and the Corsican underground to reduce communist influence in the Sicilian trades unions. This operation in Sicily was only the prelude to further United States involvement.[27]

A policy statement written by Irwin M Tobin of the Office of European Regional Affairs on 22 November 1949 stated that the leaders of moderate Italian trades unions were to 'carry out our [American] objectives in the light of local circumstances as they see them'.[28] 'Assistance' would be furnished to American trades unions and political organisations which were closely linked to Italian labour in order to influence its development. The creation of the pioneer, anti-communist labour front through the merger of the Free Italian General Confederation of Labour and FIFL on 7 February 1950 was supported by the CIA and the State Department, as was the split of the non-communist, Catholic CISL from the communist CGIL.

Simultaneously, to weld the Western unions into a more cohesive force to oppose the WFTU, a rival organisation called the International Confederation of Free Trades Unions was established in 1949. Both the CIA and British intelligence were involved, the CIA providing covert funds on a large scale to the ICFTU. Money also flowed into the hands of German miners and steelworkers who were trying to limit communist influence in their unions. In most cases the instruments of deception—the secret channels through which the United States Government money flowed—were American trades union officials. Jay Lovestone, Irving Brown and David Dubinsky were not alone. George Meany, head of the AFL, probably received the most CIA money in the first instance and he distributed it to his officials in Europe. Walter Reuther, chief of the United Autoworkers, also undertook a number of operations. On one occasion Tom Braden of the CIA travelled out to Detroit clutching a satchel containing $75,000 in hundred-dollar bills which he passed to Reuther.[29]

The West was also at a propaganda disadvantage in the field of international student politics. Cord Meyer, himself active in promoting the doomed cause of world federalism before joining the CIA in 1951, believed that:

> the Soviets and their communist allies abroad began the struggle holding nearly all the cards . . . the non-communist students in the West were badly organized, perennially broke, ill-prepared ideologically for the coming struggle, and naively innocent in comparison to the professional party activists they had to confront.[30]

In the generous and forgiving atmosphere of international co-operation existing in 1946, the Kremlin helped found the International Union of Students based in Prague. The first Soviet vice-president of the International Union was Aleksandr Shelepin, who was later to head the KGB. As with the World Federation of Trades Unions, it was the Czech coup of 1948 that led to the decisive schism in the International Union of Students. The result of discussions between the American National Student Association (NSA) and the non-communist student unions of Western Europe was a novel group which became known as the International Student Conference (ISC). A small co-ordinating secretariat was installed at Leyden in Holland in the early 1950s. ISC was to speak for the student unions of the democracies on international issues and compete with the Eastern Bloc for the allegiance of students in the emerging Third World: in other words, to fight on the student field of battle in the "Cold War".

In the short term, ISC was quite successful. By 1955, it claimed the membership of national student unions from 55 nations. Over half were from the less developed countries. The ISC's position outside the communist sphere of influence was so dominant that the Kremlin took the trouble to hold only two world youth festivals outside the Eastern Bloc—in Vienna in 1959 and in Helsinki in 1961. Both were boycotted by the ISC. But selected American students attended informally and produced a daily newspaper in several languages to challenge the orthodox, communist interpretation of world events. The 1950s and early 1960s were also special in that many future leaders of Third World countries were then prominent in student politics and the West was able to impress them favourably through the ISC by providing grants for study abroad or to attend student conferences.

Most of the funds for the ISC came from private foundations in the United States with which the American National Student Association (NSA) leaders seconded to ISC had close links. European student organisations rightly saw NSA participation and ISC finance as inseparable. An ISC Information Bulletin from the early 1950s underlined the point publicly: 'the end of the period of establishment coincided with the arrival of Mr William T Dentzer (USNSA President in 1952) to join the staff in early October. This date saw a substantial improvement in the resources of the secretariat.'[31] The two most important contributors were the Foundation for Youth and Student Affairs, founded in 1952, and the San Jacinto Foundation, founded six years later. Periodically, doubts surfaced about the almost complete reliance on sources in America for funds. ISC student leaders sought alternative finance but little was forthcoming. There was resentment at what some saw as the excessive influence and bumptious manner of some NSA leaders involved with the ISC. But with a permanent secretariat at Leyden and an extensive network of international activities to maintain, these people had no reason to wish to sever the flow of funds from the United States, so inertia reigned. All this was to change early in 1967.[32]

A series of articles in the radical American magazine *Ramparts* revealed that the leadership of the USNSA had cooperated with the CIA in founding and funding the International Student Conference and that the private foundations were simply conduits for channelling covert funds. The financing of the Conference had been a CIA deception operation from the beginning. It emerged subsequently that other youth organisations were CIA fronts to a greater or lesser extent. One was the World Assembly of Youth, based at Brussels and created about the same time as the ISC; another was the European Youth Campaign that ran from 1951 for eight years and had close links with another CIA-funded group, the European Movement; others that benefited from CIA largesse were the United States Youth Council, Young Christian Workers, Pax Romana and the International Union of Socialist Youth. The motive for deception was the same for the students as for trades unionists: overt aid would have weakened their position and so defeated the very purpose for which the money was furnished. But who exactly was misled? Undoubtedly, one or two European student leaders must have stumbled on, or been made privy to, the secret. There seems little reason, however, to doubt the protestations of former ISC or World Assembly of Youth officials who said they had no idea of the CIA's role. Their position was typified by Barney Hayhoe, who was head of the British section of WAY until 1961 and has since become a senior member of the British Conservative Party: 'I was well aware of the financial support given by the American Foundation for Youth and Student Affairs but I knew nothing of any CIA involvement at the time.'[33] European student leaders were thus effectively misled until 1967 and after them, in the baggage train of deception, came the wider European public.

The result, when the truth did finally come to light in *Ramparts*, was consternation and a bitter sense of betrayal amongst those who had been duped. ISC and WAY were emasculated as effective student organisations—a feat the Russians had been attempting unsuccessfully for almost two decades. 'In retrospect,' wrote Cord Meyer, the CIA official responsible for the student deception operations from 1954,

> the decision taken by default not to replace . . . secret CIA funding by some reliable method of open government support was an act of unilateral political disarmament in the face of a continuing Soviet challenge.[34]

The bureaucratic phlegm and bad judgement, that allowed the CIA to continue covert funding when a sea change had started in students' political attitudes in the 1960s, changed deception from being at the heart of the operation's success to being its chief liability.[35]

* * *

In the summer of 1953 the British member of parliament and political journalist, Dick Crossman, paid a visit to Munich. Crossman had spent the War in psychological warfare—first in England

and then with Harold Macmillan, who was later to become British prime minister, in exotic Algiers. Crossman became an expert in propaganda techniques and was invited to look over America's duet of psychological warfare stations at Munich, Radio Free Europe and Radio Liberty in 1953. The MP was chary of what he considered the undue freedom afforded the East European emigrés doing most of the broadcasting. But he was impressed by the 'brilliant American idea' of making Radio Free Europe independent of the State Department. 'The American Government is not responsible for it [RFE], and it is ostensibly financed by independent American business men. . . . American capitalism . . . is carrying on this private war against the Soviet Union on its own.'[36] American capitalism, however, was not alone. Crossman had been deceived, and he was not the only person duped. The uncounted grey millions behind the Iron Curtain who clandestinely tuned their radio sets to RFE and Radio Liberty, almost all Americans, and most of those who took an interest in such matters in Europe shared Crossman's misconception that the radio stations were independent—both politically and financially—of the United States Government. In fact, RFE and RL had been established with the advice and assistance of the Central Intelligence Agency which then provided the lion's share of funds.

Immediately after the War, the United States scaled down its psychological warfare to almost nothing. Only the State Department's Voice of America continued to broadcast. The unique tensions generated by the division of Berlin led to the creation of Radio in the American Sector of Berlin (RIAS) in February, 1946. It was, at that stage, simply a radio service for Germans living in the immediate vicinity. As the Cold War grew more frosty and alarm in Washington burgeoned, American politicians and the military began to take a propaganda counter-offensive seriously. In December 1946, the State–War–Navy Coordinating Committee formulated plans for American peacetime psychological warfare, considered as 'an essential factor in the achievement of national aims and objectives'.[37] RIAS stretched its broadcasting wings considerably during the Berlin blockade of 1948, as it carried the message of Allied resolution to listeners in the beleaguered city and in East Germany. But apart from RIAS there was little impetus towards an international campaign of broadcast propaganda. In fact, the organisation that was to set up Radio Free Europe officially began life outside the government machinery of Washington.[38]

One of the prime movers behind the National Committee for a Free Europe was the head of the CIA's Office of Policy Coordination, Frank Wisner. He had seen the refugee problem in Europe at first hand and understood that the huddled masses of humanity, displaced from the East and then often penned in dismal refugee camps, represented not only a source of intelligence but a means of striking back at the Kremlin. Wisner's enthusiasm for using the talents and knowledge of the refugees was shared by Allen Dulles, a former OSS executive later to head the CIA, and three other OSS colleagues, De Witt C Poole, Spencer Phenix and Frederic Dolbeare. A series of meetings took place between these men and former ambassador Joseph Grew. George Kennan, by then special adviser to Secretary of State Dean Acheson, was also consulted. All strongly supported the strengthening of psychological warfare and on 2 June 1949, the creation of the National Committee for a Free Europe was announced to a largely indifferent public. Its aim was to provide means for the emigrés to speak to their compatriots at home. The Committee was launched without any tangible source of funds but no one paid any regard because a host of similar organisations to help refugees were founded at the same time. The necessary infusions of cash in fact came from the OPC. Classified memoranda from Free Europe executives asking for funds were addressed to 'FW', 'one of our friends in the South'—a cover for Frank Wisner of truly labyrinthine complexity. Only a few members of Congress were let in on the secret.[39]

The deception over finance was for reasons which, by now, are familiar: first, to avoid unwelcome prying from Congress that might have wrecked the whole project at its inception or have revealed certain facts considered essential to Free Europe's success. One of these was that the Committee must appear to be supported solely by private American citizens, so that it could not be accused of spouting mere government propaganda. The State and Defense Departments would also be distanced from any embarrassment that might flow from Free Europe's aggressive psychological warfare. The State Department did not want the atmosphere of day-to-day diplomacy with the Eastern Bloc to be soured by needing continuously to disavow propaganda coming from another official government source. The State Department could also heave a sigh of relief as it unshouldered the burden of looking after the refugees, while still having access to them as sources of intelligence about communist subversion. Conversely, Free Europe could say and do things with which diplomats and military personnel would not wish to be tarred.

Only in 1950 did the Committee's amorphous aim assume any concrete form. Reports from Germany spoke with approval of the expanded and lively RIAS and they coincided with the adoption of

NSC 68 at the highest levels of United States government. National Security Council Document 68 epitomised America's strategy to prosecute the Cold War and contained an instruction to step up psychological warfare. It envisaged a quadrupling of spending in this area: from 34 million dollars in 1950 to 120 million in 1952. The CIA pressed for the plans of the Free Europe Committee to fit in with NSC 68. The chairman of the Free Europe Committee, considering the options early in 1950, wrote that:

> The Voice of America, because it is an arm of government, is not in a position to engage in hard-hitting psychological warfare. A Committee of private citizens, on the other hand, would suffer from no such disability. Accordingly, if such a committee succeeded in establishing adequate facilities and was provided with a flow of lively topical information, it could enter a field prohibited to the Voice.[40]

Radio Free Europe broadcast its first signal from an old and weak transmitter on 4 July 1950—probably one that the Office of Policy Coordination inherited in 1948 when it was set up. A new 135 milliwatt transmitter for broadcasts to Czechoslovakia and Hungary was opened on 1 May 1951, and the inaugural address given by Irving Brown, the European representative of the AFL who was involved with the CIA's attempts to foster non-communist unions in Europe. Free Europe executives were quite clear about their role: to engage in all shades of propaganda that would sow in the minds and hearts of the rulers of Eastern Europe 'dismay, doubt and defeatism'.[41]

What deception was practised by Radio Free Europe? Unlike some 'black' stations operated by the Allies in the Second World War, RFE never pretended that it was operating within the borders of 'enemy' territory; it was self-consciously designed to enable exiles to address their fellow citizens over the Iron Curtain. However, certain radio propaganda techniques were borrowed from the War. One was the encouragement of slowdowns at work as a protest against occupying forces. Another—which unlike the first had some element of deception in order to increase credibility—was the use of a fictional military expert to give talks on various subjects. A further weapon in the RFE armoury was to name on the air men and women who were alleged to be active collaborators with the puppet communist regimes. Addresses were also given. But deliberate deception was not inherent here; the aim was to isolate certain individuals and to subject them to the scorn of the native populations. RFE was less interested in its early days in news than in tidbits of gossip and information that could be used to undermine morale in the East. One source for these were emigrés from Russia and the satellite countries. Another, but notoriously unreliable one, was espionage, especially the debriefing of high level defectors. A colonel in the Polish secret service, Josef Swiatlo, defected to the West at CIA headquarters in Berlin in December, 1953. Swiatlo was an unparalleled fount of information about corruption and personal scandals at all levels of the Polish Communist Party and, after debriefing by the CIA, he recorded over a hundred interviews with RFE for broadcast back into Poland. Following the transmissions, the head of the Polish secret service resigned and three of his subordinates were dismissed—a powerful if isolated example of how effective RFE's psychological warfare could be. There was inevitably deception in broadcasts of this type because RFE deliberately transmitted information whose truth was one of the station's last concerns. RFE's own executives admitted that it was to broadcast effective propaganda and that effectiveness sometimes called for the distortion or the invention of facts. As RFE grew to maturity during the 1950s, however, the station toned down its bitter Cold War rhetoric and paid more and more attention to the truth of what was broadcast.[42]

Radio Liberty was founded for essentially the same reasons as RFE and was also financed by the CIA. Frank Wisner, though, wished Radio Liberty to be more malleable and to have a lower profile than Radio Free Europe. Cover would be easier to maintain, especially when the exiles were so schismatic and volatile; the OPC also resented to some extent the independence and self-confidence of the Free Europe Committee. The American Committee for the Freedom of the Peoples of the USSR was therefore much more closely under the CIA's control and the Committee's infant, Radio Liberty, began broadcasting on 1 March 1953. The fledgling station's coverage of Stalin's death on 5 March and the East Berlin riots on 17 June reached surprising levels of professionalism despite some 'news' items descending to inflammatory and simplistic calls to the security forces of the East to rise up in revolt. There was also an ill-advised change of objective by the Committee, whose aim was altered from the 'liberation of Russia' to the 'liberation from Bolshevism'. The alteration, made to foster the involvement of Ukrainians, backfired. No one on the Committee seemed to realise that 'Bolshevism' was Hitler's pet term of abuse for the Soviet system.

Ever since 1956, accusations have floated back and forth across the Iron Curtain about the role of RFE in fomenting revolt in two satellite states, Poland and Hungary. At the heart of the allegations has lurked the question: were the Poles and Hungarians deceived by RFE into revolt through promises of assistance from the West?

The tone of broadcasts to Eastern Europe in the five years from 1951 was one of 'liberation'. But transmissions to Poland in 1956—where there were a series of riots culminating in the return to power of Wladyslaw Gomulka—show RFE to have been far from a freewheeling rabble rouser. The station reported events in full but encouraged control and calm. At the height of the tension in September and October, the authorities even decided to switch off the jamming equipment which blocked RFE's signal to Poland.

In Hungary, RFE's role was more suspect. On 23 October 1956, demonstrations erupted in Budapest and within two days the whole country was in a state of insurrection. For a short period it seemed that the rebels were winning but on 6 November Russian tanks swept back into the Hungarian capital and Soviet troops ruthlessly crushed all opposition. A minute proportion of items broadcast by RFE during the revolt could in retrospect have implied that some military help would be forthcoming from the West. According to RFE's own internal investigation—which would obviously err towards exculpation—no more than 16 out of 308 fitted into this category, while another four were adjudged guilty of giving military advice to Hungarian freedom fighters. There was, for example, a broadcast by a fictional 'Colonel Bell' on how to make incendiary bombs. The overwhelming weight of evidence comes down on the side of a spontaneous rebellion, however—not one sparked off by RFE's deception.[43]

The success of RFE depended on the station being seen as separate from the American government. The station's cover could in turn only be maintained if the American public was deceived into believing RFE was funded by private citizens alone. To achieve this, a fund-raising campaign of masterful scope and style, called modestly the Crusade for Freedom, was launched. It was centred on the Freedom Bell, cast in England from a design derived from the Liberty Bell in Philadelphia's Independence Hall. The campaign was spearheaded by a contemporary American knight errant, General Lucius Clay, who organised the successful Berlin airlift in 1948. To the accompaniment of stupendous razzamatazz and a series of emotionally charged television advertisements, the Freedom Bell went on a nationwide tour of the United States before being dedicated in Berlin on 24 October 1950. The Crusade achieved its aim and hoodwinked the public both in America and Europe. Significantly, the campaign cost almost two million dollars for the eleven-month period from 1 April 1950, but garnered only about one and three-quarter million in contributions. The Crusade's objective was publicity, not fund raising. The short-term success of the Crusade must be balanced against the anger the duping caused amongst certain influential figures when they discovered the truth.

When RFE was 'exposed' as a CIA front in 1967 by journalists following up the *Ramparts* magazine article on the National Students Association, the revelations nearly led to the closure of Radio Free Europe and Radio Liberty. One of the major supporters of the drive in 1971 to stop CIA finance to the Radios was Senator Frank Church of Idaho. Church's antipathy was partly fuelled by the fact that he had been recruited to help lead the Crusade for Freedom in his home state before he had entered the Senate, but he had not been told about CIA funding. In fact, Radio Free Europe and Radio Liberty survived the legislative onslaughts of the late 1960s and early 1970s. They have subsequently been openly funded by the United States, managed and supervised by a Board for International Broadcasting.[44]

In the West, the secret of the stations' funding remained safe until 1967, even though the Soviet and East European media regularly labelled them as 'agents of the CIA'. It is difficult enough to judge the effectiveness of any propaganda radio station, let alone the extent to which its success depended on misleading people about its source of funds. The 1967 revelations had some rather feeble impact on public opinion in the United States but little measurable effect on the audience in the East. The majority of listeners, accustomed to the lies and distortions of the official media, were simply grateful for an alternative source of information in the 1950s and grew accustomed to Soviet accusations of RFE and RL being tools of the CIA, accusations which would probably have been made by Moscow whether or not they were known to be true. By the late 1960s, reports that the stations received money from American intelligence were probably treated with scepticism or a shrug of the shoulders in Eastern Europe. RFE and RL had been operating for almost 20 years and had built up a fund of credibility, certainly in comparison with the official media. This credibility showed itself in the millions of dollars the governments of the East spent on establishing a colossal network of jamming stations and the small but regular audience the stations attracted. The deception involved in the foundation and running of RFE and RL had largely fulfilled its purpose by the time it was exposed in 1967.[45]

* * *

CIA deception in post-War Europe relied on both dissimilation—activity to conceal the links between various United States agencies, and the motley European groups receiving American government assistance; and simulation—operations to make the recipients appear independent of such help—the establishment of boards of governors and fund-raising campaigns for example. Individuals involved in the simulation were often unaware that they were to a large extent puppets, deceived as to who really twitched their strings. The fact that such people were duped made them, in turn, effective deceivers. During the critical period when these operations were set up, CIA deception worked remarkably well. It may even have been an element in the political and economic recovery of parts of Western Europe. Perhaps one reason for this success was the trust that most West Europeans placed in their American ally: deception was not anticipated and defences were down. This raises wider questions.

In principle, deception and democracy are inimical. War, however, erodes democratic freedoms as survival becomes the predominant concern and as governments accept that ends justify means. Deception of an enemy in wartime is certainly legitimate and, to a lesser extent, so too is government deception of its own public, provided that it is temporary and can subsequently be explained. When peace returns, all these doubtful means are supposed to be set aside: open government, respect for the law, and regard for the conventions of international behaviour are expected to be the guiding principles of democratically elected governments.

The "Cold War", with its novel and uncertain rules of engagement, therefore posed awkward ethical as well as practical problems for the West. Was the Soviet subversive challenge to be treated as a new form of war, to be met at the same level—rather as advocated by Le Carré's 'Control'—or were the conventions of 'peace' to be upheld at all costs? How could an apathetic American public be persuaded to support a crusade to preserve democracy in Europe, and how could the vulnerable but sometimes vain Europeans be persuaded to accept financial and other help, unless the operation was disguised? Even with the advantage of hindsight, it is difficult to suggest satisfactory answers.

The danger of the course adopted was that a deception operation in peacetime develops its own self-justifying momentum and that the sponsor government does not monitor public opinion to ensure the two are in step. If a democratic government deceives its own people or those of an ally, it plays a game of double jeopardy. On the one hand, the deception may enable it to do things it would not otherwise be able to do. On the other, if the deception continues too long and is finally exposed, the results for the government engaged in the deception—and for the policy it is meant to further—can be disastrous. In fact, too many post-War

CIA deception operations were continued beyond their useful lives. Support for European groups ought either to have been suspended or assumed by some legitimate, overt agency. Exposure when it came was harmful to the CIA and, more importantly, to the trust that had made the operations possible in the first place. To be an unqualified success, deception by a democracy in peacetime must be used like a medicinal but addictive drug: little and under continuous supervision.

Endnotes

1. John Le Carré, *The Spy Who Came in from the Cold* (London, Victor Gollanz, 1963), p. 19.
2. Robert Amory Jr, Deputy-Director (Intelligence) CIA, 1952-62, interview with author, 19 March 1979.
3. Trevor Barnes (i), 'The Secret Cold War: The CIA and American Foreign Policy in Europe 1946–56, Part I,' *Historical Journal* (Cambridge, UK), Vol. 24, 1981, pp. 399–415; Barnes (ii), 'The Secret Cold War, Part II,' *Historical Journal*, Vol. 25, 1982, pp. 649–70.
4. See however Fredric S Feer, 'The Problem', Roy Godson, ed., *Intelligence Requirements for the 1980s: Analysis and Estimates* (New Brunswick, Transaction Books, 1980), pp. 126–35, 140.
5. Feer, pp. 145, 158–9. Feer's sources told him there was 'little US deception', and, in a discussion that followed, one academic said that it was a pity that 'only one side', Russia, was practising deception. Deception is in the eye of the beholder and, so far, the West has eyes only for Soviet deception. The best, although highly flawed, history of the CIA by John Ranelagh, *The Agency: the Rise and Decline of the CIA* (London: Hodder and Stoughton Sceptre Edition, 1988), typifies this blindness. 'Deception' does not even figure in the index.
6. Quoted Stephen E Ambrose, *Rise to Globalism* (Harmondsworth, Penguin, 1976), p. 150.
7. For the early history of the CIA, see United States Senate, Select Committee to Study Government Operations with respect to Intelligence Activities, *Final Report* (otherwise known as the Church Committee report) (Washington, US Government Printing Office, 1976), Vol. 1, pp. 99–106 and Vol. 4, pp. 26–30; Barnes (i), pp. 399–408; Thomas Powers, *The Man Who Kept the Secrets: Richard Helms and the CIA* (London, Weidenfeld and Nicolson, 1979), pp. 26–9; Memo, Hillenkoetter to Truman, 8 August, 1947, Leahy Files, National Archives, Washington, DC, Box 20, File 128.
8. Walter Pforzheimer, CIA legislative counsel 1946–56, interview with author, 15 March 1979; Lawrence Houston, CIA general counsel for 1946, interview with author, 16 March 1979; William R Carson, *The Armies of Ignorance: The Rise of the American Intelligence Empire* (New York, Dial Press, 1977), pp. 294–301; Ray S Cline, *Secrets, Spies and Scholars: Blueprint of the Essential CIA* (Washington, DC, Acropolis Press, 1976), pp. 97–102; Barnes (i), pp. 412–13.
9. *Ibid.*
10. *Ibid.*
11. Quoted Powers, *Secrets*, p. 31.
12. *Ibid.*
13. Kim Philby, *My Silent War* (London, Granada, 1969), p. 141.
14. Church Committee, Vol. 1, p. 106.
15. Boothe-Luce to Jackson, 19 June and 7 September 1953, quoted Barnes (II), p. 663.
16. William E Colby (i), *Honorable Men: My Life in the CIA* (New York, Simon and Schuster, 1978), p. 111.
17. William E Colby (ii), interview with author, 17 March 1979; Colby (i), pp. 111–13; Philip Agee and Louis Wolf, eds., *Dirty Work: The CIA in Western Europe* (Secausus, NJ, Lyle Stuart, 1978), pp. 168–69; Elizabeth Wiskeman, *Italy since 1945* (London, Macmillan, 1971), pp. 35–6.

18. Colby (i), p. 139.
19. The unravelling started with books such as Victor Marchetti and John D Marks, *The CIA and the Cult of Intelligence* (London, Jonathan Cape, 1974).
20. Lord Mayhew, British Foreign Office Minister 1946–50, interview with author, 23 October 1984; Barnes (ii), pp. 664–5, 667; on IRD see Lyn Smith, 'Covert British Propaganda: The Information Research Department, 1947–77', Millennium: *Journal of International Studies,* Vol. 9, No. 1, 1980, pp. 67–83; Jonathan Bloch and Patrick Fitzgerald, *British Intelligence and Covert Action* (London, Junction Books, 1983), pp. 98–9.
21. Colby (ii), p. 119.
22. Barnes (ii), p. 667.
23. Thomas Braden (1), interview with author, 21 March 1979; Braden (ii), transcript of interview in Canadian Broadcasting Corporation documentary, 'The Fifth Estate: The Espionage Establishment', broadcast 9 January 1974, p. 13; Colby (i), p. 128.
24. Braden (ii), *op. cit.*
25. Agee and Wolf, p. 192.
26. Agee and Wolf, pp. 188–93; Braden (i), and (ii), p. 15.
27. Barnes (i), pp. 404, 413; Barnes (ii), p. 662; Miles Copeland, *The Real Spy World* (London, Weidenfeld and Nicolson, 1974), pp. 235–6; Pforzheimer, *op. cit.*; Braden (i), *op. cit.*; Braden (ii), pp. 12–15; Philip Agee, *CIA Diary* (Harmondsworth, Penguin, 1974), pp. 74–5; Cord Meyer (i), interview with author, 19 March 1979; Meyer (ii), *Facing Reality: From World Federalism to the CIA* (Washington, University Press of America, 1980), p. 99.
28. Irwin Robin, 'Policy Statement and background data on unification of Italian non-communist trade unions', 22 November 1949, in *Foreign Relations of the United States* (Washington, DC, US Government Printing Office, 1976), IV, 707.
29. Note 27 *Ibid.*
30. Meyer (ii), p. 99.
31. ISC Information Bulletin, undated, ISC File, Box ISC Archives, HQ British National Union of Students, London.
32. Photocopy of document setting out sources of ISC finance, undated, International Affairs File, Box ISC Archives, HQ British National Union of Students, London; George Foulkes MP, member (1965–66) of ISC supervisory committee, interview with author, 8 November 1984.
33. Barney Hayhoe, letter to the author, 19 October 1984, quoted with permission.
34. Meyer (ii), p. 106. Ranelagh, pp. 216–19, 246–52, gives a pedestrian account of the labour and student operations and does not attempt an assessment.
35. Foulkes, *op. cit.*; *Ramparts*, March 1967, p. 29–39; Agee, pp. 72–3; Bloch and Fitzgerald, pp. 101–6; 'Report of the Executive Subcommittee investigating possible Central Intelligence Agency subversion of the International Student Conference,' October 1967, NUS Library, NUS headquarters, London. (This report concluded that 'US personnel had considerable influence over ISC policy and activities prior to 1962 but only marginal influence since.')
36. R H S Crossman, 'Psychological Warfare,' *Journal of the Royal United Services Institution*, Vol. XCVIII, No. 592, November 1953, pp. 532–3.
37. Memo, 'SWNCC: Psychological Warfare,' 10 December 1946, quoted Barnes (i), p. 405.
38. Donald R Browne, *International Radio Broadcasting: The Limits of the Limitless Medium* (New York, Praeger, 1982), pp. 132–3.
39. Sig Mickelson, *America's Other Voice: The Story of Radio Free Europe and Radio Liberty* (New York, Praeger, 1983), pp. 17–21.
40. Quoted Mickelson, p. 27.
41. Mickelson, pp. 20–33; Barnes (ii), pp. 664–6; Mayhew, *op. cit.* (The British were dubious about the possible success of RFE but they, also, were hoodwinked at the time about the source of funds: the Foreign Office in London knew some US government money was involved but not that it was provided by the CIA.)
42. Mickelson, pp. 38–40; 87–8.

43. Mickelson, pp. 59–72, 91–105; Allan Michie, *Voice Through the Iron Curtain* (New York, Doubleday, 1963), p. 258; Browne, pp. 139–40.
44. Mickelson, pp. 51–8; 121–38, 207–9; Browne, p. 147.
45. *Ibid.*

CHAPTER 13

Deception at Suez, 1956

MICHAEL HANDEL

O, what a tangled web we weave
When first we practise to deceive!
Sir Walter Scott

The circumstances which led to the Sinai and Suez Campaigns of 1956 are well known and need not be recapitulated in detail here. The Israeli Government in the period 1955–1956 came under increasing pressures to contain and counter the Fedayeen raids (masterminded by Egyptian intelligence) against its territory, as well as to eliminate the Egyptian threat created by the recent Soviet arms deal. The British and French, fighting a rear-guard action to maintain possessions and influence in the region, were interested in toppling Nasser's regime, perceived as the underlying threat to their Middle and Near Eastern interests, to the stability of the whole region, and to continued freedom of movement through the Suez Canal, which Egypt nationalised in Summer, 1956. The Western world in general was afraid that the Egyptian Government would be incapable of managing this strategically vital waterway, and was worried by the growing Soviet threat in the Canal region and the East. In short, a common interest was established between France, Britain and Israel.[1]

By mid-September, 1956, it became clear to the three governments that some military action against Egypt was almost inevitable. The problems facing the newly formed coalition were how to establish an effective military alliance, how to deceive the world as to the true extent of their co-operation and planning, and how to create an appropriate pretext for intervention. Planning proceeded through a series of secret meetings, first between Israeli and French officials, followed by contacts between French and British representatives, finally culminating in a three-sided meeting in Sèvres on 22–24 October 1956. The Sèvres meeting was attended by senior officials of the three states who had to use disguises to avoid attracting the attention of their own media or foreign intelligence services. The Israeli representatives were Prime Minister Ben Gurion, the Chief of Staff, Moshe Dayan, and Shimon Peres, the Director General of Israel's Ministry of Defence. Representing the French government were Prime Minister Guy Mollet, Foreign Minister Christian Pineau and Defence Minister Maurice Bourges-Maunoury. The British delegation included Foreign Secretary Selwyn Lloyd, and Patrick Dean, who acted as a co-ordinator for the British intelligence service.[2]

The plan of operations agreed at Sèvres was that Israel would attack Egypt through Sinai on the evening of 29 October and the following morning the British and French would issue an ultimatum to Israel and Egypt to keep clear of the Canal Zone so that the Anglo-French forces, acting as 'peace-keepers' could move in. It was anticipated that Egypt would refuse to withdraw from her own territory, and it was agreed that Israel would promptly accept the ultimatum. Directly Egypt refused, the Royal Air Force would start bombing Egyptian airfields.[3]

This bombing would be designed to eliminate Nasser's bomber force and so protect Israel from air attack. While the Israelis completed their occupation of Sinai, estimated to take 7–10 days, British and French forces could move from their bases in Malta and Cyprus, seize Port Said at the northern end of the Canal, and advance south to Suez on the Red Sea.[4] Deception would provide a *casus belli* for the French and British, who were intent on seizing the Canal and, they hoped, toppling their arch enemy, Nasser.

Anthony Nutting, who resigned from a cabinet post in the British Government because of his objection to the enterprise, described the intervention justification thus:

> . . . we were to take part in a cynical act of aggression, dressing ourselves for the part as firemen or policemen, while making sure that our fire-hoses spouted petrol and not water . . .[5]

Peter Calvocoressi, analysing the deception 10 years after the event, judged the British Cabinet's motives as follows.

> First, they wished to deceive the Arab states, who must not be allowed to think that Britain could associate with Israel; second, the British wanted to keep the United States in the dark, so that they would not thwart the operation; third, Prime Minister Eden saw the need of keeping some of his government colleagues misinformed, to prevent a major political show-

> down; and finally, Cabinet intended to hide things from the British public, which was anything but united over Suez.[6]

At the strategic level, the Israelis hid their mobilisation for the attack on Egypt behind simulated preparations for an attack on Jordan. The Israelis wanted to achieve strategic and politico-military surprise to prevent the Egyptians mobilising either their army or the diplomatic and military support of the Soviet Union, the Third World and public opinion in the West. Given the high level of incidents along the Israeli-Jordanian border, Israeli military preparations in that area would not appear unusual. Consequently, the deception was credible and successful. Not only were the Arabs and the United States taken in; the British, who should have known better, were also deceived.[7]

An additional component of Israeli deception during the initial phases of the operation was to broadcast ambiguous and incomplete information about the scope and nature of the operation. The purpose of this 'tactical' deception was not only to deceive Egypt and the Superpowers, but also to provide the Israeli Government with a fallback position in case France and Britain demurred or had second thoughts about their part in the operation. As a precaution for such a contingency, the statement by the IDF's spokesman gave the impression that, rather than being the early stages of an all-out war, the parachute operation at the Mitla Pass in the Sinai was only a routine reprisal raid of limited scope.[8]

The Israelis attacked on 29 October as planned. The ultimatum was issued next day, with results that had been foreseen. On the same day France and Britain each used their vetos in the UN Security Council twice to block ceasefire resolutions. On the 31st, Anglo-French air forces attacked Egyptian airfields. On 5 November British and French paratroops dropped at Port Said. By this time the IDF had already achieved its objectives in Sinai.

As the French and British paratroops dropped, the Soviet Union took the initiative. In messages sent by Prime Minister Bulganin to both the British and French Prime Ministers, the Soviets made vague hints about destroying England and France with nuclear missiles.[9] The Soviet note sent to the Israeli government was much more threatening. Bulganin suggested that the actions taken by the Israeli government endangered Israel's very existence. He added that the 'Soviet Government is taking at this very moment appropriate measures to stop the war and restrain the aggressors'.[10] To lend weight to these words, the KGB elaborated on a hint dropped by Khrushchev in August, and initiated an extensive disinformation rumour campaign 'leaking' to various newspapers and intelligence agencies in the world information concerning the sending of Soviet 'volunteers' to Egypt to serve as tank crews, pilots and infantrymen. The KGB campaign was backed in addition by a TASS anouncement making similiar threats. The goal of the Soviet deception campaign was to achieve a complete and unconditional withdrawal of the Israelis, British and French and present the Soviet Union as the most reliable ally of the Arab world. These rumours from 'unknown sources' increased in their intensity as the Suez campaign continued. The Israeli intelligence service, Mossad, even received information from what were considered to be 'reliable sources' that Soviet pilots had arrived in Egypt and were ready to attack targets in Israel.[11]

Peter Wright, the former British Security Service (MI5) officer, has described an operation that enabled the British to read the Egyptian cipher in that country's London Embassy throughout the Suez Crisis. The Russians, however, discovered what was going on and turned the arrangement to their own advantage. Aware that the British would probably accept covert intelligence at face value, the Soviets fed 'a continuous account of Egyptian/Soviet discussions in Moscow, details of which were relayed into the Egyptian Embassy in London direct from the Egyptian Ambassador in Moscow'.[12] The messages were of course, part of the disinformation designed to convince the British that the Soviet Union was indeed serious in its threat to become involved in the Suez Crisis on the Egyptian side. According to Wright, this ploy was effective, creating 'panic' and prompting Eden to withdraw.[13]

Reality, however, was more prosaic. With limited power projection capabilities at that time and with large forces already committed to suppressing unrest in Hungary and Poland, unilateral Soviet intervention in the Suez conflict was not likely. In fact, according to Stephen Kaplan, Soviet and Czechoslovak military advisers were withdrawn from Egypt to Sudan during the fighting. So, the intervention threat was only a deception.[14]

In the short run, all the deceptions of the Sinai-Suez campaign were successful. By the time of the Israeli paratroop drop at the Mitla pass, neither the Egyptians, United States nor Soviet intelligence had suspected the nature, scope, timing and destination of the attack. As a result, the IDF caught the Egyptians off their guard, and were able to achieve their objectives faster than the plan anticipated.[15] Soviet deceptions may or may not have been accepted at face value by the British, French and Israeli Governments, but the threats and sabre rattling contributed to an atmosphere of imminent major war thus strengthening French and particularly British domestic opposition to the adventure. Even in Israel, reports of Soviet 'volunteer' fighters were accepted as credible. Undoubtedly, it was American rather than Soviet pressure that forced the Allied decision to stop the fighting and eventually to withdraw. But in their propaganda to Egypt and the Arab and Third Worlds, the Soviets could at least claim to have saved Egypt.[16]

As for the Anglo-French collusion with Israel, this too enjoyed short-term success of a sort. It was not so much that the peace-keeping story was believed, rather that the secrecy and duplicity in the preceding weeks and days misled target audiences. Nutting has described how the Foreign Office deliberately kept their men in Washington in the dark, on the grounds that no one deceives better than one who is himself deceived. The operation went beyond mere secrecy: United States Ambassador Winthrop Aldrich was fobbed off with lies about British efforts to restrain Israel, and these lies were from people in London who did know better.[17] The terms of the ultimatum were so obviously loaded against Egypt that only Mr Eden's staunchest supporters took them seriously as a peace-keeping effort.

The key deception, concealing the collusion with Israel, however, did hold up quite well, although it required some brazen statements to hold it together. Asked in Parliament on 31 October whether there had been collusion, Selwyn Lloyd, the Foreign Secretary, replied that 'there was no prior agreement between us about it [Israel's attack]'.[18] And two months later, Eden would argue that to say 'that Her Majesty's Government were engaged in some dishonourable conspiracy is completely untrue . . .'[19]

In the longer run, only the Israelis and the Soviets benefited by their deceptions; the Israelis from their initial tactical, operational and strategic success, the Soviets from the opportunity to pose as the saviour of the Egyptian underdog. France and Britain, particularly the latter, paid the price for becoming involved in a project they were unprepared to see through to the end. A last imperial fling, having failed, contributed to the disintegration of colonial patterns.

Their deception also backfired on them. President Eisenhower and Secretary of State Dulles were personally outraged at having been cheated by their allies. The trust on which alliances are built was temporarily, but severely, shaken. Perhaps more serious was the distortion of policy brought about by the simulation that presented a false picture of 'peace-keeping'. Once the French and British had justified their aggression by this means, it became impossible for their leaders to resist proposals that their troops be replaced by a United Nations Emergency Force, who were real peace-keepers. London and Paris were 'hoist with their own petard'.

Israel discovered how problematic and unreliable alliances could be, and how even a close friend, America, could be instrumental in forcing the relinquishment of territorial gains. On the credit side, Israel had established a formidable military reputation and she obtained in the war's aftermath a decade of tranquillity based on deterrence.[20] In retrospect, this has been the longest period of tranquillity in Israel's history so far.

Endnotes

1. The major sources on the Suez Crisis include: A J Barker, *Suez: The Seven Day War* (London: Faber and Faber, 1964); André Beaufre, *The Suez Expedition—1956* (New York: Praeger, 1967); David Ben-Gurion, *Israel: A Personal History* (New York: Funk and Wagnalls, 1971); Moshe Dayan, *Diary of the Sinai Campaign* (London: Weidenfeld and Nicolson, 1966); Abba Eban, *An Autobiography* (New York: Random House, 1977), p. 211; Dwight D Eisenhower, *Waging Peace: 1956–61* (Garden City, N Y: Doubleday and Co., 1965); Derman Finer, *Dulles Over Suez* (Chicago: Quadrangle, 1965); Kennet Love, *Suez: The Twice-Fought War* (New York: McGraw-Hill, 1969); Anthony Moncrieff, ed., *Suez: Ten Years After* (London: British Broadcasting Corporation, 1966); Anthony Nutting, *No End of a Lesson: The Story of Suez* (London: Constable, 1967); Nadav Safran, *Israel, the Embattled Ally* (Cambridge: Belknap Press of Harvard, 1978); Hugh Thomas, *Suez* (New York: Harper and Row, 1967); Ezer Weizman, *On Eagles' Wings*, (Tel Aviv: Steimazky's, 1976).
2. Yosef Evron, *Be'yom Sagrir: Suez Me'akhorei Haklain*, (On a Rainy Day. Suez behind the Scenes), [Hebrew], (Tel Aviv: Otpaz 1968). This book is based on interviews with Shimon Peres, who accompanied Ben-Gurion to Sèvres. Chaim Herzog, *The Arab-Israeli Wars: War and Peace in the Middle East* (New York: Methuen, 1982), pp. 113–14 suggests that the initiative to involve Israel in Anglo-French planning came from the French.
3. Nutting, p. 105.
4. *Ibid.*
5. *Ibid.*, p. 94.
6. Moncrieff, p. 107.
7. Herzog, p. 117, describes the 'Jordanian' deception in considerable detail. Nutting, p. 110, asserts that the 'Jordan ploy' was, in fact, worked out with the British in advance, both to reinforce the Israeli strategic deception and to strengthen Britain's position as a peace-keeper. He does not substantiate this argument with evidence however.
8. Herzog, pp. 117–18.
9. Ephraim Karsh, 'Soviet Arms for the Love of Allah', *US Naval Institute Proceedings*, April 1984, pp. 44–50; see also, this volume, Chapter 4.
10. Quoted in *Maariv*, 6 November 1956.
11. Private information. On the Khrushchev hint, see this volume, Chapter 4. See also *TASS*, 6 November 1956, and subsequent .
12. Peter Wright, *Spycatcher: The Candid Autobiography of a Senior Intelligence Officer* (Toronto: Stoddart Publishing, 1987), p. 85.
13. *Ibid.*
14. Stephen S Kaplan *et al*, *Diplomacy and Power: Soviet Armed Forces as a Political Instrument* (Washington, D C: Brookings, 1981), pp. 154–5. Owing to obvious force structure, logistic and deployment limits, particularly in the airborne forces and air transport, Western analysts did not take the Soviet threats seriously. See Kenneth Allard, 'Soviet Airborne Forces and Pre-emptive Power Projection', *Parameters*, Vol. 10, No. 4 (December 1980), p. 43; Dennis M Gormley, 'The Direction and Pace of Soviet Force Projection Capabilities', *Survival*, Vol. 24, No. 6 (November/December 1982), p. 270.

15. For a detailed account of the Mitla pass operation, and the other battles of the Sinai campaign, see Herzog, pp. 117–38.
16. Kaplan, p. 155, and notes 15, 17; see also Joseph L Nogee and Robert H Donaldson, *Soviet Foreign Policy Since World War II* (New York: Pergamon, 1981), p. 105.
17. Nutting, pp. 110–13.
18. Quoted in Moncrieff, p. 84.
19. Quoted in Moncrieff, p. 85.
20. On this issue, see Zvi Lanir, 'Political Aims and Military Objectives—Some Observations on the Israeli Experience', in Zvi Lanir, ed., *Israeli Security Planning in the 1980s: Its Politics and Economics* (New York: Praeger, 1984), pp. 21–4; and Gunther E Rothenberg, 'Israeli Defence Forces and Low-Intensity Operations', in David A Charters and Maurice Tugwell, eds., *Armies in Low-Intensity Conflict: a Comparative Analysis* (London: Brassey's, 1989), p. 59.

CHAPTER 14

Nine Days in May: the U-2 Deception

DAVID A CHARTERS

Does an American, when he represents all Americans, have to tell the truth at any cost? The answer is yes . . . We can be dramatic, even theatrical; we can be persuasive; but the message . . . must be true.

Edwin Land[1]

On Sunday, 1 May 1960 a U-2 reconnaissance aircraft piloted by Francis Gary Powers took off from an airfield at Peshawar, Pakistan to begin a 3,788 mile flight that was supposed to take Powers over the heartland of the Soviet Union before landing at Bodo, Norway. It was Powers' first overflight of the Soviet Union; it was also to be his last, and the last for the U-2 programme. About half-way through the flight, over Sverdlovsk—some 1,300 miles inside Soviet airspace—a surface-to-air missile (SAM) brought down the U-2. Once the aircraft was overdue beyond its known endurance and assumed to be down somewhere in the Soviet Union, Washington was notified, and the cover story which had been prepared for such circumstances was approved for release. Unknown, however, to Washington, which had no means to determine the location or circumstances of the downing of the aircraft, the Russians had captured Powers alive. Furthermore, they had salvaged significant portions of his aircraft. Thus the SAM that downed the U-2 did more than just bring to an end a remarkable intelligence effort; it set in train a series of events that unravelled the deceptive cover which had protected it, in a manner that caused considerable embarrassment to President Eisenhower, and contributed to the subsequent collapse of the forthcoming Paris Summit.[2]

The U-2 programme had its roots both in the "Cold War" and in the collective American trauma over Pearl Harbor. Eisenhower was deeply concerned about the paucity of hard information about the Soviet Union and the danger of surprise attack arising out of such ignorance.[3] In the mid-1950s, the Soviet Union was, in intelligence parlance, a 'denied area'. Its strict system of internal population controls, hostility to and constant surveillance of foreigners inside Russia, and an almost impenetrable frontier, made it extremely difficult for Western intelligence services using traditional means (diplomats, attachés, and agents) to collect information about the Soviet Union, and thus to make accurate assessments and estimates about Soviet intentions, and equally important, about military developments, particularly in the fields of nuclear weapons, long range bombers and missiles.[4] And, as Hannes Adomeit observes elsewhere in this book, Nikita Khrushchev did not make the Western task any easier by his deliberate use of bluff and deception over the size and capabilities of Soviet forces during this period.[5]

To deal with this problem, Eisenhower convened a secret team of scientific advisers, the Technological Capabilities Panel (TCP), to explore new and innovative means to improve American intelligence collection on the Soviet Union. The Chairman of the Intelligence sub-committee was Edwin H Land, President of Polaroid Corporation, who drew to their attention the potential opened up by recent advances in camera technology. All that was required was a stable aerial platform with sufficient range and altitude, and the new high resolution cameras would do the rest. The U-2, a glider-like jet, designed to fly at extreme altitudes, offered just such a platform. After consulting CIA Director Allen Dulles and his very able Special Assistant, Richard Bissell, as well as the Assistant Secretary of the Air Force for Research and Development, Land and TCP Chairman James Killian took their recommendation to the President. He approved it in December, 1954. The U-2 was rushed into production in 1955 and the first operational flights were made in July, 1956 under the code-name IDEALIST.[6]

At the time the programme represented a technological revolution in intelligence collection by aerial photographic reconnaissance. During the period of nearly four years over which the programme functioned, vast areas of the Soviet Union were opened to inspection and the intelligence gained gave the United States valuable data and insights on Soviet military developments; it allowed the CIA, amongst other things, to revise downwards, by a considerable margin, its estimate of the Soviet strategic bomber programme, thus laying to rest—inside the intelligence community at least—the myth of the 'bomber gap'.[7] But overflying the Soviet Union—even relatively small portions of it, as was the case with many flights–involved considerable risk, and no one was more keenly aware of this than the President himself. Clearly impressed by the capabilities the system demonstrated in the testing phase, he was worried nonetheless about the consequences of an 'incident' should one of the aircraft be shot down. He admonished his intelligence advisers in this respect, prophetically as it turns out: 'Well boys, I believe the country needs this information, and I'm going to approve it. But I'll tell you one thing. Someday one of these machines is going to get caught and we're going to have a storm.'[8] As the programme continued, his fears increased. At a February, 1960 meeting of the National Security Council (NSC) the President opined that if a U-2 were lost at a time when the United States was engaged in 'apparently sincere deliberations', it could be put on display in Moscow and ruin his effectiveness.[9] The programme would have to be protected, and its intelligence function disguised. From the moment that the project was handed over to the CIA,the Agency went to great lengths to conceal the programme beneath several layers of deceptive cover.

The effective security and cover of the programme owed its initial success to the efforts of Richard Bissell who, with Dulles' blessing, ran the entire project 'out of his hip pocket'. A graduate of Yale and one-time acting administrator of the Marshall Plan, Bissell exhibited a genius for administration which manifested itself in its best light in his direction of the U-2 programme. He arranged for covert financing out of the CIA's Reserve Fund. Although the project ultimately involved several hundred people in the CIA, the U-2 squadrons, and others in the intelligence and defence community, Bissell did his utmost to limit knowledge to those who really had a 'need to know'. He kept his own staff very small. Communications were largely by word of mouth, thus keeping the number of documents to a minimum. The NSC, under whose authority the CIA operates, issued a general policy directive for the Agency to conduct the overflight programme but, as a body, was excluded from the project. The flights were rarely discussed at NSC meetings. Instead, Bissell created an informal inter-agency committee to co-ordinate intelligence collection requirements, consisting initially only of representatives of the three Services and the CIA; subsequently, it was expanded to include the State Department, the Joint Chiefs of Staff, and the National Security Agency. Similarly, authorisation of specific flights was closely held; in this, Bissell or Dulles consulted directly with the President and a few of his closest advisers.[10] The plane itself was constructed at a maximum security Lockheed plant in Burbank, California. No one who was not working on the aircraft was allowed near the site, which led the plant to acquire its now famous nickname: the 'skunk works'.[11]

It was one thing to enforce strict physical security at the production site, or to limit knowledge of the project and the source of its output. It was quite another to disguise an aircraft of such unusual design once it became operational. Sooner or later, it was going to be seen, and its existence would have to be acknowledged and explained; after all, only one or two countries could have built such an aircraft. What was required was a plausible cover story and, as Allen Dulles later told the Senate Foreign Relations Committee, it was recognised from the outset that the unusual nature of the project made plausible denial extremely difficult.[12] But the project's unusual nature also afforded a credible cover. General James Doolittle, then Chairman of the National Advisory Committee on Aeronautics (NACA—the predecessor of NASA), was approached in this regard. He persuaded Dr Hugh Dryden, then Director of NACA, to accept a CIA proposal that NACA provide the operational cover for the U-2 programme. In return, the CIA would allow NACA to use U-2s for genuine scientific research in conjunction with the United States Air Force Air Weather Service (USAF AWS).

Consequently, during the next four years, U-2s flew 264,000 miles on 200 flights for the benefit of NACA's scientific research. Valuable data was collected on upper atmospheric turbulence—essential knowledge for modern aircraft design. This programme (which was paid for by the CIA) provided an almost perfect cover for the U-2's intelligence gathering flights and explained its presence around the world. The secret was known to only Dryden and three other NACA/NASA officials.[13] To launch the programme and its deceptive cover, NACA issued a press release on 30 April 1956, which read in part:

> Tomorrow's jet transports will be flying air routes girdling the earth. This they will do at altitudes far higher than presently used except by a few military aircraft. The availability of a new type of airplane, the Lockheed U-2, makes possible obtaining the needed data (on gust-meteorological conditions) in an economical and expeditious manner . . . .[14]

The press release, which went on to praise the role of the USAF AWS, was a classic example of misdirection. Everything it said was true; but the most important truth—the covert intelligence operation—was hidden behind the cloak of overt meteorological flights. Consistent with the cover story for the programme as a whole, the two U-2 formations were designated respectively the

Weather Reconnaissance Squadron (Provisional) and the 2nd Weather Observational Squadron (Provisional) in NACA press releases announcing the extension of the weather programme to Europe. The pilots, all military fliers, had been 'sheep-dipped'—provided with new professional credentials to cover their assignments. Powers, for example, was made a civilian employee of the Air Force during the selection phase, then later was assigned to Strategic Air Command. Upon the acceptance to the U-2 programme, Powers was instructed to resign from the Air Force, and to sign a (secret) contract with the CIA. During the training phase he was a Lockheed employee on loan to NACA. Before deployment overseas, he and the others were given yet another identity, using their real names, but once again as civilian employees of the Air Force working for NACA. They were allowed to tell family and friends that they were going overseas as part of NACA's weather study project and, if necessary, could hint that it was in conjunction with the forthcoming International Geophysical Year.[15]

In the event that a U-2 crashed while on a mission, the general policy was to describe the missing aircraft as NACA/NASA plane. The CIA had also drafted pre-packaged cover stories to suit the geography and circumstances of any foreseeable incident. These consisted of 'innocuous, short announcements to be plucked out of the files and issued by Air Force public information officers' at the bases concerned. In the case of overflights of the Soviet Union, the cover story 'was intended to be of use particularly if the plane had come down in the border areas where one could plausibly have said that the plane had gone over, strayed over the frontier without meaning to'.[16] Until Powers was shot down, the cover stories had never been used.

This appears somewhat surprising since, as early as the training phase and with a certain regularity thereafter, the U-2 attracted an unseemly amount of attention through a series of crashes and other exposures. One of the most spectacular and unsettling of these occurred in Japan in September 1959, when a U-2 on a test flight ran out of fuel and made an emergency landing at a glider club 10 miles short of its base at Atsugi. Curious locals took many pictures until American military police arrived and forced everyone away at gunpoint. Naturally, the Japanese press commented on the peculiarity of the weather plane and its pilot, since neither carried any identification and occasioned such extreme security precautions. A startling disclosure had come more than a year earlier from a most unlikely source; the March 1958 issue of *Model Airplane News* carried a short article on the U-2, complete with three simplified diagrams, and reported that 'An unconfirmed rumour says that U-2s are flying across the Iron Curtain taking aerial photographs.'[17]

That secret was supposed to be very closely held, and its exposure in print—even as an unconfirmed rumour—must have caused some concern among those 'in the know'. Yet, it is clear that the reconnaissance programme was no secret to the Russians. After the first flights over the Soviet Union in July 1956, the Soviet Foreign Ministry quietly lodged a protest with the United States concerning the intrusions. The United States Government did not reply to the note, which was then followed up by a TASS bulletin on 10 July mentioning a Soviet protest over violations of its airspace. Four days later, at a French Embassy reception in Moscow, Premier Nikolai Bulganin questioned United States Ambassador Charles Bohlen about the matter. Bulganin, who claimed to have indisputable evidence from radar tracking, emphasised the serious nature of such violations. Bohlen had general knowledge of plans for the flights, but no details, and gave nothing away. On 23 July, Marshal R A Rudenko, Chief of Staff of the Soviet Air Force, told Western reporters that 'all necessary measures' would be taken if the intrusions continued. The Americans halted the intrusive flights temporarily, after the first six had been completed.[18] Early in 1958, Pyotr Popov, a GRU (Soviet Military Intelligence) officer and CIA 'mole' stationed in East Germany, told his Agency 'controller' that a visiting senior officer had bragged that the KGB had full technical details on 'a special high-altitude aircraft CIA had been flying over the USSR'.[19] That Soviet Intelligence had indeed learned something about the U-2 was confirmed a few months later. In May 1958, *Soviet Aviation*, the official daily newspaper of the Soviet Air Force, published a series of articles about the U-2, which it dubbed 'the black lady of espionage'. Powers implies that much of the information the newspaper published was either incorrect or outdated.[20]

Taken together, the two reports do not necessarily add up to a major security breach inside Bissell's project team. After all, the Soviets had made it clear that their radars could track the U-2 from the start, and that would have given them some data on the aircraft's performance. Eastern Bloc diplomats had frequently observed U-2s taking off from their bases in Germany. The Soviets would have been able to gather scraps of information from official statements (such as the NACA press releases) and from the open literature, perhaps even including *Model Airplane News*. That might go some way to explain the apparently doubtful accuracy of the data published in the Air Force newspaper. On the other hand, it is not in the nature of the Soviet system to be very forthcoming with important intelligence in large circulation publications, even inside their armed forces. If they had genuine and accurate information on the U-2, they would not publish all that they had if only to avoid alerting the Americans to the extent of their information. And if that information had come from a highly placed 'mole' of their own, they would have had even greater incentive to be circumspect or, more likely, deliberately misleading. In any case, the important point is that the Soviet military and their political leadership were aware of the American overflights from the outset.[21] This fact helps to place in perspective Khrushchev's reaction to the Powers incident; if nothing else, *he* was not taken by surprise.

Rather, it was the United States Government that was taken by surprise when the Soviet Government announced that it had captured the pilot alive and had recovered large portions of the plane intact. There were several reasons for this. First, the CIA believed that Soviet SAMs were likely to score only a near miss against a U-2 at its normal operating altitudes—68,000 feet or higher—and that this would not be sufficient to bring the plane down. Second, the Agency assured President Eisenhower that the U-2's fragile airframe could not withstand a direct hit by a SAM; rather, it would disintegrate into pieces too small to be useful in identifying the plane's origins or function. Nor was the pilot expected to survive under such circumstances. Third, the aircraft was equipped with a destruct mechanism, a small explosive charge which was intended to destroy the sensitive intelligence equipment (particularly cameras and film) if the pilot had to bail out over hostile territory. Finally, some of those involved in the programme—although apparently not Bissell—may have shared the view of Secretary of State John Foster Dulles that the Soviets would never admit to having captured a U-2, since to do so would require them to admit that the United States had been conducting such flights for years while the Soviets had been unable to do anything about it.[22]

In retrospect, these assumptions appear to have rested on rather flimsy foundations, in view of past Soviet protests about the intrusions, genuine and realistic concerns about the effective range of the SA-2 SAM, and doubt's about the ability of the destructive device to destroy rolls of film. Nonetheless, they undoubtedly explain why the preparation of the pilots for these eventualities was so completely at odds with the carefully constructed cover story for the programme as a whole. Although provided with new operational identities, there was no mistaking the pilots for who they were. Their survival kit and clothing was unquestionably

American; in fact, it included a large silk American flag poster bearing the following message in 14 languages: 'I am an American and do not speak your language. I need food, shelter, assistance. I will not harm you. I bear no malice toward your people. If you help me, you will be rewarded.'[23] The very existence of a survival kit would have been enough to arouse suspicion that this was no ordinary 'innocent civilian'; that the kit included a silenced pistol and a curare-tipped 'suicide' pin was a dead giveaway for a covert intelligence operation. It is hardly surprising then that when Powers asked the squadron intelligence officer what he should say if forced down over the Soviet Union, he was told: 'You may as well tell them everything because they're going to get it out of you anyway.'[24] A CIA guidance to U-2 pilots on the subject of capture, made public in 1962, indicates clearly that pilots were to adopt a co-operative attitude toward their captors, and to feel free to tell the truth about the mission, including their employment by the CIA.[25] The complete disjunction and contradiction between the pilots' preparation and the cover story seems to have escaped everyone. Consequently, when the Russians captured Powers alive, and recovered large sections of the aircraft and its intelligence equipment intact, but then concealed for a week what they really knew, the stage was set for an American deception operation to boomerang on its initiators.

On the afternoon of Sunday, 1 May, Hugh S Cumming, Chief of Intelligence and Research at the State Department, was notified that Powers was overdue at Bodo and was presumed down inside Russia. Cumming met with Bissell and others at the U-2 project office. They decided to release the pre-planned cover story, modified to show an intended flight plan that would have kept the plane inside Turkish airspace. Allen Dulles was told upon his return to Washington later that afternoon. The CIA contacted the NASA Administrator and his deputy Hugh Dryden to warn them. General Andrew Goodpaster, Eisenhower's Staff Secretary, called the President at Camp David and advised him that a Turkish-based American reconnaissance aircraft was overdue and possibly lost. Eisenhower realised immediately that he was referring to a U-2, probably over Russia.[26]

With Eisenhower's approval, the cover story was issued the following day by the information office at the USAF base at Incirlik, Turkey, Powers' home station. The story, first carried in the Turkish newspaper *Yeni Istanbul* on 3 May, was that a U-2 weather reconnaissance aircraft based at Incirlik was reported missing on Sunday, that the pilot had reported a breakdown in his oxygen equipment, and that nothing further had been heard from the plane thereafter. The story went on to say that the plane might have made a forced landing in the Lake Van area, but that search planes had failed to confirm this. The story was also picked up by the Istanbul bureau chief for United Press International, who cabled the story to New York. On 4 May it appeared as a three inch item on page one of the *Washington Post*, but was dropped in the final street edition in favour of a baseball headline.[27]

In the meantime, efforts were underway among the agencies concerned in Washington to co-ordinate the response to inquiries anticipated from the American news media. On the afternoon of 2 May, Major James Smith of Headquarters USAF AWS called Harry Press, Chief of Loads and Structures Division at NASA and one of the four cleared on the U-2's real mission. Smith told Press the details of the cover story which was being released in Turkey and Press then passed on the information to Dryden and to Walter Bonney, NASA's director of public information. Since the operation was not NASA's, they checked with the CIA as to what they should say in response to questions. During the next two days NASA and the CIA jointly drew up a list of likely questions and plausible answers, to be used when required. Bissell himself apparently was directly involved in this process. Simultaneously, the CIA was consulting with the State Department about the information to be given to NASA to help them answer questions. At no time, however, did the State Department and NASA work together on this matter,[28] a fact which was to cause some difficulties in the days to come.

On Thursday, 5 May, the cover story began to unravel. In a speech to the Supreme Soviet, Premier Nikita Khrushchev announced that an unmarked American plane had been shot down after violating Soviet airspace. He added that an earlier American flight had violated the Soviet frontier, but had been permitted to depart without military action or protest. In the course of his long harangue he denounced the act as an 'aggressive provocation' and suggested the militarist forces in the United States might have carried out the operation on their own initiative, behind the President's back, in order to wreck the forthcoming Paris summit.[29]

News of Khrushchev's speech reached the key American officials at an inopportune moment. They were taking part in a nuclear emergency simulation exercise at an underground command post in Virginia. After the formal NSC meeting the President convened a smaller group, including Dulles, Goodpaster, Under-Secretary of State Douglas Dillon, and Secretary of Defense Thomas Gates. Raising the possibility that the Soviet Government might have physical evidence, Gates expressed concern that the President could be caught out in a lie. In spite of these reservations, however, the group decided to maintain the existing cover story. It was also decided that the State Department would handle all questions relating to the matter. Unfortunately, this decision was not communicated to NASA; nor was the President's Press Secretary James Hagerty, informed of this until the President returned to the White House.[30]

In the event, confusion prevailed. Shortly after noon, Hagerty announced to the press corps, who had beseiged him all morning, that the President had asked for a complete inquiry, and that the results of this inquiry would be made public by NASA and the State Department. It is not clear whether the reference to NASA was intentional or inadvertent, but NASA spokesman Walter Bonney soon found himself hounded by reporters expecting him to make some kind of statement. He fended them off until NASA had time to take the previously agreed list of questions and answers and redraft them into the form of a prepared text. Dryden later told the Senate Committee that this approach was thought to have been a better alternative than to engage in a question and answer 'free for all'. He was given the impression by the CIA that the statement had been cleared with the State Department, but he was misinformed. While the Agency had advised State of the general guidelines of the original question and answers given to NASA, State was unaware that NASA was going to make a formal and rather complete statement. When it was issued, at 1.30 pm on 5 May it came, in Dillon's words, as 'somewhat of a surprise'.[31] For in the meantime, the State Department, operating under the assumption that it was the 'lead agency' on this matter, had issued its own brief to reporters 45 minutes earlier. At 12.40 pm, State Department spokesman Lincoln White, who was not cleared for the truth on the U-2, read the following statement:

> The Department of State has been informed by NASA that as announced May 3 an unarmed plane, a U-2 weather research plane based at Adana, Turkey, piloted by a civilian, has been missing since May 1. During the flight of this plane, the pilot reported difficulty with his oxygen equipment. Mr Khrushchev announced that a US plane has been shot down over the USSR on that date. It may be that this is the missing plane. It is entirely possible that having a failure in the oxygen equipment, which could result in the pilot losing consciousness, the plane continued on automatic pilot for a considerable distance and accidentally violated Soviet air space. The United States is taking the matter up with the Soviet Government, with particular reference to the fate of the pilot.[32]

White then went on to describe the U-2 as a test plane used for upper atmospheric research, with a ceiling of 55,000 feet and an endurance of three to four hours. U-2 pilots, he said, were Lockheed employees on contract to NASA. In response to a question about the earlier (9 April) intrusion, White replied, 'We have absolutely no—N-O, no—information on that at all'.[33]

White's account was an almost total fabrication. There had been no contact with Powers, no report of difficulty with oxygen equipment. He stuck to the NASA cover story, and threw in an understatement of the U-2's altitude and endurance capabilities. Bonney's four page NASA press release, which received very full coverage in the *New York Times*, embellished this fabrication with a notional triangular flight plan of some 1,400 miles, all within Turkish airspace. He provided elaborate descriptions of the legitimate NASA weather research programme and the instruments used, and denied that the plane carried any military equipment. He added that a search was still being carried on in the Lake Van area of Turkey, where the aircraft was thought to have disappeared.[34]

Later that afternoon, with the cover story fully launched and the United States Government publicly committed to it, the administration received unsettling news. The Moscow Embassy reported that rumours were circulating there that the pilot had been captured alive. According to Dillon, 'it gave us pause'.[35] On the following day *Trud* published a photo of aircraft wreckage, which it claimed was the U-2. American experts quickly identified the wreckage as being from a Soviet-made aircraft. Nevertheless, it was argued in Washington that unless the Soviets actually had U-2 wreckage, they would not have made the claim. Dillon later testified that, from that point on, the Administration acted on the assumption that the Soviets did indeed hold Powers and probably had parts of the U-2.[36]

If this was the case, the Administration's subsequent actions displayed stubborn inflexibility. Clearly, this was the moment when the President might have recovered the initiative by abandoning the cover story and justifying the U-2 programme on grounds of national security. Instead, the Administration blundered deeper into the trap that was being set for it. NASA announced that U-2s were being grounded while their oxygen systems were checked. The unfortunate Lincoln White, innocent of the conspiracy and lacking direction from his superiors in the State Department, stuck grimly to a line which was rapidly losing credibility. 'There was no deliberate attempt to violate Soviet air space', he said, 'and there never has been'.[37]

The next day, 7 May, Khrushchev sprang his trap. After paraphrasing the State Department's 5 May statement, he told the Supreme Soviet:

> Comrades, I must tell you a secret. When I was making my report [on May 5th] I deliberately did not say that the pilot was alive and in good health and that we have got parts of the plane. We did so deliberately because had we told everything at once, the Americans would have invented another version. And now just look how many silly things they have said—Van Lake, scientific research and so on and so forth. Now that they know that the pilot is alive they will have to invent something else and they will do it.[38]

He went on to state that the U-2 had been shot down over Sverdlovsk, more than 2,000 kilometres inside Soviet air space. It was clear from Khrushchev's speech that Powers had been very thoroughly interrogated and much of the plane's intelligence-gathering instrumentation had been captured intact. Powers' CIA employment, his unit and commander, his flight path and destination were described accurately, and the falsehood about oxygen difficulties was demolished by Powers' own confession. Khrushchev also displayed aerial photographs of Soviet installations supposedly developed from the film in the U-2's cameras. And he ridiculed the American cover story with a personal dig at the CIA chief: 'The whole world knows that Allen Dulles is no great authority on meteorology.'[39]

Clearly some American response was called for. White again found himself deluged by calls from newsmen asking for a State Department reply to Khrushchev's charges, but he couldn't locate anyone to give him guidance. The key officials—Secretary of State Christian Herter, Dillon, Hugh Cumming, Goodpaster, Foy D Kohler and Charles E Bohlen—were closeted in Herter's office trying to reach a consensus on how far the Government should go in admitting the truth. Herter and Kohler felt that the cover story had been demolished and that American responsibility could no longer be denied. Others were uncertain. When the President was consulted he apparently wished to accept full responsibility for ordering the flight. Herter however, dissuaded him and, according to Wise and Ross, the outcome was the insertion into the public statement of a sentence dictated by Eisenhower denying that the overflight had been authorised by Washington,[40] the worst compromise imaginable.

At 6.00 pm Lincoln White appeared before the media to read a statement which contradicted all his earlier remarks. Before offering a justification 'after-the-fact' for intelligence collection behind the Iron Curtain, White's statement made two important assertions:

> As a result of the inquiry ordered by the President, it has been established that insofar as the authorities are concerned, there was no authorisation for any such flights as described by Mr Khrushchev. Nevertheless, it appears that in endeavouring to obtain information now concealed behind the Iron Curtain a flight over Soviet territory was probably undertaken by an unarmed civilian U-2 plane.[41]

The statement was deliberately evasive, and at least partly untrue. The President had authorised the flight, and Allen Dulles later confirmed that he had given the go-ahead himself. So White's statement compounded the Administration's problems by unnecessarily raising the spectre of an intelligence service out of control. The *New York Times*' James Reston concluded that if the flight had not been authorised in Washington then 'it could only be assumed that someone in the chain of command in the Middle East or Europe had given the order'.[42]

Overnight, Eisenhower had a change of heart, feeling that he had to 'come clean' with the American people. He consulted with Herter, and a new statement was drafted. It was issued by White mid-afternoon on Monday, the 9th, under Secretary Herter's name. After a lengthy justification of the requirement for such intelligence, the statement stated that the President had put into effect 'directives to gather by every possible means the information required to protect the United States and the Free World against surprise attack and to enable them to make effective preparations for their defence'.[43] The statement added that the programme included 'extensive aerial surveillance . . . normally of a peripheral character but on occasion by penetration'.[44] It emphasised, quite incorrectly, that the President did not authorise specific flights himself.

With that the U-2 deception operation was effectively at an end. Khrushchev's skilful concealment of the extent of Soviet knowledge had neatly entrapped the President of the United States. The Americans were encouraged to destroy their own credibility through a cover story that could only have worked, and then only up to a point, had the U-2 been totally destroyed and the pilot been killed. Eisenhower's gut feeling all along had been to rely on secrecy for so long as flights were successful, and to rely on the security benefits of the programme to justify the flights if a U-2 were to be lost over Russia. His offence was to allow his sound instincts to be overruled by short-sighted and not very competent advisers, all of whom allowed their obsessions with the short-term to blind them to long-term considerations.

Poor deception planning can be detected in failures to heed the evidence that the Soviets were aware, at an early stage, of the existence of U-2 and its probable true mission. The first Soviet protest provided ample warning that the dissimulation part of the deception—the hiding of the real—had collapsed. Without dissimulation, the chances of successful simulation—the presentation of a false picture—are virtually nil. Additionally, the cover story was designed for U-2 overflights close to the Soviet borders; it was deficient for deep penetrations. Yet no one accepted the implications: it was trundled out and used in a quite reckless manner. Finally, when it was abundantly clear that the deception had failed and was about to be exposed, the President was persuaded to confound the folly with the 'no authorisation' nonsense, implying that he was not in control of his own intelligence establishment.

The result of Eisenhower's errors and Khruschev's trap was a major American diplomatic embarrassment. The sole intended victim of the United States deception had been Moscow. But the Administration's fixation with a flawed cover story and its very public exposure by Moscow created the unwanted secondary effect of an American public angered and dismayed by government mendacity. In Congress, these reactions were bipartisan, heightened no doubt by the occasion of a Presidential election year. Republicans and Democrats alike expressed concern about the wisdom of the timing of the flight—just before the Paris Summit, the apparent lack of co-ordination evidenced by the issuing of statements by two government agencies—State and NASA, the former ultimately repudiating the latter—and most important, the damage done to the moral stature of the President—and the United States itself—by being caught out in a transparent lie, thereby handing the Soviets a propaganda victory. Editorial comment in the major American newspapers and views of foreign leaders, especially American allies, reflected similar concerns and anxieties. NASA officials Dryden and Bonney disavowed any knowledge of the U-2's real mission, and pointed to the USAF AWS as the source of the original deception. 'I thought I was telling the truth', Bonney protested. He left government service six months later, apparently anxious to put the U-2 mess behind him.[45]

It has since become an article of faith among diplomatic historians and foreign policy analysts that the U-2 incident 'sabotaged' the Paris Summit which convened on 16 May.[46] Certainly, Eisenhower's worst fears were realised. Khrushchev came to Paris in a belligerent mood. He accused the United States of acts of aggression, and demanded an apology, the suspension of overflights, and the punishment of those responsible as pre-conditions for proceeding with the Summit. Eisenhower replied that the overflight programme had been suspended, but he refused to apologise. The two leaders traded charges at their respective press conferences, and the Summit broke up without ever really getting started.[47] It is clear that the meeting would have been a difficult one at best, owing largely to the intractable issue of Berlin. Khrushchev clearly recognised this well before the U-2 incident occurred, as his public statements on the issue and his hopes for the Summit became increasingly pessimistic. That he seized upon the U-2 incident as a pretext for disrupting a meeting which was unlikely to produce the results he desired, appears quite plausible. Eisenhower, too, had come to have doubts beforehand, about the likelihood of its success. Nevertheless, on reflection in retirement, he came to realise that the U-2 incident had made the failure more likely and that the fault was largely his own. 'The big error', he concluded ruefully, was 'the issuance of a premature and erroneous cover story. Allowing myself to be persuaded on this score is my principal personal regret . . . regarding the whole affair.'[48]

In realising this he was more honestly perceptive than many of his colleagues, but not everyone has found the President's *ex post facto* candour convincing. Historian James A Nathan argued in a 1975 article that the anomalies in the case—including the timing of the flight, the wholly inadequate cover story, and the Administration's behaviour during the incident—tend to support the hypothesis that the Eisenhower Administration staged the incident deliberately in order to sabotage a détente that would not serve American interests at that time. This 'revisionist' interpretation posits a deception of considerably greater proportions than that of a mere cover story.[49] More recent scholarship, however, has challenged Nathan's hypothesis. Michael Beschloss, whose full-length treatment of the affair must be regarded as definitive, dismisses the allegation and points out at some length the benefits that would have accrued to Eisenhower from a successful Summit.[50] Nevertheless, the impact of the exposed cover story on American attitudes towards their own government was significant and clearly negative. Victor Marchetti and John Marks suggested in their 1974 book that 'for much of the American press and public it was the first indication that their government had lied, and it was the opening wedge in what would grow during the Vietnam years into the "credibility gap".'[51] Intelligence critic Morton Halperin echoed these sentiments the following year when he told the Church Committee, 'We should not forget that the erosion of trust between the government and the people in this Republic began with the U-2 affair . . .'[52] It may be fair to suggest that it was this erosion of trust, as much as the existence of hard evidence, that would create the climate of opinion in which conspiracy theories such as Nathan's would gain credibility. At the time, however, the practitioners' concerns were purely operational. Characteristic perhaps of official perceptions at the time was the view of Secretary of State Herter. Asked by the Chairman of the Foreign Relations Committee what lessons the Administration had learned from the experience, he replied, 'Not to have accidents?'[53]

Endnotes

1. Commencement Address, MIT, June 1960, quoted in James R Killian Jr, *Sputnik, Scientists and Eisenhower: a Memoir of the First Special Assistant to the President for Science and Technology*. (Cambridge, Mass.: MIT Press, 1977), pp. 84–5.
2. The definitive scholarly account of the incident and its consequences is Michael R Beschloss, *Mayday: Eisenhower, Khrushchev and the U-2 Affair* (New York: Harper and Row, 1986).
3. *Ibid.*, p. 74; see also, Dwight D Eisenhower, *The White House Years: Waging Peace 1956–61* (Garden City, N Y: Doubleday, 1965), pp. 544–5; and Stephen Ambrose with Richard Immerman, *Ike's Spies: Eisenhower and the Intelligence Establishment* (Garden City, N Y: Doubleday, 1981), p. 267.
4. The difficulties the Americans encountered using human sources to collect intelligence in the USSR are discussed in: Jeffrey Richelson, *American Espionage and the Soviet Target* (New York: Morrow, 1987), pp. 42–55; John Prados, *The Soviet Estimate: US Intelligence Analysis and Russian Military Strength* (New York: Dial Press, 1982), pp. 25–6; William Hood, *Mole* (New York: Norton, 1982), pp. 181–4.
5. See Chapter 4: see also, Prados, pp. 41–3.
6. Prados, pp. 30, 33; Killian, pp. 81–2; Ambrose and Immerman, pp., 267–8; Jeffrey Richelson, *The US Intelligence Community* (Cambridge, Mass.: Ballinger, 1985), p. 318.
7. Allen Dulles, Director of Central Intelligence, Testimony to US Senate, Committee on Foreign Relations, 31 May 1960, Report of Proceedings, 'Hearing Regarding Summit Conference of May 1960 and Incidents Relating Thereto', Vols. I, II microfilm (Frederick, Md.: University Publications of America), Reel 1, frames 760–61, 763. Beschloss, pp. 314–16, 410 warns historians against taking any portion of the Hearings testimony entirely at face value, but also points out that it does comprise the basic primary source on the incident. It is noteworthy, for example, that Dulles' assessment of the value of the programme has since been supported by the independent analysis. See for example, US Senate, Select Committee to Study Government Operations with Respect to Intelligence Activities, *Book 1: Foreign and Military Intelligence* (Washington, DC: US Government Printing Office, 1976), pp. 112–13; see also, Prados, pp. 42, 62–3, 87; Mark Lowenthal, *US Intelligence: Evolution and Anatomy*, The Washington Papers, No. 105 (Washington, DC: Center for Strategic and International Studies, Georgetown University, 1984), p. 27; John Lewis Gaddis, *Strategies of Containment: A Critical Appraisal of Postwar American National Security Policy* (New York: Oxford University Press, 1982), pp. 186–7; and Walter Laqueur, *A World of Secrets: the Uses and Limits of Intelligence* (New York: Basic Books, 1985), p. 145.
8. Quoted in Herbert S Parmet, *Eisenhower and the American Crusades* (New York: Macmillan, 1972), p. 538. These may or may not have been the President's exact words, but they express his sentiment on the subject. See Beschloss, pp. 8–9.
9. Richelson, pp. 304, 310, n. 30. Note the discrepancy between the date cited in this text and the date on the document as cited as the source.
10. William R Corson, *The Armies of Ignorance: the Rise of the American Intelligence Empire* (New York: Dial Press, 1977), pp. 374–5; Killian, p. 82; David Wise and Thomas Ross, *The U-2 Affair* (New York: Random House, 1962), pp. 42, 45, 48, 54–5; see also, correspondence, Edward McCabe to Jerry Morgan and Gordon Gray, 16 June 1960, from Official File 225-G, Eisenhower Papers, in Robert L Branyan, and Lawrence H Larsen,

eds., *The Eisenhower Administration 1953–1961: a Documentary History* (New York: Random House, 1971), Vol. 2, p. 1253. On the question of flight authorisations, Ambrose and Immerman, pp. 272–3, drawing on an interview with Bissell, have him consulting directly with the President and his advisers. Prados, p. 35, who does not give a source, says flight requests were cleared first by the Secretary of State; Bissell would then take the request to General Andrew Goodpaster at the NSC, who served as Eisenhower's Staff Secretary. The President himself apparently discussed the risks and opportunities of each flight programme with the Secretaries of State and Defense. Likewise, Beschloss, pp. 6, 140, places the President at the centre of the flight approval process.

11. Wise and Ross, p. 45; Ambrose and Immerman, p. 269.
12. Dulles, testimony, 31 May 1960, frame 782.
13. *Ibid.*, frame 776; Dr Hugh Dryden, Deputy Administrator of NASA, testimony, 31 May 1960, frames 946–7, 949, 951, 989, 991; Wise and Ross, pp. 50–52. Doolittle, along with Killian and Land, was also a member of the President's Board of Consultants on Foreign Intelligence Activities.
14. Quoted in Wise and Ross, p. 51. According to Dryden's testimony, frame 968, the first NACA press release on the U–2 was issued 7 May 1956, one week later than the date given by Wise and Ross.
15. Francis Gary Powers with Curt Gentry, *Operation Overflight: the U-2 Spy Pilot Tells his Story for the First Time* (New York: Holt Rinehart and Winston, 1970), pp. 21, 24, 26, 33, 38; the term 'sheep-dipping' is in Prados, p. 31. Powers' squadron (the 2nd) was officially known as 10–10 Detachment.
16. Dulles, Dryden, testimony, frames 789, 957; Wise and Ross, p. 31.
17. Wise and Ross, pp. 53–4; Powers and Gentry, pp. 65, 67–8. Dulles, in his testimony, frame 755, claimed that there had not been any damaging disclosures, and that the U-2's full capabilities had not been exposed. Beschloss, pp. 56–7, 234, points out that this was due largely to voluntary restraint on the part of the US media, a number of whom had discovered the truth about the U-2 missions.
18. Prados, pp. 33–4.
19. Hood, p. 251.
20. Wise and Ross, pp. 53-4; Powers and Gentry, p. 65.
21. Powers and Gentry, p. 61; Christian Herter, Secretary of State, and Dulles, testimony, 27, 31 May 1960, frames 530, 665–6, 673–4, 797–8. Both men emphasised that the US Government was aware that the Soviets knew about the over-flights. Beschloss, pp. 236–7, suggests that Lee Harvey Oswald, a radar controller at Atsugi before his defection to the USSR, probably did not know enough to have passed on information of exceptional intelligence value. However, James Bamford, *The Puzzle Palace: America's National Security Agency and its Special Relationship with Britain's GCHQ* (London: Sidgwick and Jackson, 1983), p. 144, states that two NSA analysts who defected in 1960 did have access to information on U-2 flights, including Powers'.
22. Eisenhower, p. 546; Ambrose and Immerman, pp. 279–82.
23. Ambrose and Immerman, p. 284; Prados, pp. 96–7. Dulles, Testimony, 31 May 1960, frame 769, confirms that even if the destruct device had been used, sizeable and identifiable parts of the plane might be expected to survive the explosion. See also, Powers and Gentry, p. 45.
24. Powers and Gentry, pp. 46, 69, 76.
25. *Ibid.*, p. 315, wherein the document is quoted. Dulles, testimony, frame 774, stated that pilots were to delay as long as possible the revelation of any damaging information, but to tell the truth about those matters obviously within the knowledge of their captors; and if some attribution was necessary, a pilot could say he was working for the CIA.
26. Douglas Dillon, Under-Secretary of State, and Dryden, testimony, 27 and 31 May 1960, frames 522, 955-6; Eisenhower, p. 543; Wise and Ross, pp. 28–31, 35–7; Beschloss, pp. 32–3, 427–8. Wise and Ross have Cumming meeting with Dillon, but Beschloss cites evidence indicating that Cumming informed Dillon of the group's decision, after the 1 May meeting. This may explain an apparent discrepancy in Dillon's testimony—his assertion that it was decided that the previously prepared cover story would be released, without mentioning any modifications.

27. Dillon, Dryden testimony, frames 522, 956–7; Beschloss, p. 37; Wise and Ross, pp. 61–2, 64–5.
28. Dillon, Dryden testimony, frames 525–6, 956–8, 974; Wise and Ross, pp. 62–3.
29. *New York Times*, 6 May 1960; Ambrose and Immerman, pp. 284–5.
30. Dillon, Dryden testimony, frames 524, 970–71; Prados, pp. 98–9; Wise and Ross, pp. 78–80.
31. Dillon, Dryden testimony, frames 649–50, 662, 958–60; Wise and Ross, pp. 80–81, 84. Eisenhower, p. 549, gives the impression that he told Hagerty to make reference to both NASA and State, while Dillon testified that he didn't know who decided that NASA should make a statement. According to Beschloss, pp. 50–51, Goodpaster later felt he was to blame for the confusion, since it was he who called NASA to suggest that the CIA-inspired questions and answers be redrafted as a statment.
32. Dillon, testimony, frames 586–8.
33. *Ibid.*, frames 547–9 'Excerpt from Press and Radio News Briefing'.
34. *New York Times*, 6 May 1960; Wise and Ross, pp. 84–5.
35. Dillon testimony, frames 585–6. Wise and Ross, p. 87, say the warning from the embassy reached Dillon at 6.00 pm on Thursday. Beschloss, p. 54, sets the time at 1.34 pm, but still too late to prevent release of NASA's statement.
36. *Ibid.*, frames 528, 586. Copy of the *Trud* photograph in the *New York Times*, 7 May 1960.
37. *New York Times*, 7 May 1960; Dillon, testimony, frames 587–9; Beschloss, p. 58. The wording of the statement in the *Times* differs slightly from the transcript in the presidential papers.
38. *New York Times*, 8 May 1960.
39. *Ibid.*; the quote is in Wise and Ross, p. 98.
40. Wise and Ross, pp. 104–7. Ambrose and Immerman, p. 286; Eisenhower, p. 550 makes no mention of the statement denying that flight had been authorised in Washington. See also Beschloss, pp. 243–4, 246–8.
41. 'Text of the US Statement on Plane,' *New York Times*, 8 May 1960.
42. Dulles, testimony, frame 758; *New York Times*, 8 May 1960.
43. 'Herter Statement on U-2 Flight,' *New York Times*, 10 May 1960. See also Eisenhower, p. 550; Wise and Ross, pp. 113, 115.
44. *New York Times*, 10 May 1960.
45. *New York Times*, 9 and 10 May 1960; see also, 'Senate Speech by J William Fulbright (Chairman, Senate Foreign Relations Committee) on the U-2 Crisis and the Paris Summit Meeting' and 'Report of the Committee on Foreign Relations on Events Relating to the Summit Conference' 28 June 1960, in Branyan and Larsen, pp. 1254–77; and comments by Senators Gore and Fulbright in Senate, Report of Proceedings, frames 738, 1011–12. See also, Wise and Ross, p. 120.
46. See, for example, Gaddis, pp. 196–7; John Spanier, *American Foreign Policy Since World War II*, 8th ed. (New York: Holt, Rinehart and Winston, 1980), pp. 102–3; Alexander De Conde, *A History of American Foreign Policy*, 2nd ed. (New York: Scribners, 1971), pp. 797–9; and Hannes Adomeit, *Soviet Risk-Taking and Crisis Behavior: A Theoretical and Empirical Analysis* (London: George Allen and Unwin, 1982), p. 199; and Beschloss, pp. 376, 380–81.
47. Elmo Richardson, *The Presidency of Dwight D Eisenhower* (Lawrence, Kansas: Regents Press, 1979), pp. 174–5; Robert A Divine, *Eisenhower and the Cold War* (Oxford and New York: Oxford University Press, 1981), p. 150–51.
48. See Herter, Testimony, frames 460–62, 475–8; Ambrose and Immerman, 287; Divine, pp. 140, 147; Adomeit, pp. 198–9; Eisenhower, p. 558; Gaddis, pp. 197, 392, n. 83.
49. James A Nathan, 'A Fragile Détente: the U–2 Incident Re-examined', *Military Affairs*, Vol. 39, No. 3 (October 1975), pp. 97–104.
50. Ambrose and Immerman, pp. 289–92 take issue with Nathan, calling his hypothesis 'absurd'. See also Beschloss, pp. 358, 374–5.
51. Victor Marchetti and John D Marks, *The CIA and the Cult of Intelligence* (New York: Dell, 1974; 1980), p. 306.
52. *Foreign and Military Intelligence*, p. 521.
53. Herter, testimony, frame 742.

CHAPTER 15

Breaking Cover: The Bay of Pigs Intervention

DAVID A CHARTERS

Anything that can go wrong, will go wrong, and at the worst possible time.
Murphy's Law

Shortly before dawn on 17 April 1961 a force of some 1,500 Cuban exiles began to disembark from landing craft on to the beaches of the Bay of Pigs on the southern coast of Cuba. They were quickly engaged and contained by the regular Cuban Army and Air Force under the direct command of President Fidel Castro. After three days of combat, the exile force, deprived of air cover, reinforcements and supplies, surrendered; some 1,100 were marched off into captivity. From the outset, the United States Government disclaimed any responsibility for or association with the invasion. But this disclaimer was a deception, one that collapsed almost as quickly as the invasion itself. The invasion force was Cuban, but in every other respect the attack was an American operation. The CIA had planned it, and had recruited, organised, financed, and trained the exile force. The Joint Chiefs of Staff had examined and endorsed the plan, and the new American President, John F Kennedy, had given the final 'go ahead' for the invasion. The American role was supposed to have been 'plausibly deniable'; during more than a year of preparation, the CIA went to great lengths to conceal the American hand, all to no avail. The proposed operation and its American sponsorship had been an open secret throughout Latin America, apparently even in Cuba, for months before the landing. In the wake of the disaster 'plausible deniability' was shattered. For the second time in less than a year an American President was forced to admit that his administration had lied about its involvement in an international incident.[1]

The operation had its origins in the Cuban revolution which

resulted in the overthrow of the corrupt dictator Fulgencio Batista at the beginning of 1959. The revolution led by Fidel Castro was initially very popular, both in Cuba and in the United States. Castro himself was perceived as a genuine national liberator in the tradition of Simon Bolivar. But opinion in the Eisenhower Administration changed rapidly as Castro expropriated American property, executed or jailed his political opponents, developed close ties with the Cuban Communist party and the Soviet Union, and appeared to be preparing to export his brand of social revolution to other parts of Latin America.[2]

On 11 December 1959 Colonel J C King, Chief of the CIA's Western Hemisphere Division recommended a programme of activities intended to 'eliminate' Castro.[3] On 13 January 1960, Director of Central Intelligence (DCI) Allen Dulles took King's recommendations to the '5412 Committee', the 'Special Group' of presidential advisers 'charged with reviewing and approving covert action programmes intiated by the CIA'.[4] At the meeting, Dulles suggested that 'covert contingency planning to accomplish the fall of the Castro government might be in order . . . actions designed to enable responsible opposition leaders to get a foothold'.[5] But when, in February, Dulles presented some proposals in this regard to the President, Eisenhower was unimpressed. Dulles then directed the 'Cuba Task Force', established in January under his capable Deputy Director of Plans Richard Bissell, to develop further options. The Task Force concluded that the Cuban leadership probably could not be eliminated by a single assassination operation, only by a prolonged programme involving the use of force. On 10 March, the National Security Council (NSC) discussed policy to 'bring another government to power in Cuba'.[6] The NSC directed the 'responsible departments and Agencies' to keep current their plans to deal with the contingencies which might develop with respect to Cuba.[7] Four days later Bissell presented to the Special Group a policy paper entitled 'A Program of Covert Action Against the Castro Regime.' It called for a four-part programme:

> **a.** The creation of a responsible and unified Cuban opposition to the Castro regime located outside Cuba.
> **b.** The development of means for mass communication to the Cuban people as part of a powerful propaganda offensive.
> **c.** The creation and development of a covert intelligence and action organisation within Cuba which would be responsive to the orders and directions of the exile opposition.
> **d.** The development of a paramilitary force outside of Cuba for future guerrilla action.[8]

Bissell's plan envisaged first the selection and training of a cadre of leaders. In the second phase they would serve as instructors for a number of paramilitary cadres, who would be trained at secure locations outside the United States. He believed it would be six to eight months before the trainees would be ready for deployment into Cuba, where they would organise, train, and lead resistance forces. On 17 March 1960, three days after Bissell's presentation, the President approved the programme.[9]

The ease with which the programme was accepted requires some explanation. First, as Trevor Barnes has shown,[10] the political climate of the "Cold War" era favoured covert action. A 1954 report to Eisenhower on covert action argued that the United States faced an implacable enemy, and thus the American tradition of 'fair play' had to be reconsidered. Urging an aggressive strategy and organisation for covert action more effective, more sophisticated, 'and if necessary, more ruthless than that employed by the enemy . . .' the paper stated that 'There are no rules in such a game. Hitherto acceptable norms of human conduct do not apply'.[11] Second, by 1960, the Clandestine Service (as the Directorate of Plans had come to be known) was in the ascendancy in the CIA. Dulles gave it a high priority, and it commanded the major share of the Agency's budget, personnel and resources. Covert action was perceived to be an 'essential service' that only the CIA could provide. Finally, the Agency's covert action successes in Iran and Guatemala had given both the CIA and its political masters confidence in the Agency's ability to produce victory 'on the cheap'.[12]

This last point may be the most telling one. According to Peter Wyden, 'The Guatemala model was on everyone's mind, especially Bissell's'.[13] This reaction was understandable; the June 1954 operation which removed Jacobo Arbenz Guzman, the allegedly-leftist president of Guatemala, from power made the covert *coup d'état* appear deceptively easy. The CIA had achieved this 'victory' with low United States visibility and at minimum cost. Arbenz had not been defeated militarily; the CIA's force of some 300 Guatemalan exiles and mercenaries, supported by a motley handful of aircraft, actually performed poorly. Instead, he was ousted largely by a deception: a CIA radio disinformation campaign that persuaded him and—more important—the Guatemalan Armed Forces that they confronted a large invading force. The bluff worked. Arbenz panicked, and ordered the military to arm the population. The officers, however, probably fearing the loss of their privileged status if Arbenz prevailed, refused and forced him to resign.

Shortly thereafter, the CIA's man, Colonel Carlos Castillo Armas, was installed as the new president of Guatemala.[14]

Bissell's proposals for Cuba incorporated some of the Guatemala plan's main features: an exile paramilitary force and a radio propaganda offensive. It also involved some of the same Agency personnel in key positions: Tracy Barnes, David Atlee Phillips and E Howard Hunt. There was an important difference, initially at least. Bissell's original Cuba plan called for the gradual infiltration and creation of a guerrilla army. By the Autumn of 1960, however, the operational concept had changed to a more traditional amphibious 'invasion'. While this might be attributed in part to the influence of the 'Guatemala model' and its architects, there were at least arguable tactical grounds for making the change. Cuba was acquiring large quantities of Soviet Bloc weapons for its armed forces. Castro was demonstrating increasingly effective control of the population, and air drops to the Cuban 'underground' were not proving effective. The size, equipment, tactics and training of the exile force were modified accordingly. By the end of the year, the guerrillla warfare option had been dropped entirely.[15] What is less certain is that the planners appreciated the implications of these changes for the visibility of the operation. It would be much harder to keep an amphibious operation covert and to maintain the fiction that America was not involved.

Nonetheless, deception planning proceeded apace. The intention was to minimise the American role and profile in the operation, so as not to alert the Cuban Government or to upset friendly Latin American regimes who were understandably sensitive about American intervention. Deception planning attempted, therefore, to make the operation appear as 'Cuban' as possible. This was reflected in virtually every aspect of operational planning, organisation, and implementation. In the proposed plan, Bissell's first priority was the creation of a responsible, unified Cuban opposition in exile. There were many Cubans opposed to Castro in exile in the United States, but they were a politically fractious lot. There was no single self-motivated, organised group ready and able to take the initiative in forming a viable political opposition. Consequently, the CIA had to create one where none had existed before. This proved to be more difficult than originally anticipated, owing to the divisions and mistrust within the American-based Cuban community. Thus, the Revolutionary Democratic Front and its somewhat broader successor, the Revolutionary Council, which was supposed to become the rebel 'government' in the wake of the landings, was never more than a creature of the CIA, imposed on a reluctant and suspicious paramilitary force. It did not, in fact, command the force which acted in its name; indeed, it wielded no real political influence or leadership at all. It was not privy to the exact details of the operation (such as the location and timing of the landing), and during the landing operation itself was kept 'incommunicado' at an abandoned naval air station at Opa Locka, Florida, on the outskirts of Miami. The communiqués issued in its name during the operation were drafted by the CIA and released by Lem Jones Associates, a New York public relations firm.[16] The exile leadership, therefore, was not even a genuine 'front'; it was a fraud.

The propaganda operation was organised and directed by David Atlee Phillips, who had been involved with the 'Voice of Liberation' radio deception in Guatemala. The Cuban operation, however, was to be of a rather different character. Unlike the Guatemalan scenario, where radio deception was central to the success of the operation, radio broadcasts to Cuba were apparently intended (in conjunction with leaflet drops) to prepare the ground in advance of the intervention, by creating a psychological climate among the Cuban people that would favour the exile force. Phillips estimated that this would take six months. His main weapon was 'Radio Swan', a medium-wave transmitter established on Great Swan Island off the coast of Honduras. Phillips had wanted to disguise the station as a commercial operation based in the Florida Keys, but Bissell advised him that the State Department would never agree. As it was, however, the studios remained in Miami, while the transmitter broadcast from the island under the ostensible ownership of the Gulf Steamship Company, which was actually a CIA commercial front or 'proprietary'.[17]

A number of experienced exile broadcasters and technicians were recruited to run the station, which went on the air in May, 1960, 30 days after Bissell had given Phillips his mandate. But Bissell, ever concerned with maintaining the cover story, concluded that the broadcasts sounded too smooth to be credible. 'Too professional', he told Phillips, 'too American. Go back and put some rough edges on it. Make it more Cuban.'[18] This did not prove to be difficult. The on-air announcers were told not to worry about their diction, and rugs were removed from the studios so that listeners would hear chairs scraping. Later, the leading exile groups were permitted to buy airtime (with CIA funds provided by Hunt and others). They used their time to push their own political views, rather than the CIA's official line, and to engage in on-air squabbles with rival groups. The broadcasts were in no danger of sounding too American; if anything, they were almost too authentically Cuban. Phillips also created a secondary broadcast capability by having CIA agents buy time on other radio stations around the Caribbean. During the pre-intervention period, these broadcasts were low key and not explicitly anti-Castro. Had the invasion succeeded, it was intended that they would become activist voices designed to persuade Cubans to join the winning side. In the event, Radio Swan played only a minor deception role once the invasion began. Phillips and Hunt composed, and Radio Swan broadcast, a series of cryptic, nonsense messages which were supposed to confuse and misdirect Castro's security intelligence service about the extent to which internal resistance forces were expected to rise up in support of the landings. There were, in fact, no such resistance forces, and no risings were planned.[19]

A more important aspect of the cover plan was the concealment of American 'material' involvement. Cuban exile recruits were told that wealthy Cuban and American 'industrial interests' were sponsoring and funding the operation, and that the United States Government was not involved. The recruits knew better, and treated that part of the cover story as an 'in joke'.[20] The CIA acquired military equipment, vehicles, ships, and aircraft for the Cubans by covert means. The Department of Defense provided the CIA with weapons out of its own stocks in exchange for funds paid from Agency accounts to the Pentagon. The weapons were 'sanitised'—stripped of identifying marks—so that they could not be traced back to official American sources. For shipping, the CIA acquired the services of the Garcia Line Corporation, a Cuban company run by a family opposed to Castro. Their small fleet of slow, tiny, creaky freighters was just the sort of low profile operation that suited the Agency's needs. The ships were leased for a modest sum. Other vessels were acquired in a similar manner. To mount the naval side of the operation the CIA had taken over the premises of a defunct ferry service in Key West. There, the naval component operated under commercial cover, Mineral Carriers Limited, as a company supposedly drilling for oil in the Marquesas Keys (there was, in fact, a legitimate oil company working there). To run the 'Cuban' air force, Bissell appointed United States Air Force Colonel Stanley W Beerli, who had worked under him previously as commander of the U-2 Detachment 10-10 at Incirlik, Turkey. Bissell told Beerli his mission was to assure air superiority over the landing site, with the proviso that the operation had to look 'Cuban'—in other words, amateurish and shoddy. Beerli decided to keep the air force small, and to equip it with the Second World War vintage B-26 bomber. This aircraft, in service with the real Cuban Air Force, was available in considerable quantities in the United States and elsewhere. He recruited flight instructors for the Cubans from the Alabama Air National Guard, which had recently flown the B-26, and some of whose personnel had previously been security cleared to work on other CIA operations. Transport aircraft—again, only a handful—were acquired by similar roundabout means. Beerli based the aircraft and their Miami-recruited Cuban crews, at Retalhuleu, in Guatemala, not far from the training base used by the exiles' landing force. Although small numbers of specialists were trained in the United States, the bulk of the landing force was trained at a coffee plantation in Guatemala. Recruitment and selection of personnel, however, was carried on openly in Miami.[21]

To ensure secrecy, Bissell ran the Bay of Pigs operation in much the same manner as he had managed the U-2 project as his own personal fiefdom. The Cuba Task Force was 'compartmentalised' and isolated from the rest of the Agency. Those inside the CIA like Robert Amory, who had managed to learn the truth about the operation but who were not part of the team, were told politely to keep their noses out of it and their mouths shut. Even Bissell's deputy in the Plans Directorate, Chief of Operations Richard Helms, was excluded from the project. The Joint Chiefs of Staff (JCS), the military body with actual expertise in amphibious operations, were not permitted to review the plans until late January, 1961; at that time the CIA expected to launch the operation no later than 1 March. Dulles briefed the Senate Foreign Relations Committee on American involvement in training Cuban exiles, but did not tell them about plans for an invasion. United Nations Ambassador Adlai Stevenson and United States Information Agency (USIA) Chief Edward R Murrow were kept in the dark until the last moment. This degree of internal security was only partly effective. But it led to an anomalous situation in which most of the Agency's intelligence professionals and the American diplomatic and military establishment knew nothing about the operation, while it was common knowledge among the Cuban exiles in Miami, the American news media, and apparently in Cuba itself.[22]

In retrospect, this hardly seems surprising. The operation was a counter-intelligence nightmare, and Bissell's security efforts notwithstanding, the operation leaked like a sieve almost from the outset. This can be attributed in part to the scale of the operation. Outside the carefully compartmentalised Task Force at CIA headquarters, thousands of people were involved. Secretary of State Dean Rusk, Allen Dulles, even Bissell himself all conceded later that the scale of the operation meant that the cover plan was inadequate both to preserve the secrecy of the operation and to conceal the American hand.[23] Location and character may also have been factors. The Cuban exiles seemed incapable of doing anything quietly or subtly. The flood of refugees into Florida had transformed Miami, and scores of rival exile groups formed in this milieu, gossiping and politicking. The headquarters of the CIA's 'Front' on Biscayne Boulevard was a hive of highly visible activity, hardly the appropriate profile for a clandestine headquarters. Recruitment for the exile force was carried on openly, as was a certain amount of training in nearby localities. The Opa Locka airfield was unusually active for a supposedly abandoned air base. Biscayne Bay was busy with a variety of craft belonging to the CIA and exile groups. The organising of an exile army and the prospect of its return to Cuba were open secrets among the Cuban community in Miami.[24]

At this time the FBI estimated that Cuban intelligence agents in Miami numbered in the hundreds. But in September, 1960, there was only one CIA counter-intelligence officer assigned to the project. Helms and James Jesus Angleton, the CIA's Chief of Counter-Intelligence, were alarmed by several egregious security lapses. One of Howard Hunt's couriers lost a briefcase full of incriminating documents (including a list of agents and contacts within Cuba) in Mexico City. In a second incident, project officer Gerry Droller (alias Frank Bender) was overheard discussing sensitive matters with one of the Cuban exiles in a motel room near Miami airport. The eavesdropper, a zealous stenographer, took notes and sent them to her brother in the FBI. The FBI eventually turned the report over to the CIA and Bender was reprimanded. Bender was identified as a 'Mr B' in a *Time* Magazine article on the operation on 27 January 1961. At least one event subsequent to the invasion suggested that Cuban intelligence had penetrated the operation. Two days after the collapse of the operation, two unidentified men withdrew funds from a CIA dummy account at a Washington bank. A week later, the *Washington Star* reported a speech by Castro, in which he alluded to CIA funding of plots against Cuba, citing the cover name of the Agency bank account.[25]

Thomas Powers, in his study of former DCI Richard Helms, takes the view that while Bissell and his assistant Tracy Barnes were not indifferent to the counter-intelligence aspects of the operation, Bissell's relative lack of operational experience may have left him without a sufficiently keen appreciation of the risks posed to a leak-prone operation by a hostile intelligence service.[26] Bissell's record with respect to the security of the U-2 project neither supports nor undermines this assertion. It has since been suggested, however, that the elaborate secrecy surrounding the Cuba Task Force, as well as its monopoly on information, made a more significant contribution to the failure of the operation than any of the security lapses, by discouraging independent assessment of the feasibility of the plan.[27]

With the election of a new President, John F Kennedy, in November 1960, the plan underwent yet another significant change, related to maintaining the deception. The new administration was very concerned about the 'visibility' of the operation and the possible exposure of American involvement. Existing plans had the brigade landing in the vicinity of Trinidad, on the south coast of Cuba. The 'Trinidad Plan' had several advantages: the local population were expected to be 'friendly' to the exile force and, in the event that the landing encountered strong military opposition, it could melt away into the nearby Escambray mountains—the seat of Castro's revolution and ideal guerrilla territory. Even so, the JCS evaluation concluded that the operation stood only a 'fair chance' of success, subsequently defined as about 30 per cent—hardly a glowing endorsement. But in a meeting about the operation on 11 March 1961, President Kennedy objected that the plan was 'too spectacular', like a Second World War invasion. He made it clear that he preferred a quieter operation, a landing by night at a less visible location. Assistant Secretary of State for Inter-American Affairs Thomas Mann made probably the most persuasive argument against Trinidad: the airfield was too short to take B-26 bombers. Thus, it would be impossible to sustain the fiction that the aircraft were operating from within Cuba. These arguments carried the day. Trinidad was dropped and the Bay of Pigs, which had an adequate airstrip, was selected in its place. The new location was less than ideal. It was closer to Havana, and was too far from the mountains to permit exploitation of the 'guerrilla option' if the landing was opposed. Nor would the local terrain sustain guerrilla operations. Apparently unknown to the CIA, Castro himself was personally very familiar with the Bay of Pigs as a recreation area. Moreover, the bay itself was obstructed by coral reefs which were not marked on American maps, but which would wreak havoc on a landing operation. Oblivious to these limitations, Bissell assured the President that the amphibious landings could be completed there before first light, thus preserving the covert nature of the operation. On 16 March Kennedy told him to proceed.[28]

Early in April the Task Force amended the plan yet again, in a further attempt to increase the chances of success and also to enhance 'no US involvement' deception theme. The Task Force decided that on D minus 2 the exile's air force would carry out limited air strikes against Cuban Air Force bases. The raids were supposed to achieve two objectives; neutralise Castro's air force before the landings, and also to create the impression that it had defected to the opposition. On the morning of Saturday 15 April, six B-26's, in Cuban Air Force markings, flew from Nicaragua and raided three military airfields in Cuba. A seventh B-26 flew directly to Miami, where the pilot Mario Zuniga, posed as a defector from the Cuban Air Force. Prior to Zuniga's take off, a CIA team had removed one of the engine covers, fired some shots through it and then replaced it on the plane, so that the aircraft would appear to be battle-damaged. Zuniga added to this shortly before landing in Miami; he fired his pistol at one of the engines until it was feathered. Upon landing he recited the story he had been instructed to tell: that he and three fellow pilots, all disenchanted with the Castro regime, had been plotting to defect for some time. The opportunity came during a routine patrol that day. He returned to his base, strafed it, was hit by ground fire, and was forced to land at Miami because he was low on fuel. Once he had told this story to reporters without revealing his name, Zuniga was spirited away by the Immigration and Naturalisation Service, who turned him over to the CIA. That night he was flown back to Nicaragua. Unknown to Zuniga, however, a genuine defector had landed his plane at Jacksonville, Florida the previous day and was seeking asylum. Then, shortly after the air strikes, one of the participating aircraft developed engine trouble and landed at Key West.[29] The United States now had more 'defectors' than the plan called for, and the business of explaining these fortuitous incidents dragged the United States Government even deeper into the murky waters of deception. In doing so, it added to the suspicions of the many newsmen who had been gradually uncovering the plot against Cuba.

During the days following the air raids and the landings, the entire deception came unglued. This was the result not only of the operation's failure, but was also the end product of a series of leaks and exposures which had rendered the covert operation almost transparent even before the first exile soldier waded out of the surf on to the shores of the Bay of Pigs.

According to Wyden's account, the first serious breach of security which attracted media attention occurred in August, 1960. Some American youngsters tossed firecrackers into a Cuban training camp near Homestead, south of Miami. The Cuban trainees, thinking they were under attack, burst out of the camp, shooting and wounding an American youth. Several Cubans were arrested but the case against them was dropped as a result of pressure from Washington. The *Miami Herald* sent its Washington reporter,

David Kraslow, to ferret out the story. By meticulously checking his Washington sources over several weeks, Kraslow put together a remarkably accurate account of the CIA's operation. The *Herald* editors then agonised over its national security implications. Eventually, Kraslow and his bureau chief confronted Dulles with the story. He told them publication would be 'harmful to the national interest', and the *Herald* dropped the story.[30] But the story would not go away.

On 30 October the Guatemalan newspaper *La Hora* published an account describing the CIA's base near Retalhuleu and the training of the Cuban exiles there. The story caught the attention of Dr Ronald Hilton, Director of Stanford University's Institute of Hispanic American Studies and editor of the highly-regarded *Hispanic American Report*, who happened to be in Guatemala at the time. He published a report on the CIA story, which was picked up by the left-wing political weekly *The Nation*. On 19 November, it published an editorial entitled 'Are We Training Cuban Guerrillas?' The editorial urged the mainstream media to take up the story in order to put pressure on the Administration to abandon the operation. *The Nation* sent proofs of its editorial to the major American papers and wire services, and the *New York Times* (20 November) carried a report on the training and the Guatemalan President's denial. Two days earlier, CIA Inspector General Lyman Kirkpatrick, speaking before the Commonwealth Club in San Francisco, was asked whether Hilton's report was true. The questioner quoted Hilton to the effect that it would be a black day for the United States if the attack on Cuba took place. Kirkpatrick's reply was, 'It will be a black day if we are found out.'[31]

Professor Hilton was not the only American outsider in on the secret. Republican Senator Bourke Hickenlooper was in Guatemala in November and learned about the camps. He later told his Foreign Relations Committee colleagues, 'apparently everybody in Guatemala knew about it. It was not any secret. It was supposed to be very covert, except that everybody in Central America knew about it . . .'[32]

On 10 January 1961, the *New York Times* carried a front page headline story by the paper's Central American reporter Paul Kennedy. He had managed to penetrate the Guatemalan training area; his report included a map, and a detailed account of the training and American involvement. The *Miami Herald*, which had suppressed its knowledge at the request of the government, resented the 'scoop' by the *Times*, and followed up with stories on the camp and the air traffic between Florida and Guatemala. *Time* magazine on 17 January included a photograph of the CIA's aircraft at Retalhuleu. A *New York Daily News* account, however, preserved at least part of the cover story, by reporting that the camps were sponsored by 'American and Cuban industrial interests'.[33]

The CIA created some of its own problems. In order to boost recruiting for the brigade, Howard Hunt got Bissell's permission to publish photographs of the force in training. Published in Spanish-language papers across the United States', they had the desired effect—a significant increase in volunteers—but they were also published in Cuba. If Castro previously had harboured any doubts about the existence of the force, the photographs must have dispelled them. While Hunt's actions appear to have been unwise from a security standpoint, they were at least consistent with NSC policy, which directed the CIA to seek the maximum publicity for the exile political leadership, 'especially those who may be active participants in a military campaign of liberation.'[34]

Around the end of March, the Administration received more unwelcome news. A visit from a *US News and World Report* reporter, and galley proofs of an article for *The New Republic* sent by the publisher, indicated that both publications had substantial, accurate knowledge of the plan. On President Kennedy's request *The New Republic* withheld its article from publication.[35] Worse was to follow. Tad Szulc, a *New York Times* correspondent, stumbled on to the story while in Miami en route to New York for reassignment. He was told to pursue it, while James Reston, the *Times* Washington Correspondent, followed up leads in the capital. Szulc sent a detailed account and after an acrimonious debate among the *Times* editorial staff, a watered-down version of the story ran on 7 April. It overestimated the size of the force, but emphasised that training had been completed, and that the invasion was imminent. CBS news also reported that an invasion was at hand. President Kennedy reportedly was enraged by the stories.[36] The speculation continued. On 9 April, the *New York Post* reported that Cuba was about to get 'the Guatemala treatment' and that the CIA 'had cast Captain Manuel Artime in the Role of Colonel Castillo Armas'.[37] If the accuracy of that observation was lost on most readers, those 'in the know' could hardly have failed to realise that the news media knew enough of the truth to render any further attempts at plausible denial of American involvement an exercise in self-delusion.

This spate of stories may have prompted President Kennedy to attempt to clear the air about Cuba at his next press conference, on 12 April. But his efforts to clarify the official United States position in respect of Cuba still amounted to deception. In retrospect, his remarks appear to have been deliberately evasive.

> First I want to say that there will not under any conditions, be an intervention in Cuba by the United States Armed Forces. This Government will do everything it possibly can, . . . to make sure that there are no Americans involved in any actions inside Cuba . . . The basic issue in Cuba is not one between the United States and Cuba, it is between the Cubans themselves. I intend to see to it that we adhere to that principle. As I understand it, this Administration's attitude is so understood and shared by the anti-Castro exiles from Cuba in this country.[38]

When asked if the Neutrality Acts or the Rio Treaty barred the United States from giving aid or arms to anti-Castro elements, Kennedy replied,

> Well . . . there is a revolutionary committee which is, . . . extremely anxious to see a change of government in that country. I am sure that they have—that they are very interested in associating with all those who feel the same way.[39]

Given the amount of information by then in the public domain, Kennedy's remarks were scarcely credible. It is doubtful that anyone in the American news media who had looked seriously into the invasion story—and their numbers were legion and growing—took the President's denials seriously.

Ironically, those in the Administration who were not cleared on the operation either knew nothing about it, or knew only what they had read in the papers or picked up on the Washington 'cocktail circuit'. It was only Tad Szulc's inquiries in early April that alerted the United States Information Agency to the fact that something was about to happen. USIA director Ed Murrow confronted Allen Dulles with Szulc's story, but the DCI refused to discuss it. McGeorge Bundy, the President's Adviser for National Security Affairs, later briefed Murrow about the operation. Murrow was appalled by the plan and by the lack of guidance given to the government's official 'voice'. He felt that USIA had no choice but to play along, to act as if it knew nothing, and pick up the 'news' of the invasion from the wire services.[40]

The 15 April air strikes and 'defections' raised the stakes and the profile of the American deception effort. It moved from the back alleys of covert action to the centre stage of high-level diplomacy. In doing so, these events set in train the process which rapidly unravelled the deception and discredited the Administration and its spokesmen. At the United Nations on 15 April Cuban Foreign Minister Raul Roa charged the United States with responsibility for the attacks. He persuaded the United Nations Political Committee to advance its planned debate on the Cuban situation (scheduled for Monday 17 April) to the afternoon of the

15th.[41] The first American responses to the Cuban charges came shortly after noon at a news conference given by Presidential Press Secretary, Pierre Salinger. Salinger, who apparently had not been cleared on the operation at that time, told reporters,

> Our only information on the situation in Cuba comes from the wire service stories that we have read. We have . . . no first-hand information. We are trying to determine what the situation on it is . . . These two planes which have landed in the United States reportedly have been interned, as I understand it.[42]

Meanwhile, at the United Nations, United States Ambassador Adlai Stevenson was trying to collect enough information to permit him to make a credible reply to Roa's charges. Stevenson had been briefed a week earlier on the Bay of Pigs operation. Kennedy had told Arthur Schlesinger Jr, who assisted CIA officer Tracy Barnes in briefing Stevenson, that he wanted Stevenson fully informed and that nothing he said at the United Nations 'should be less than the truth, even if it could not be the full truth. "The integrity and credibility of Adlai Stevenson, . . . constitute one of our great national assets. I don't want anything to be done which might jeopardise that".'[43]

Tracy Barnes, however, had a reputation for being 'the soul of vagueness', and he discussed the operation only in very general terms. He told Stevenson that it was a strictly Cuban affair, that the United States was involved only in training and financing; there was no direct American military involvement. The operation would appear to happen from inside Cuba, with some outside (Cuban) participation. He said nothing about the form, size, timing, or location of the operation. Nor did he say anything about air strikes.[44] Consequently, the air strikes and the 'defector' came as a complete surprise to Stevenson and sent his staff scrambling for details.

Harlan Cleveland, Assistant Secretary of State for International Organisation Affairs, called the State Department's Bureau of Inter-American Affairs. They in turn called the CIA, who assured them that the air strikes had been the work of genuine defectors. Joseph Sisco, Clevelands deputy, then drafted a statement denying American involvement, and dictated it over the telephone from Washington to Stevenson's political officer, Richard Pederson, in New York. Stevenson got Sisco to confirm that the charges had been carefully checked, then delivered the American response to the United Nations committee. Stevenson's statement made four points: first, that there would not be any intervention in Cuba by

American armed forces; second, that the United States would do everything possible to ensure that no Americans participated in actions against Cuba; third, that it would consider the defecting pilots' requests for asylum, and finally, that the pilots who landed in Florida were 'to the best of our knowledge' Castro's own, and that they were apparently genuine defectors. To support his statement Stevenson read into the record what he believed was a corroborative account from a press wire service (it was in fact the CIA's cover story), and displayed a wire service photograph of Zuniga's B-26. He also pointed out the Cuban Air Force markings on the tail as proof that the plane was that of a genuine defector.[45]

Stevenson's account was believed, up to a point. The *New York Times* of 16 April reported the air strikes, the Cuban Government charges against the United States, Zuniga's story, Salinger's responses, and Stevenson's explanation to the United Nations. It also quoted Miro Cardona, leader of the CIA 'front', the Revolutionary Council, to the effect that the Council had been in contact with and encouraged the defection of B-26 pilots. The Council claimed to have had advance notice of the attack, which of course was true, but not for the reasons they implied. The *Times* also carried a curious report that the Cuban Air Force was rationing fuel to prevent further defections.[46] This report may have been based on rumours picked up in the crisis atmosphere in Cuba. Yet, given that at least one of the defectors was genuine, the story might have been true. Obviously, the CIA would have been delighted had the 'defections' truly paralysed Castro's air force in the same manner that the Agency had hamstrung Arbenz in Guatemala. So the possibility that the story was planted in Cuba to sow confusion cannot be discounted.

By this time, however, even Stevenson was aware of the deception to which he had been an unwitting party. He was 'dismayed', and felt that he had been 'deliberately tricked'. He sent a telex to Rusk at the State Department indicating that he knew that he had misled the United Nations about the truth of the bombing raids, and warning that 'There is the gravest risk of another U-2 disaster in such uncoordinated action.'[47] It was too late. The next day, with the invasion underway, Lem Jones Associates began issuing the Revolutionary Council's deceptive communiqués, and the *New York Times* had started to question Zuniga's and Stevenson's accounts. Reporters had noticed that the guns on Zuniga's plane were taped, and thus could not have been used for strafing. It was also pointed out that Cuban Air Force B-26's were equipped with a plexiglass nose; Zuniga's was opaque. In Cuba itself, the Cuban government had displayed physical evidence which indicated that the raids had not originated in Cuba. The most damning was the long range fuel tanks jettisoned by the exile B-26's; aircraft originating in Cuba would not have required them, or at least would not have jettisoned them so soon. On Monday, at the United Nations, Roa charged that the CIA had organised, armed and financed the invasion. Stevenson, in his biographer's words, 'could only reply in weasel words that "The United States has committed no aggression against Cuba and no offensive has been launched from Florida or any other part of the United States".'[48] Stevenson had been humiliated publicly in just the way Kennedy had hoped to avoid.

Pierre Salinger fared little better. By the second day of the invasion, no one was buying his denials. At the daily press conference, a question about the extent to which the President was being briefed on the Cuban situation led to the following exchange:[49]

Salinger: 'I think he is being kept apprised of the situation.'
Question: 'By whom, Pierre?'
Salinger: 'Through our normal channels.'
Question: 'Still getting all your information from the news dispatches, Pierre?'
Salinger: 'Yes.'

In the years since, most details of the operation and the deception have become public knowledge, the subject of polemic, exposé, memoir and serious study. The incident remains controversial. But there is a degree of consensus in the assessment of the event; it was a foreign policy disaster of the first order. Some have called it 'the perfect failure'.[50]

Deception planning contributed directly to that disaster. At the strategic level, the Kennedy Administration forced upon the CIA operational changes which were intended to preserve 'plausible deniability'. But, as Secretary of Defense Robert McNamara told the Taylor Commission, this requirement was given precedence over the requirements of operational success. The result was that 'the plan was compromised in order to reduce the chances of attribution'.[51] At the tactical level, deception planning had been very thorough, even if not leak-proof. It was probably sufficient for Bissell's original proposal: a small-scale infiltration of guerrilla fighters. Once the plan was changed to a large-scale invasion, the cover story was insufficient, but deception planning was not altered accordingly. Instead, the invasion plans were changed to protect the cover story. The end result was an operation that, in Peter Wyden's words, was 'too large to remain secret and too small to succeed'.[52]

As in the case of the U-2, the inability of the Americans to 'hide the real' made it impossible to 'show the false' credibly. Again, ineffective dissimulation precluded effective simulation. McGeorge Bundy remarked to the Taylor Commission, 'As I reflect on the covertness of this operation, I'm amazed that we thought there was a chance of deniability'.[53]

In a manner similar to that of the U-2 incident, the Americans destroyed their own credibility by adhering to a cover story which was untenable. Unlike the U-2 case, the story was blown before the operation began. That should have been a signal either to call off the operation or to abandon the compromised cover story. For political and operational reasons, however, neither course was adopted. If the operation had no enthusiastic supporters outside the CIA Task Force, neither did it have vocal, influential opponents. It went ahead for lack of such opposition and of palatable alternative plans.[54] Moreover, once committed to the operation, the United States could hardly admit to a blatant act of aggression. So it was committed to sustaining the deception, although even the President was aware that it was no longer credible. 'How could I have been so far off base?' Kennedy commented later. 'All my life I've known better than to depend on the experts. How could I have been so stupid to let them go ahead?'[55] Peter Wyden suggests, with the advantage of hindsight, that Kennedy and his advisers, many relatively new to Washington, were unduly impressed by the CIA as a 'can-do' organisation. They lacked the experience or knowledge to assess objectively the Agency's capabilities and plans. The CIA's secrecy and lack of candour compounded this problem. It left the President and his staff poorly briefed, and thus contributed to inadequate policy analysis and decision-making. Consequently, like Eisenhower, Kennedy let the experts overrule his own best political instincts. The Agency, concerned only with the operational short term, had not given sufficient thought to the long-term consequences should the operation and the deception fail. But the fault was not the Agency's alone; few of Kennedy's own advisers had considered these matters properly, let alone expressed their reservations.[56]

The consequences of the Bay of Pigs operation and the accompanying deception were serious and wide-ranging. First, it achieved the exact opposite of its intended goals; it consolidated the Castro regime by providing a convenient external threat which rallied most of the Cuban people to defend the Government. It provided Castro with the perfect opportunity and excuse to clamp down on the internal dissenters, which he did with vigour and ruthlessness. It also cemented the growing Cuban–Soviet relationship and fixed the Cuban regime firmly in the Marxist camp.[57]

Some of these processes were already underway before the invasion, but it may be fair to suggest that the invasion hastened the pace.

The failure clearly, if only temporarily, made Kennedy look incompetent as a national leader and statesman. Many analysts have suggested since that Soviet leader Nikita Khruschev may have taken this as a sign of weakness and sought to exploit it. This line of argument posits a direct connection between the Bay of Pigs and the Cuban Missile Crisis in the following year. If Kennedy's fumbling emboldened Khrushchev in this way, they argue, then the operation may be said to have contributed to a significant deterioration in East–West relations, even to the point of encouraging reckless Soviet risk-taking. More recent scholarship, however, has cast considerable doubt on this thesis.[58]

Through the Bay of Pigs operation the United States inflicted damage on its international moral standing which has yet to be repaired. In Latin America and the rest of the Third World, the name 'Bay of Pigs' was to become synonymous with 'Yankee Imperialism'.

Finally, the operation widened the 'credibility gap' opened the year before by the U-2 incident. The Bay of Pigs stood as testimony to the view that the United States Government could not be trusted to tell the truth. The consequences of that credibility gap were most evident during the Vietnam War, when a significant proportion of American public opinion refused to accept as credible the Administration's explanations of its involvement in and conduct of the war. It is a moot point whether disillusionment with their own government's apparent lack of commitment to truth may have left some Americans unwittingly predisposed to accept deceptive information propagated by the other side during that war.[59] If so, then the Bay of Pigs deception ultimately boomeranged on the United States with serious political impact.

Endnotes

1. The literature on the Bay of Pigs operation is extensive. Two works bear special mention. Trumbull Higgins, *The Perfect Failure: Kennedy, Eisenhower, and the CIA at the Bay of Pigs* (New York: Norton, 1987) is the most complete account, based on access to private papers, recently declassified documents, interviews, and the secondary literature. However, it suffers from convoluted writing and somewhat tortuous analysis. As such, it adds little to the more readable Peter Wyden, *Bay of Pigs: the Untold Story* (New York: Simon and Schuster, 1979). A journalist's account based largely on interviews with individuals who had played a part in the operation, and to a lesser extent on documentary sources, it has been widely and authoritatively praised, although it falls short of being a proper history.
2. On the Administration's perceptions of and responses to Castro's policies, see John Spanier, *American Foreign Policy Since World War II*, 8th ed. (New York: Holt, Rinehart

and Winston, 1980), pp. 106–7; Alexander DeConde, *A History of American Foreign Policy*, 2nd ed. (New York: Scribners, 1971), p. 733; Elmo Richardson, *The Presidency of Dwight D Eisenhower* (Lawrence, Kansas: Regents Press, 1979), p. 178.

3. US Senate, Select Committee to Study Government Operations with Respect to Intelligence Activities, *Alleged Assassination Plots Involving Foreign Leaders: an Interim Report* (New York: W W Norton, 1976), p. 92.
4. Stephen Ambrose with Richard Immerman, *Ike's Spies: Eisenhower and the Intelligence Establishment* (Garden City, NY: Doubleday, 1981), p. 308; US Senate, Select Committee, *Book 1: Foreign and Military Intelligence* (Washington, DC: US Government Printing Office, 1976), pp. 50–51. The group consisted of the DCI (Dulles), the Special Assistant to the President for National Security Affairs, a Deputy Under Secretary of State, and Deputy Secretary of Defense, and got its name from NSC Directive 5412/2 (November 1955), which brought it into being.
5. US Senate, Select Committee, *Alleged Assassination Plots*, p. 93.
6. *Ibid.*; Ambrose and Immerman, p. 309. The Directorate of Plans was responsible for CIA clandestine operations.
7. NSC Action No. 2,191, 'US Policy Toward Cuba', Record of Actions by the National Security Council, 436th Meeting, 10 March 1960, Documents of the National Security Council, 1947–77 (Washington, DC: University Publications of America, 1980), Reel 5.
8. Quoted in *Operation ZAPATA: the 'Ultrasensitive' Report and Testimony of the Board of Inquiry on the Bay of Pigs*, ed., Thomas F Troy (Frederick, Md.: University Publications of American, 1981), pp. 3–4. This is a 'sanitised' version of the report; the complete document was declassified in 1985.
9. *Ibid.*, p. 4; Dwight D Eisenhower, *The White House Years: Waging Peace 1956–1961* (Garden City, NY: Doubleday, 1965), p. 533.
10. See Chapter 12.
11. US Senate, Select Committee, *Foreign and Military Intelligence*, p. 50.
12. *Ibid.*, pp. 108–9, 111–12; Wyden, p. 21. Between 1953 and 1961, clandestine intelligence collection and covert action consumed an average of 54 per cent of the Agency's annual budget.
13. Wyden, p. 20.
14. The most thorough scholarly account is Richard Immerman, *The CIA in Guatemala: the Foreign Policy of Intervention* (Austin, Texas: University of Texas Press, 1982), esp. chapters 6–8.
15. *Ibid.*, pp. 190, 194–7; Wyden, pp. 20, 22; *Operation ZAPATA*, pp. 4–6.
16. Wyden, pp. 119, 166, 290–91; *Operation ZAPATA*, pp. 5, 11–12, 203, 58, 66, 104, 127, 295, 337, 339–40, 356. E Howard Hunt, *Give Us This Day* (New Rochelle, NY: Arlington House, 1973), pp. 45–7, 59–62, 143–4, 188, 192–3, 207 provides some interesting insights on the internal political feuding of the Front and the Council. The reader should bear in mind, however, that this 'first hand' account by one of those deeply involved in the operation was written from memory, without notes, long after the event. In Wyden's words (p. 333), it is 'not distinguished for accuracy of detail'. Copies of the Council's invasion press releases are reproduced in Karl E Meyer and Tad Szulc, *The Cuban Invasion: The Chronicle of a Disaster* (New York: Preager, 1962), pp. 130–32, 139–40. Higgins, p. 68, says there were 184 different Cuban exile groups.
17. David Atlee Phillips, *The Night Watch* (New York: Ballantine, 1977), pp. 108, 110–14. Wyden, p. 118, and others have referred to it as the Gibraltar Steamship Corporation. 'Proprietaries' are defined in *Foreign and Military Intelligence*, pp. 205, 208, as 'business entities, wholly owned by the Central Intelligence Agency, which either actually do business as private firms, or appear to do business under commercial guise'. They provide cover, attribution for funding and other services to clandestine activities not available through normal commercial channels.
18. Quoted in Wyden, p. 23.
19. *Ibid.*, pp. 23, 118; Philips, p. 122; Hunt, pp. 200–201; *Operation ZAPATA*, pp. 19, 20, 41, 86, 111–12 and *passim.* The latter refers to the use of 11 CIA-controlled radio stations. These may have been the secondary stations to which Phillips refers. The CIA had small teams of agents and infiltrators in Cuba, but although internal opposition to the Castro regime was thought—probably over-optimistically—to be considerable, little

was done to stimulate a viable internal resistance movement; the exile landings were not intended to be dependent upon internal uprisings.

20. Haynes Johnson, *et al*, *The Bay of Pigs: the Leaders' Story of Brigade 2,506* (New York; W W Norton, 1964), pp. 26–8, 30–31, 33, 37.
21. *Ibid.*, p. 44; Wyden, pp. 45, 57, 70–71, 77–8, 85–6; US Senate, Committee on Foreign Relations, 'Briefing on Cuban Situation', 2 May 1961, Reel 2, frame 860, Reel 3, frames 198–9; The Bissell plan had anticipated the creation of a small air supply capability 'under deep cover as a commercial operation in another country': *Operation ZAPATA*, p. 14. Wyden and other sources indicate that the CIA operated a 'black' (clandestine) air transport capability in connection with the operation but it is not entirely clear whether this was integral to or separate from Beerli's Cuban exile air force. In testimony to the Senate Committee, Dulles and Bissell confirmed that small groups of the exiles had received training in the United States.
22. US Senate, Committee on Foreign Relations, Sub-Committee on American Republics Affairs, 28 April 1961, Reel 2, frames 669–70; see also frame 902; US Senate, Select Committee, *Foreign and Military Intelligence*, pp. 112, 134; *Operation ZAPATA*, pp. 6–7, 237–8; Thomas Powers, *The Man Who Kept the Secrets: Richard Helms and the CIA* (New York: Alfred Knopf, 1979), pp. 107, 109; Wyden, pp. 45, 98, 142–6, 152–3, 156–8, 190; Joseph Burkholder Smith, *Portrait of a Cold Warrior* (New York: Ballantine, 1976), p. 316. See Higgins, pp. 68–71, 89 on Cuban and others' knowledge.
23. See Rusk testimony to the sub-committee on American Republics Affairs, 1 May 1961, and by Dulles and Bissell to the full Foreign Relations Committee, 2 May 1961, Reel 2, frames 716, 818, 888–9.
24. Wyden, pp. 45, 118–19; Hunt, pp. 59–60; Warren Hinckle and William W Turner, *The Fish is Red: the Story of the Secret War Against Castro* (New York; Harper and Row, 1981), pp. 47–8; *Operation ZAPATA*, p. 105.
25. Hunt, p. 60; Powers, pp. 107, 109–10.
26. Powers, pp. 109–10.
27. Wyden, pp. 310, 320, 323; U.S. Senate, Select Committee, *Foreign and Military Intelligence*, pp. 156, 159; Tyrus G Fain, *et al*, *The Intelligence Community: History Organization and Issues*, Public Documents Series (New York: R R Bowker, 1977), p. 122; William R Corson, *The Armies of Ignorance: the Rise of the American Intelligence Empire* (New York: Dial Press, 1977), pp. 384–5; Lucien S Vandenbroucke, 'Anatomy of a Failure: the Decision to Land at the Bay of Pigs', *Political Science Quarterly*, Vol. 99, No, 3 (Autumn 1984), pp. 471–91; *Operation ZAPATA*, p. 181.
28. Wyden, pp. 100–103; Higgins, pp. 82–4, 91–2; *Operation ZAPATA*, pp. 8, 11-15, 67, 69, 75–90, 103–11, 144–5, 177–8, 202, 206, 210, 213, 215, 262–70 *passim.* The Taylor Commission testimony dealt extensively with the reasons for the change of location from Trinidad to the Bay of Pigs, and the operational disavantages that accrued therefrom. The JCS preferred the Trinidad plan, even though its own evaluation team rated the chance of success as low as 15 per cent. When it was dropped, the JCS acquiesced in the Bay of Pigs location as the best of the proposed alternatives. *National Security Action Memorandum*, No. 31, 11 March 1961, Documents of NSC, Reel 5, para 4 makes it clear that as of the 11th, Kennedy did not believe the CIA had produced 'the best possible plan, from the point of view of combined military, political and psychological considerations. . . .'
29. *Operation ZAPATA*, pp. 15–18; Wyden, pp. 163, 173–7; see also Arthur M Schlesinger Jr, *A Thousand Days: John F Kennedy in the White House* (Boston: Houghton Mifflin, 1965), p. 271; Kenneth E Davis, *The Politics of Honour: a Biography of Adlai E Stevenson* (New York: G P Putnams, 1967), p. 457.
30. Wyden, pp. 45–6.
31. *Ibid.*; Hinckle and Turner, pp. 67–9; see also Higgins, pp. 66–7, 186 n. 15.
32. Quoted in transcript of testimony, US Senate Committee on Foreign Relations, Subcommittee on American Republics Affairs, 28 April 1961, Reel 2, frames 691–2.
33. Wyden, pp. 46–7; Hunt, p. 115; Smith, p. 338–9; Hinckle and Turner, p. 169.
34. Hunt, pp. 144–5, 153; Johnson, p. 63; *NSAM* No. 31, 11 March 1961.
35. Wyden, pp. 142–3.
36. *Ibid.*, pp. 143–4, 153–5; *New York Times*, 7 April 1961.

37. Meyer and Szulc, p. 115.
38. Press Conference No. 9, 12 April 1961, Transcripts of Press Conferences, President Kennedy and Press, (Frederick, Maryland: University Press of America, 1982), Reel 19.
39. *Ibid.*
40. Wyden, pp. 144–6.
41. Schlesinger, p. 271; Davis, pp. 456–7.
42. News Conference No. 120 with Pierre Salinger, 12.15 pm, Saturday 15 April 1961, in 'Press Conferences with Presidential Press Secretary Pierre Salinger and Assistants', President Kennedy and the Press, Reel 1.
Reel 1.
43. Quoted in Schlesinger, p. 271; see also Davis, pp. 456–7; and Walter E Johnson, ed., *The Papers of Adlai E Stevenson Volume VIII Ambassador to the United Nations 1961–1965* (Boston: Little, Brown, 1979), p. 53.
44. Wyden, pp. 156–7. Both Schlesinger, p. 271, and Johnson, p. 53, n. 87 emphasise the vagueness of the briefing.
45. Schlesinger, pp. 271–2; Johnson, p. 53, n. 90; Wyden, pp. 186–8. According to Wyden, the use of the photograph was cleared through the Assistant Secretary of State for Public Affairs, who assured one of Stevenson's assistants that the photograph was 'clean'. Wyden does not say whether the Assistant Secretary knew the truth about Zuniga's plane.
46. *New York Times*, 16 April 1961; Johnson, p. 91.
47. Quoted in Wyden, pp. 189–90.
48. Wyden, pp. 176, 183–4; *New York Times*, 17 April 1961; Davis, p. 458; Johnson, pp. 93, 129, 146, 172. The Revolutionary Council's bulletins described the landings as support for an internal resistance movement, and later claimed that large numbers of Cubans were defecting to the rebel forces.
49. News Conference No. 123 with Pierre Salinger, 18 April 1961, President Kennedy and the Press, Reel 1.
50. Higgins, p. 13.
51. Quoted in *Operation ZAPATA*, p. 182.
52. Wyden, p. 310.
53. Quoted in *Operation ZAPATA*, p. 182.
54. Various sources refer to the 'disposal problem'—the risks attendant in cancelling the operation and dispersing the exiles, who would then be free to talk. This was one of the key factors in keeping the plan moving forward. See *Operation ZAPATA*, pp. 12, 18, 41, 176, 201; Schlesinger, pp. 239–42; Higgins, pp. 99, 102–3, 106, 107–9, 167–9.
55. Quoted in Wyden, p. 310.
56. Wyden, pp. 94–5, 31, 316–20; Corson, pp. 28–9; Schlesinger, p. 241. See also, Schlesinger quote in Gregory F Treverton, *Covert Action: the Limits of Intervention in the Postwar World* (New York: Basic Books, 1987), p. 91.
57. Jorge I Dominquez, ed., *Cuba: Internal and International Affairs* (Beverly Hills: Sage Publications, 1982), pp. 12–13, 168; see also Sandor Halebsky and John M Kirk, eds., *Cuba: Twenty-Five Years of Revolution, 1959–84* (New York: Praeger, 1985), p. 99.
58. Those making the case for Khrushchev's misjudgement of Kennedy include: Spanier, pp. 107–9; Herbert S Dinerstein, *The Making of a Missile Crisis: October 1962* (Baltimore: Johns Hopkins University Press, 1976), pp. 130–31, 134–5; and Jacques Lévesque, *The USSR and the Cuban Revolution: Soviet Ideological and Strategical Perspectives, 1959–77*, trans. Deanna Drendel Lebouef (New York: Praeger, 1978), pp. 29–30. For the contrary view, see Richard Ned Lebow,'The Cuban Missile Crisis: Reading the Lessons Correctly', *Political Science Quarterly*, Vol. 98, No. 3 (Autumn 1983), pp. 438–46.
59. See Chapter 7; see also David Wise, *The Politics of Lying* (New York: Random House, 1973), and John M Orman, *Presidential Secrecy and Deception: Beyond the Power to Persuade* (Westport, Conn.: Greenwood Press, 1980).

CHAPTER 16

'The Uncounted Enemy: A Vietnam Deception' Revisited

MICHAEL A HENNESSY

If there were a verb meaning 'to believe falsely', it would not have any significant first person, present indicative.

Ludwig Wittgenstein

The 'credibility gap' opened by the U-2 and Bay of Pigs incidents[1] grew to alarming proportions during and after the Vietnam war.[2] Official United States Government secrecy or apparent misrepresentation surrounding events such as the Gulf of Tonkin incident and the secret bombing of Cambodia attracted charges of deceit.[3] The most recent and perhaps most controversial of these arose in 1982 with the airing of the *CBS Reports* television documentary 'The Uncounted Enemy: A Vietnam Deception'. The programme accused General William C Westmoreland, Commander of the United States Military Assistance Command Vietnam (MACV) from 1964 to 1968, of deliberately conspiring to deceive President Lyndon Johnson and the American public. In particular, it was argued that General Westmoreland and others at MACV falsified the Viet Cong order-of-battle and under-reported the rate of North Vietnamese infiltration prior to the 1968 Tet Offensive. This conspiracy, it was further alleged, although intended to strengthen both the President's re-election drive and America's determination to stay the course in Vietnam, backfired and actually heightened the disillusionment of the President and the nation resulting from the offensive.[4] Westmoreland sued CBS for libel, but the case ended inconclusively with an ambiguous out-of-court settlement. Presiding Judge Pierre Leval commented at the termination of the case that it was perhaps 'best that the verdict will be left to history'.[5] It has been said that journalism is the first draft of history. Several accounts have taken to task 'The Uncounted Enemy' for its biases and journalistic failures. Rehearsed testimonials, erroneous attribution of quotes, failures to disclose contradictory evidence and the knowing incorporation of false information, are among the well-documented ethical failures of this documentary. Yet, despite those deficiencies, the central thesis of the documentary has been widely accepted in the secondary literature on the war—CBS' first draft has held sway.[6]

This acceptance has largely resulted from the absence of a thorough, critical evaluation of the substantive allegations of the documentary. Its producer, George Crile, and the programme's paid adviser, former CIA analyst Samuel A Adams, advanced the ambiguous, and inevitably misleading thesis that MACV had misrepresented the enemy threat in Vietnam to the public, President, and to the United States Armed Forces. This chapter examines one central question of 'A Vietnam Deception', namely, the debate within the United States intelligence community over the nature of the threat posed by the irregular forces arrayed against the United States and South Vietnamese forces. It demonstrates that CBS was right on one count of deception, wrong on another, and missed at least one other important deception altogether. The Crile–Adams thesis woefully failed to explore other explanations of the intelligence failures surrounding the Tet Offensive. Chief among such alternative explanations is that the American planners deceived themselves about the nature of their enemy. In contrast to the argument of the CBS documentary, it is suggested here that it was self-deception, aided by mirror-imaging and rigid thinking, that resulted in a classic form of intelligence failure. This self-deception, rather than a 'conspiracy', compounded the disillusionment born of the 1968 Tet Offensive.

The documentary took as its point of departure America's apparent unpreparedness for the Tet Offensive. By the close of 1967 the situation on the ground had greatly improved for American forces over the previous years. The Unites States intelligence community estimated that the population support base of the guerrillas had been eroded by about 30 per cent.[7] In part motivated by these gains, and in an effort to bolster the resolve of a discontented, war-weary America,[8] the President, General Westmoreland, and other senior administrative representatives embarked on a public-speaking campaign in September 1967. As the year drew to a close, however, ambivalence over the real successes and high costs of the war was heightened by the knowledge that the Viet Cong and the North Vietnamese Army would soon launch their annual Winter–Spring Offensive. There were reports that this year's assaults would prove unprecedented in size and scope. The President had been warned in November that the war was 'probably nearing a turning point and the outcome of the 1967–68 Winter–Spring campaign . . . [would] . . . in all likelihood determine the future direction of the war'.[9] As well, equivocal warnings of an impending 'Kamikaze' attack on South Vietnam's cities had been circulating since late in the year. So Westmoreland clearly was forewarned.

Forewarned, however, did not prove forearmed. There were conflicting reports on the form the offensive might take, and overt North Vietnamese actions, especially the placing under siege of the United States Marine base at Khe Sanh, complicated American estimates and planning. On 29 January a solid warning was issued of an impending attack on an undetermined number of urban areas. Westmoreland soon cancelled the Tet holiday leaves for American personnel and prevailed upon the South Vietnamese to do similarly, at least in I Corps. Despite this alert, an interagency review of the intelligence available before the offensive later concluded, the 'fact that the attacks would involve near simultaneous assaults against over three quarters of the province capitals and other major cities, however, was not anticipated, nor was it probably possible to do so on the basis of the available intelligence'.[10] Consequently, the enemy achieved a strategic *coup de main*.

Nothing had prepared the public or the Johnson administration for the Tet attacks that began on 30–31 January 1968. The American Embassy in Saigon, 44 provincial capitals, five of six autonomous cities and many district capitals, villages and hamlets were either overrun or attacked by fire.[11] Although American and South Vietnamese forces had little difficulty breaking the Viet Cong offensive, the ability of the supposedly weak Viet Cong to infiltrate to the heart of South Vietnam's urban centres, evading more than one-million American and South Vietnamese troops, brought American domestic dissatisfaction with the war to a head. Gallup polls recorded the sharpest decline in American domestic support for the war since early 1966: dropping between January and April from roughly 42 per cent in favour of the war to 34 per cent.[12]

Shock and dissatisfaction was most pronounced at the White House. As sustained enemy offensive actions continued into March, President Johnson ordered a complete 'A–Z' reassessment of America's objectives. While immediate reinforcements were despatched, the forces necessary for a prolonged counter-offensive were not authorised. Instead Johnson ordered a unilateral reduction of the strategic air campaign against the North, announced his withdrawal from the Presidential race, and offered

Hanoi the olive branch—which was grasped with alacrity. This retraction of the American war effort resulted not simply from the offensives of 30–31 January but, rather from the whole protracted North Vietnamese/Viet Cong Winter–Spring Offensive.[13] Yet, the events of 30–31 January struck the key blow, from which the Johnson administration never recovered.

The unpreparedness of the Administration and general public for those events was the centrepiece of the CBS documentary. It focused on the failure of the United States Government to anticipate such enemy assault capabilities as those demonstrated at Tet. CBS alleged that the surprise of Tet resulted, as George Crile (the series producer) would contend, from a process whereby the 'US Military Command in Vietnam entered into an elaborate conspiracy to deceive Washington and the American public as to the nature and size' of the enemy in Vietnam. Furthermore, the programme intended to demonstrate that 'a number of very high officials—General Westmoreland included—participated in a conspiracy that robbed this country [USA] of the ability to make critical judgements about its most vital security interests during a time of war'.[14] While, in the build-up to the libel trial, CBS would waffle on the 'conspiracy' charge against Westmoreland, yet, the documentary clearly alleged, as *Newsweek* explained, that General Westmoreland 'and his aides' suppressed 'accurate estimates of growing enemy troop strength in South Vietnam before the 1968 Tet Offensive'.[15] The programme asserted that the shock of Tet was directly attributable to this conspiracy.[16] Several writers have challenged the programme on ethical grounds instead of challenging the assumptions of the conspiracy theory. It is the latter course that this essay follows.

At the heart of the CBS conspiracy theory was a debate which raged within the American intelligence community[17] on the eve of Tet. At issue was the specific composition of the enemy Order-of-Battle (OB) produced within MACV. These OB estimates provided the most important data base on the enemy's human resources. In turn they were employed in constructing National Intelligence Estimates which were utilized for war planning by the entire American policy, intelligence, and defence establishment. Consequently, the OB debate which took place during 1967 was directly related to the production of Special National Intelligence Estimate (SNIE) 14.3 (November 1967) and subsequent policy decisions derived from it. The MACV OB included three main categories of enemy combat forces. 'Main' and 'Local' force designations were applied to both North Vietnamese and Viet Cong forces which operated in formed units of platoons, or larger, and under direction from regional, front, province, or district commands. The third category, 'guerrillas' applied to all forces subordinate to village or hamlet party commanders. However, there were also district and province guerrilla units, of platoon and company size, recorded within this category.[18]

Although the internal intelligence debate over SNIE 14.3-67 concerned a wide number of forces, such as administrative personnel and the 'Village Assault Youth',[19] CBS focused on one category in particular. At issue was the fate of the Viet Cong 'Self-Defence' and 'Secret Self-Defence' militias,[20] which operated at the village and hamlet level. These had been included in the third (guerrilla) category until late 1967. CBS chose to use the decision to drop them from the OB at that time—just prior to Tet—as a major buttress to its allegation that Westmoreland, and others at MACV, purposely misled the Administration and the public on the size of the enemy thereby ensuring that MACV's forces, the President, and the American public were, in the words of one observer, 'unprepared for the all-out attack the enemy launched on Vietnam's Tet holiday . . .'[21]

The documentary's allegation that the suspected size of the Viet Cong irregular forces was purposefully hidden from the public has been substantiated. As was made clear at the trial, anticipation of a negative reaction from the American press induced General Philip Davidson, head of MACV's J-2 intelligence section in 1967, to keep MACV's total enemy military strength substantially around the 298,000 figure. Part of an August internal memo from General Davidson to Colonel Godding, also of MACV J-2 stated:

> **2.** Further consideration reveals the total unacceptability of including the strength of the self-defence forces and secret self-defence forces in any strength figure released to the press.
>
> **3.** The figure of about 420,000 which includes all forces including SD and SSD [Self-Defence and Secret-Self-Defence-Forces], has already surfaced out here. This figure has stunned the embassy and this headquarters and has resulted in a scream of protest and denials.
>
> **4.** In view of this reaction and in view of General Westmoreland's conversations, all of which you have heard, I am sure that this headquarters will not accept a figure in excess of the current strength figure carried by the press.[22]

In a later cable to the Joint Chiefs of Staff on the subject of this decision, General Creighton Abrams, Westmoreland's military deputy and successor, explained that MACV had rightfully presented a picture of success in the war but continued:

> All available caveats and explanations will not prevent the press from drawing an erroneous and gloomy conclusion as to the meaning of the increase. All those who have an incorrect view of the war will be reinforced and the task will become more difficult.[23]

Similarly, several days later Ambassador Ellsworth Bunker communicated this point directly to Presidential adviser Walt Rostow. Indeed, in the documentary, Westmoreland conceded the fear of press reaction when he argued that neither the press nor many people in Washington would understand the different categories of enemy; they would simply see the dramatic increase in numbers.[24] Given the growing criticism of the war, a fact of which both the policy-makers and the press were well aware, these men presumably thought they were merely being politically prudent, not deliberately deceptive. That the decision to exclude the self-defence and secret self-defence forces from the OB represented a consensus decision between the Ambassador, his deputy Robert Komer and the two senior American military commanders in Saigon cannot be explained away as simply a conspiratorial coincidence. Rather, it suggests, as will be shown, a convergence of opinion, that while perhaps wrong, was premised on a shared set of assumptions extending far beyond any clear concerns over 'press reaction.' Nevertheless, efforts to keep bad news from the press cannot be denied.

Does this mean that MACV kept two sets of books on the enemy OB, thereby allowing it to deceive policy-makers in Washington with false statistics? The evidence suggests that it did not,[25] and it is this point which clearly undermines that part of the CBS' deception conspiracy thesis that relates to Washington decision-makers. Further, it is the key to unlocking a far more plausible explanation of MACV's conduct.

Before offering that more plausible explanation, however, it is essential to dispose of the allegation that MACV conspired to deceive Washington. This requires an exploration of the genesis of the debate over the OB and the SNIE, which itself must begin with a brief explanation of American intelligence resources and processes in Vietnam.

Until mid-1965, when MACV's 'J-2' (Joint-service intelligence staff section) began to produce its own independent, detailed OB, American intelligence was dependent on the South Vietnamese Government's estimates of Viet Cong strength. These were notoriously unreliable.[26] Because General Westmoreland was also the commander of the United States Army, Vietnam, his J-2 section, initially headed by Major General Joseph A McChristian, also functioned as the Army's 'G-2' operational intelligence branch, responsible for generating tactical estimates.[27] By 1967 all estimates were generated through the office of Deputy J-2 Production. This section's appreciations were based on the historical listings compiled by the Combined Intelligence Center, OB section, and on the daily tactical and long-range forecasts provided by the joint Warning, and Current Indications and Intelligence Branch.[28]

While there were many intelligence consumers within the US government with agencies or representatives operating in Vietnam, only the Central Intelligence Agency (CIA) had resources even approaching those of MACV. Strategic intelligence, in the form of national estimates, was produced by the Office of National Estimates, under the supervision of the Director of the Central Intelligence Agency (CIA). MACV, on the other hand, primarily generated operational and tactical estimates. Hence the Defense Intelligence Agency, National Security Agency, and State Department remained largely dependent on intelligence garnered by either MACV or the CIA; however, the distinctions between their respective roles can prove misleading. This is especially the case in the generation of an accurate estimate of the Viet Cong political, administrative, and military forces that were purely indigenous to South Vietnam. These factors could not be so readily traced by trail watching, photo-reconnaissance, and signals intelligence as were the Northern infiltrators. While the CIA was particularly strong in tracing the political intrigues of Saigon, MACV held a virtual monopoly of sources on the Viet Cong administrative and military organisation, especially in the countryside. Indeed, MACV's dominance in this field is clearly substantiated by the fact that the bulk of the information scrutinised by Sam Adams, the CIA's chief irregular forces analyst, came to his desk in Langley, Virginia, via MACV channels.[29] Why the CIA allowed such dominance has never been examined publicly nor has the dependency of the entire United States intelligence community on MACV sources been explained. That explanation is beyond the scope of this chapter but it is necessary to emphasise MACV's unchecked dominance in this field.

America's intelligence community proved particularly remiss in failing to generate an accurate estimate of the Viet Cong strength and composition during the initial years of United States combat. This is not to suggest that this would have been an easy task. The Viet Cong was a clandestine organisation, well-disciplined and securely compartmentalised, prone neither to easy penetration nor to large numbers of defections. It would take time for American intelligence to become familiar with the organisational structure, let alone to obtain an accurate count for each component. After all, even the Viet Cong itself did not have a definitive count.[30] The least readily verified component was the irregular, administrative and political forces, which included the Self-Defence and Secret Self-Defence units. The significance of these forces to the Tet Offensive would become a central issue in the CBS documentary.

The omission or under-reporting of the Self Defence forces in the OB and SNIE military-threat categories had long been the particular concern of CIA analyst Adams. Indeed, the issue became an *idée fixe* which eventually cost him his position within the CIA. Adams played a central role in the debate from the first. Responsible for reviewing intelligence on Viet Cong irregular forces, Adams concluded in August 1966 that MACV's OB included many categories whose strength estimates had remained virtually unaltered from those adopted in the first MACV OB generated the previous year. Beyond the Main and Local force figures, the original OB listed 103,573 irregulars, broken into: 34,318 guerrillas; 35,661 Self-Defence; and 33,597 Secret Self-Defence militia troops. Adams grew alarmed that neither these figures nor those for the political cadre or service-administrative forces had been updated to reflect either growth or attrition since their incorporation into the OB. In a series of memos he protested this oversight.[31] Reports prepared by Adams, bolstered by his personal canvassing of MACV field intelligence personnel, supported his contention that the Viet Cong irregular forces were, in fact, under-reported. In a failure that condemns not only MACV but the entire intelligence community, Adams revealed that as late as November, 1966 sub-district and sector intelligence sections responsible for appraising the situation at the district, village, and hamlet levels for MACV had no 'criteria' spelling out 'categories of irregulars IE, guerrilla self-defense force or secret self-defense force [sic].'[32] A 'fast and dirty' estimate of irregulars completed by the MACV OB section at that time arrived at a figure in the order of five times greater than that then listed in the official OB.[33] A protracted debate within the American intelligence community ensued.

Throughout 1967, Sam Adams was a member of an analysis team responsible to Dr George Carver, the DCI's Special Adviser on Vietnamese Affairs, tasked with finalising the CIA's contribution to SNIE 14.3-67. As a member of Carver's staff, Adams attended a February 1967 meeting in Honolulu called at the behest of the Chairman of the Joint Chiefs of Staff, General Earle G Wheeler. Wheeler, concerned with the discrepancies between the MACV and CIA estimates, wished to see a reconciliation so that the military and CIA would 'play off the same sheet of music'[34] and agree how and what to count. Joseph McChristian and his deputy in charge of the OB section, Lieutenant Colonel Gains Hawkins, attended for MACV. According to Adams, Hawkins concurred with his revised aggregate OB figure of 500,000 including main, local and irregular forces, which was almost twice the figure of approximately 270,000 currently listed by MACV's OB. McChristian, however, would later write that the meeting simply substantiated MACV's criteria.[35]. Although Adams, and apparently Hawkins, believed MACV accepted the higher estimates for inclusion in the OB, in fact it had not. While the February meeting did not resolve the issue Hawkins and McChristian continued to refine their estimates of irregular forces. By May, 1967, Hawkins had completed a far-reaching review of the OB which later contributed to the proposed SNIE 14.3-67. His new estimate was summarised in a cable prepared by McChristian, who chose to inform Westmoreland of its contents before forwarding it to Washington. Hawkins had concluded the irregular strength should read some 185,000, which was an aggregate well short of Adams' estimated 250,000 but well above MACV's current 112,000. Furthermore, his estimate of 'political cadre' read some 90,000 while MACV's remained at 39,000. Westmoreland delayed forwarding the cable.[36]

That decision became a central point of the CBS documentary—the argument being that Westmoreland kept the figures from his superiors. However, Westmoreland did inform his direct superior, Ambassador Ellsworth Bunker and his deputy, the newly arrived Robert Komer. According to Komer, the higher figures were made known to Washington through 'back channels';[37] moreover, the CIA had not acquiesced in the debate, but chose to maintain a figure of roughly 500,000. Yet, all these estimates of 'irregular forces' were derived, inductively, from extremely small data bases. In the later debate between MACV and the CIA it became clear that each would admit some logic to the other's position, because both were premised on scanty evidence and different assumptions. MACV's estimate was derived from an in-depth study of only three of South Vietnam's 234 districts, while Adams' data base was several times greater; neither data base was sufficient to guarantee a statistically relevant representation.[38] So Washington was not stuck with only a single estimate of uncertain accuracy, but rather with several. CBS would virtually ignore these caveats raised during the protracted negotiations over the exact composition of the enemy forces that occurred between May and November 1967.

With the order of battle debate still deadlocked in September and the Special National Intelligence Estimate coming due, DCI Richard Helms despatched Carver, Adams and fellow analyst William Hyland to MACV for a marathon eleventh-hour negotiating session. The negotiations ended in impasse until, in a *tête-à-tête*, Carver suggested a compromise to Westmoreland. Carver's compromise was to accept the reclassification of the Self Defence and Secret Self-Defence as non-offensive military troops while increasing their estimated size to 150,000, and increasing the cadre estimate to a range of 75–85,000.[39]

In November MACV's new OB was released to the press. Though the omission of the self-defence units was acknowledged, the overall increase in numbers was not. Completed within days of the new OB, SNIE 14.3-67 'Capabilities of the Vietnamese Communists for Fighting in South Vietnam', confidentially acknowledged and explained the higher figures.[40] Given the problem of accurately determining the size of the self-defence and other irregulars, the total was included in prose form, described within broad ranges, as an addendum to the numerical OB. The Special National Intelligence Estimate read: 'Our current evidence does not enable us to estimate the present size of self-defence, secret self-defence, the 'Assault Youth,' or other similar Viet Cong (VC) organisations with any measure of confidence.'[41] Hence, in line with the Carver compromise, no clear aggregate estimate of these forces was given; nevertheless, anyone who took the time to tally the actual figures contained in the estimate would have arrived at a total enemy force of between 518,000 and 568,000, a figure that would seem to vindicate Adams' estimates and demonstrates they were available to the policy-makers in Washington, and implicitly, if reluctantly, acknowledged by MACV.[42]

But Adams refused to accept this outcome. Following Tet, he restated his objections on numerous occasions. This led eventually to an enquiry by the inspector general of the CIA, but Adams was never completely satisfied.[43] After resigning from the Agency, he aired his objections in a 1975 article for *Harper's*, 'Vietnam Cover-up: Playing War with Numbers'.[44] CBS producer George Crile, who had been an editor of *Harper's* at the time, grew convinced he'd found the makings of a damning exposé. He would later adopt Adams' thesis and hire him as the documentary's only paid consultant.[45]

The decision to drop the Self-Defence and Secret Self-Defence units from the numerical order-of-battle was partly explained during the trial. CBS had alleged that it was all part of the conspiracy of silence masterminded by MACV. However, the trial clearly demonstrated that senior officials within the CIA knew of the problems with the OB figures, and that moreover, it was George Carver who suggested to General Westmoreland the final compromise whereby the Self-Defence and Secret Self-Defence units were removed from the numerical OB but retained in the addendum. Carver had explained this to Crile, but the point was deleted from the broadcast.[46]

Nor did CBS elaborate on the need for and inherent problems of consensus building which plays a major role in generating all Special National Intelligence Estimates. Intelligence is at best a grey art. Two men reviewing the same information may well disagree over its significance and basic interpretations. This is both necessary and unavoidable. Reasonable differences are generally solved through some form of compromise unless everyone agrees to a more extreme interpretation of the data. Dissenting views are often relegated to footnotes[47] but SNIE 14.3-67 had not employed this evasive approach; it had simply stated in prose the problem of actually estimating the size of the irregular forces. CBS did not explore this and the wider problems inherent in producing estimates. Of that process the former Deputy Director of Intelligence Ray Cline has observed:

> Intelligence estimators are always wrong, and there are always plenty of people around to tell them so. The questions that are brought up for public scrutiny and even scrutiny in the high levels of the government are never the simple questions on which intelligence can give clear and precise answers. . . . your judgement is all that you have to go on.[48]

With respect to generating estimates on Vietnam, all estimates were complicated both by saturation of the intelligence system and dubious data bases. Even Sam Adams conceded that it was physically impossible for him to analyse fully the 4,000 pages 'or more' of summaries and reports which crossed his desk each month.

Beyond the trivialisation of the complexities of the intelligence process and its down-grading of the ever present problems surrounding the human assessment of intelligence, CBS never established the actual influence that Special National Intelligence Estimate 14.3-67 played in setting up the senior members of the executive branch for the psychological blow of Tet.[49] No real light can be shed on that issue here either, except to examine what the White House did know before Tet.

Although it is clear the CIA knew of the internal OB debates at MACV, CBS further contended that information was withheld from the White House. Walt Rostow, a former special adviser to the President, has repeatedly argued that the White House was not dependent solely on MACV intelligence.[50] Just how much the White House knew of the OB debate has long been in question. Recently declassified material, however, clearly establishes that Robert Komer, then the President's special adviser on Vietnam, charged with monitoring progress in the war, was informed of the debate at least as early as 28 September 1966.[51] So the issue was never kept from the White House staff. Knowledge of the debate is clearly traceable to the President's senior staff. But this fact was not acknowledged in the documentary.

In a further failure to test this thesis, CBS did not explain that internal debates over the SD and SSD units continued well after the Tet Offensive. In fact, although the issue was re-examined after Tet, the intelligence community never resolved the dispute over the strength of the irregular forces and who should be included in the OB. MACV's position in the debate proved unwavering; but, as demonstrated, it was also a position well known to the entire intelligence community. Although the Defence Intelligence Agency accepted MACV's approach, it was rejected by the National Security Agency, the State Department and the CIA. But the CIA was the only national level intelligence agency with resources that could have surpassed MACV's field intelligence capacity and obtained a truly representative count of the village level irregular forces.[52] Hence it was clearly imprudent for CBS to claim that a conspiracy existed when those who were supposedly misled were openly informed of the issues—as was the case of the White House—and continued to debate the issue before and well after the events of Tet. These shortcomings of the CBS thesis are even more surprising considering that Sam Adams was present at several of the post-Tet meetings examining the irregular force issue.[53]

The foregoing suggests that CBS' charge of deception of Washington elites does not stand up to scrutiny. While unable to provide Washington with a precise statement of the size of the Viet Cong, MACV did concur in a compromise estimate which reduced uncertainties to a range, but did not delete them entirely. In any event, SNIE 14.3-67, which was generated for the President and his advisers, more closely approximated the CIA's figures than those of MACV. Moreover, the discrepancies in the rival estimates were widely known, so MACV can hardly be accused of keeping two sets of books which might have allowed it to mislead the President. Indeed, the fact that it did not do so points clearly to the issue that CBS overlooked; simply put, American military doctrine and MACV's mission in Vietnam did not construe the Viet Cong irregular forces to be the United States Army's responsibility. As the MACV press release announcing the decision to drop the Self-Defence and Secret Self-Defence forces explained: 'they are not a fighting force comparable to the guerrilla . . . they do not represent a continual or dependable force and do not form a valid part of the enemy's military force'.[54] This same reasoning led MACV to argue for their deletion from the OB. It raises two important, interwoven questions: was it a sound assumption, and did MACV really believe it?

As will be explained, the answers are, respectively, no and yes; it was a faulty assumption, but MACV's Commander did believe it. Together these answers indicate the scale of the American military's misappreciation of the type of war it was fighting. They also demonstrate why the Viet Cong was able to achieve strategic surprise with the Tet offensive.

In theory a distinction exists between deception engendered by enemy actions and that which results from self-induced misunderstanding of the threat. In practice, where self-deception leaves off and active enemy deception operations begin, is often difficult to distinguish. Although CBS did not examine this problem, there was sufficient evidence available to suggest the shock of Tet resulted from the combined effects of an active enemy deception operation[55] and an incorrect conception of the enemy threat. It is the latter which is perhaps least acknowledged or understood, but it is arguably the most important factor in explaining the decision to drop the SD/SSD and other irregulars.

Michael Handel has advanced the proposition that the 'major causes of all types of surprise are rigid concepts and closed perceptions'.[56] The historical record indicates quite clearly that the United States Army in Vietnam was burdened in just this way. Marshalling impressive evidence in support of his case, US Army Major and historian Andrew Krepinevich argues that a rigid 'Army Concept' of war-fighting closed the Army's eyes to the nature of the war in which it was engaged. Further, guided by a mission statement, the 'Overseas Internal Defense Plan', that kept military and civilian roles in the war distinct and separate, the Army resisted adaptation to 'counter-insurgency'. Instead, it redefined the war in its own terms, and this oriented American forces to the conventional, large-unit threat, rather than against the guerrilla forces.[57]

The mission that Washington assigned to the American forces was to drive the enemy main-forces out of South Vietnam and thereby allow the South Vietnamese Army and National Police to neutralise the guerrillas and village and hamlet militias. The tactical role of MACV's army dictated that the OB, though receiving wide distribution, was really intended for Westmoreland, wearing his second hat as Commander United States Army, Vietnam, to plan operations against the main-force enemy. Significantly, the chief of the Current Intelligence, Indications and Estimates branch, Lieutenant Colonel Daniel Graham, who ranked next to Westmoreland in the list of 'conspirators' vilified by CBS, was one of the chief opponents of adopting Adams' estimates for the OB. As Graham saw it, their inclusion would have confused the issue for American forces. Consequently, given MACV's explicit mission of driving out the enemy's main forces, the irregulars did not, by definition, constitute a major concern. There were of course a number of inconsistencies and problems with such an approach—many of which Sam Adams pointed out, and the Tet offensive confirmed.

Nevertheless, under the functional mission given to MACV, keeping the Self-Defence and Secret Self-Defence units off the OB was fully consistent with the role of the American forces. Explicitly, that role warrants further explanation than can be provided here, but it is important to note that the CBS programme did not address at all the mission the Joint Chiefs of Staff had given to MACV. With this oversight, however, CBS missed the real 'Big Story'. The MACV model caused many commanders to give little weight to the warnings sounded on 29 January for several reasons. First, from the previous pattern of enemy actions during periods of truce, and from the observation that the enemy had never attempted a general offensive, few commanders or troops paid much attention to the warnings. Second, the threat of a 'general uprising' was down-graded because the MACV analysts believed that the enemy would not risk the 'potential disaster to their military machine', and did not possess the 'capability' for launching such a sweeping offensive.[58] Tet, of course, proved this wrong.

The Army Concept, the conventional orientation of the Army's mission, and the consequent deletion of the irregulars from the OB demonstrate that Army planners were engaged in 'mirror imaging'; they assumed that the enemy saw the war in the same terms as themselves. Since the Americans could not conceive of mounting such a high risk operation as that of Tet 68, with the forces believed available to the enemy, they assumed that the North Vietnamese and Viet Cong could not do so either.

The Vietnamese Communists laboured under no such illusions. They saw their war as a revolutionary insurgency, encompassing politics and violence, wherein control of the social structure and of perceptions of power and legitimacy were more important in the long run than control of territory or winning set-piece battles.[59] As such, they did not draw the same kind of clear-cut distinctions as the Americans did between political and military roles, or between the forces which would carry them out. Of the three types of forces, main force units, local force units and guerrilla forces, and self-defence militia, each clearly had its own tasks, but shared the purpose of achieving a political victory. Hence all three forces were seen as complementary. North Vietnamese leaders stated that:

> If we fail to combine the three categories of troops, ou[r] capability will not be adequate. . . . All our three kinds of armed forces have been fostered and developed in a planned way, evenly, proportionally, and in conformity with the practical conditions on the battlefield and with the combat tasks of each kind of armed forces. Thus by closely co-ordinating our three kinds of forces . . . and closely co-ordinating the armed struggle with the political struggle . . . [the] armed forces and people . . . are fully in a position to defeat enemy forces far superior in number.[60]

After the event, the CIA concluded that more than one-third of the 400,000 or more North Vietnamese and Viet Cong troops who took part in the 1968 Winter–Spring Offensive were Viet Cong irregulars which had been discounted by MACV because, according to United States military doctrine, they did not represent a 'military threat'.[61]

There was no 'conspiracy' to deceive America's national security policy-makers, as CBS alleged. There was a serious misunderstanding on the part of the United States military and civilian policy-makers of the nature of the war that was being fought—one that seems to have generated an institutional self-deception about the potential importance of irregular troops. This significant failure of military intelligence and politico-military strategy deserved critical analysis, but was missed entirely by CBS in their haste to pillory General Westmoreland on a charge of deception. A more balanced documentary would have acknowledged the serious ambiguities of its evidence, thus allowing the viewer to apprehend the very real complexities of the isssues, and to question the Adams–Crile deception thesis.

On the other hand, it did demonstrate that there was a deliberate and sustained effort by senior MACV commanders and staff officers to present the American public with enemy strength figures that would reinforce confidence that the war could indeed be won. The explanations outlined above provide honourable motivations for the decisions of busy and harassed men, burdened with intelligence from all quarters, whose overriding concerns were those of all good military commanders: to pursue the war to a successful conclusion. In any case, such motives need not be condemned out of hand. After all, democratic governments regularly refrain from making public damaging news about the security of financial institutions in order to forestall panic selling. The moral dilemmas of such matters are less acute than those of war.

If the war had been won, presumably America would have forgiven the deception worked upon it, just as the British public after the Second World War forgave the lies perpetrated by officialdom to fool the Nazis, which inevitably fooled the public too.[62] As it was, the deception was exposed in the cruellest manner possible, by the impact of the Tet Offensive. By raising false expectations, the optimistic figures left the United States public totally unprepared for what was to come. The subsequent disillusionment may have been deepened at least in part, by the attempt to avoid it.

Endnotes

1. See Chapters 14 and 15.
2. George C Herring, *America's Longest War: the United States and Vietnam 1950–1975*, 2nd ed. (New York: Alfred Knopf, 1986), pp. 175, 191, 203. Portions of this chapter are based on the author's thesis 'Divided they Fell: America's Response to Revolutionary War in I Corps, Republic of Vietnam, 1965–1971'. (MA Thesis: University of New Brunswick, December 1987.) The author is indebted to the editors and Lieutenant Colonel Alexander ('Sandy') Cochrane, PhD, of the US Army Center for Military History, for comments on a previous draft of this chapter.
3. Herring, pp. 119–22, 225; see also John M Orman, *Presidential Secrecy and Deception: Beyond the Power to Persuade* (Westport, Conn.: Greenwood Press, 1980), pp. 98–9, 142; and Edward J Epstein, 'The Pentagon Papers: Revising History', in his *Between Fact and Fiction: The Problem of Journalism* (New York: Vintage, 1975).
4. 'CBS Reports', 'The Uncounted Enemy: A Vietnam Deception' (New York: CBS Inc., 1982), telecast 23 January 1982.
5. Judge Leval quoted in Jonathan Alter, 'The General's Retreat', *Newsweek*, 4 March 1985, p. 59.
6. Military accounts of the war have virtually ignored the intelligence factors of the confrontation, and histories of America's intelligence agencies have generally accepted, unaltered, the conclusions of the CBS documentary by relying largely on the comments of Sam Adams. Bruce Palmer's *25 Year War: America's Military Role in Vietnam* (Lexington, Ky.: University of Kentucky Press, 1984), is an exception to the military histories. Of the intelligence histories, compare John Ranelagh, *The Agency. The Rise and Decline of the CIA*, rev. ed. (New York: Simon & Schuster, 1987), pp. 454–70, and Thomas Power's, *The Man Who Kept Secrets* (New York: Alfred A Knopf, 1979), p. 187. See also, Walter Laqueur, *A World of Secrets. The Uses and Limits of Intelligence* (New York: Basic Books, 1985), p. 171-passim. See also, Bob Brewin and Sydney Shaw, *Vietnam on Trial: Westmoreland vs. CBS* (New York: Atheneum, 1987); cf. the accounts of two former MACV intelligence officers, T L Cubbage II, 'Westmoreland vs. CBS: Was Intelligence Corrupted by Policy Demands?', *Intelligence and National Security*, Vol. 3 (July 1988), pp. 118–80; and, Bruce E Jones, *War Without Windows* (New York: The Vanguard Press, 1987).
7. 'Missions and Activities of the VC Irregular Forces', 24 April 1968, p. 7 [CIA Research Reports, Supplement, (Frederick, MD: University Publications of America, 1986), Reel 6, 00075]. Hereafter cited as [CIARR Supp.].
8. Robert Pisor, *The End of The Line* (New York: W W Norton), pp. 22, 69; 'We Are Definitely Winning', and 'End of the Vietnam War in Sight?' *US News & World Report*, 12 August and 11 September 1967; Peter Braestrup, *Big Story* (New York: Anchor

Press, 1978), p. 50; excerpts from President Johnson's 17 November news conference as reproduced in the *New York Times*, 18 November 1967; William C Westmoreland, *A Soldier Reports* (Garden City, NY: Doubleday, 1976), p. 22; and, Lyndon B Johnson, *The Vantage Point* (New York: Holt, Rinehart & Winston, 1971), p. 370.

9. Johnson, p. 371.
10. For example, see the conflicting expectations advanced at the time, as noted in: Braestrup, *Big Story*, p. 54; Walt W Rostow, *The Diffusion of Power* (New York: Macmillan, 1972), pp. 462–3; 'Situation in Vietnam', CIA Memo, 8 December 1967, [Declassified Documents Reference System: CIA 37C (Carrollton Press, 1978)]. Hereafter cited as [DDRS]. And CIA Post Tet internal report, 15 February 1968, pp. 1–5 [DDRS:CIA 38A 1978].
11. MACV Command History 1968, pp. 376, 386, 390, 397, 884, 890, 894–5 and 902, cited in Herbert Schandler, *The Unmaking of a President: Lyndon Johnson and Vietnam* (Princeton, NJ: Princeton University Press, 1977), p. 74.
12. See Mark Lorrell and Charles Kelley Jr 'Casualties, Public Opinion, and Presidential Policy During the Vietnam War', R-3060-AF (Santa Monica, CA: Rand Corporation, March 1985), pp. 17–19, and John E Mueller, *War, Presidents and Public Opinion* (New York: John Wiley & Sons, 1973), passim.
13. Herring, p. 191; Schandler, pp. 99–120, and 158; Johnson, pp. 504–6.
14. As cited from producer George Crile's pre-production 'bluesheet' in Don Kowet and Sally Bedell, 'Anatomy of a Smear', *TV Guide*, 29 May 1982, pp. 5-6.
15. Charles Kaiser, 'Who Broke the Rules', *Newsweek*, 14 June 1982, pp. 81–2. See also Kowet and Bedell, p. 3.
16. Kowet and Bedell, p. 3. See also, Don Kowet, *A Matter of Honor* (New York: Macmillan, 1984), presents a very detailed, if not completely reliable, account of the production of the CBS Reports and the internal intelligence debate.
17. See Peter Braestrup, 'The Uncounted Enemy: A Vietnam Deception, A Dissenting View', *Washington Journalism Review* (April 1982), pp. 46–8; and, Renata Adler, *Reckless Disregard: Westmoreland v. CBS et al., Sharon v. Time* (New York: Alfred A Knopf, 1986).
18. Col. Hoang Ngoc Lung, *Intelligence*, Indochina Monograph series (Washington, DC: US Army Center of Military History, 1982), pp. 87–8. *A Study of Strategic Lessons Learned In Vietnam*, Vol I. *The Enemy* (McLean Va: BDM Corp., 1979), AD-A096-424, p. 5–24.
19. The many thousands of Village Assault Youth policed battlefields, acted as porters and provided other support functions but because many were so young, without uniforms, and few had weapons, they were not listed in the OB, see, 'Memo for Chief of Indochina Division', from Sam Adams, 1 December 1966 [CIARR supp. reel, 00986].
20. Both terms reflect an imprecise taxonomy incompatible to divisions employed by the VC. Compare Douglas Pike's description of paramilitary forces with those of Edwin Moise; Pike, *Viet Cong* (Cambridge, Mass.: Massachusetts Institute of Technology, 1966), pp. 234–6; Moise, 'Why Westmoreland Gave Up', *Pacific Affairs*, Vol. 58, No. 4, (Winter 1985–6), p. 664; George Carver admitted to many of the empirical and epistemological problems of quantifying the irregulars, and argued that perhaps the standard criteria of the OB were really too narrow or inappropriate for this type of enemy, Message, Carver to Helms, 3 November 1967, [CIARR Supp. reel 5, 00261].
21. Kowet and Bedell, p. 5.
22. Message: Gen. P Davidson (MACV J-2) to Col. G Godding, (Washington), 19 August 1967, cited from trial transcript, in Moise, p. 670.
23. Abrams' further observation reveals that MACV was not unaware of the VC subversive infrastructure which United States combat forces were directed largely to ignore: 'They [Self-Defence & Secret Self-Defence forces] are no more effective in the military sense than the dozens of other non-military organizations which serve the VC cause in various roles'. See Memo from Abrams, acting commander MACV, to General Wheeler, copy forwarded to George Carver, re. controversy over inclusion of SD-SSD in OB estimate, 21 October 1966 [CIARR supp. reel 5, 00218]. See also, Eleanor Randolph, 'The Military Misled Us In Vietnam', *Washington Post*, 23 December 1984, p. B-1.
24. Abrams' Memo 21 October 1967 [CIARR supp. reel 50, 00218]. Westmoreland's explanation is also notable: '. . . the people in Washington were not sophisticated enough to understand and evaluate this thing, and neither was the media'. Westmorland com-

ments in 'The Uncounted Enemy', transcript cited in Kowet, p. 77. On similar dilemmas faced by democratic societies see Orman, pp. 8–9, 97–119.

25. Moise, p. 670.
26. Palmer, p. 25; Joseph H McChristian, *The Role of Military Intelligence* (Washington, DC: Department of the Army, 1974), p. 13; the weakness of GVN intelligence resulted from many factors, see Hoang, *Intelligence* passim. See also William Colby and Peter Forbath, *Honorable Men* (New York: Simon and Schuster, 1978), pp. 266–8 describe many of the failures to conduct and co-ordinate operations at the village level. Further indication of this failure is the fact that J-2's estimates of SD-SSD strength were developed through induction from a database of only three districts out of a national total of 234, see, Message: George Carver to DCI, 10 September 1967 [CIARR Supp. Reel 5, 00021–00024].
27. On the propensity of MACV to produce operational and tactical estimates suited primarily for their conventional force opponent, see the remarks of former OB Chief Lieutenant Colonel Gaines Hawkins, cited in James J Wirtz, review article in *Intelligence and National Security*, Vol. 2 (October 1987), pp. 180–83.
28. The Combined Intelligence Center was a joint US-GVN operation, see Hoang, p. 86, and McChristian, p. 17.
29. MACV's dominance in this field, however flawed, is discernible from the pivotal role it played in the production of SNIE 14.3.-1967. See, for instance, Carver's acknowledgements of Komer's criticisms of the CIA's analysis and its levels of staffing which were dwarfed by MACV J-2, in Message Carver to DCI, 12 September 1967 [CIARR Supp. Reel 5, 00014].
30. See Note 25 above, and Hoang, p. 86.
31. Palmer, p. 39; 'The Strength of the Irregulars', December 1966 [CIARR supp. reel 3, 00821]; Kowet, p. 39.
32. Cable, 'Saigon 9165', for DCI, subj. J-2 OB, 9 November 1966 [CIARR supp. reel 3, 00814]
33. *Ibid.*; 'Memo: for Chief, Indochina Division', 27 November 1966 [CIARR supp. reel 6, 0472]; 'Draft Working Paper "The Strength of the Viet Cong Irregulars" ' [CIARR supp. reel 6, 0472]; Moise, p. 664.
34. Memo from Gen Wheeler, cited in M A Faber, 'CBS Westmoreland Trial: A Reprise', the *New York Times*, 31 December 1984, p. 27.
35. McChristian, p. 129. Kowet, p. 40.
36. Specifically why Westmoreland did not immediately forward McChristian's estimate remains a moot question, given the subsequent wide ranging debate over the OB.
37. Eleanor Randolph, 'Troop Estimates Supported. Pacification Chief Denies Westmoreland "Cooked the Books" ', *Washington Post*, 17 October 1984, p. 12. According to one source, the memo was only held for three days, see Wirtz, p. 181.
38. See message Carver to DCI, 10 September 1967 [CIARR supp. reel 5, 00021–00024]; cf. Jones, p. 103.
39. See message Carver to DCI, 13 September 1967 [CIARR supp. reel 5, 00025].
40. 'Capabilities of the Vietnamese Communists for Fighting in South Vietnam', SNIE 14.3-67, see especially pp. 10–16 [CIARR supp. reel 5, 00281]; both Kowet and Ranelagh, previously cited, provide sound summations of the debates; on Carver's 11th hour meeting with Westmoreland see, cable, Carver to DCI 13 September 1967; Carver apparently had little difficulty convincing Westmoreland to accept a compromise, and no mention of an arbitrarily low figure being maintained by Westmoreland was made—an allegation which Carver had made in a series of previous cables to Helms. Further, Carver attributes two important comments to Westmoreland; namely, 1) Westmoreland, stated he could see 'clear logic behind both sets of figures' and 2) he requested that the DIA, CIA, and J-2, continue to consult on estimates to 'see if we could resolve our differences'. Moreover, Carver admitted that he could see legitimate intelligence reasons for reducing the CIA figure [CIARR supp. reel 5, 00025].
41. Ranelagh, p. 461.
42. See, SNIE 14.3-1967 [CIARR supp. reel 5].
43. See Ranelagh, pp. 456–69, and Powers, pp. 187–9.
44. *Harper's* (New York), May 1975, pp. 41–4, 62–73.

45. See the previously cited works of Kowet, Kowet and Bedell, Kaiser, *et al.*
46. '1. We would count hard hats when we could count—like main-force regulars. 2. We would use ranges where intelligence was soft—like guerrillas. Where we couldn't really determine how many, we would just stop counting them. Carver's comments to George Crile, reproduced in Kowet, p. 146.
47. Even footnotes, however, carry much weight. As one insider has written, the 'power of footnotes is formidable, since there is a deeply rooted view that the right way to read estimates is simply to find the split judgements'. Quoted from G Paul Holeman Jr, 'Estimate Intelligence', in Gerald W Hopple and Bruce W Watson, eds., *The Military Intelligence Community* (Westview Press: Boulder, 1986), p. 137.
48. Ray Cline comments in, Roy Godson, ed., *Intelligence Requirements for the 1980's: Analysis and Estimates* (Washington, DC: National Strategy Information Center, Inc., 1980), p. 77.
49. On saturation see Moise, p. 665, while for the failure to dissent see, Lt Col Evan H Parrott, 'CBS News, General Westmoreland, and the Pathology of Information', *Air University Review*, September–October 1982, p. 102. On the problems with the data base see note 26 above. See also the top level memo produced for the President's new OB file which detailed the areas of 'potential controversial judgements' within SNIE 14.3-67, 'Memo: Potentially Controversial Judgments or Data Holding Changes in 14.3-67, 15 November 1967 [CIARR supp. reel 5, 00357].
50. Kowet, pp. 12 and 88; Rostow, pp. 462–3.
51. Memo for Lt Col Robert M Montague, military assistant to Robert Komer, from Carver, 28 September 1966 [CIARR supp. reel 3, 00713]. See also Kowet, p. 86.
52. 'Briefing Materials: Conference on Assessment of Enemy Strengths', 10–16 April 1968 [CIARR supp. reel 6, 0001]; 'Report on Conference on DCI Assessment of Enemy Strengths', April 1968 [CIARR supp. reel 6, 00030]; and 'Missions and Activities of the VC Irregular Forces', 24 April 1968 [CIARR supp. reel 6, 00075]. But it is not clear that the CIA's assets in Vietnam were ever directed towards measuring the full extent and detailed composition of the village/hamlet level insurgency.
53. See, 'Briefing Materials', cited note 52 above.
54. MACV briefing on OB, November 1967, pp. 5–6 [DDRS-DOD 1985. 1585].
55. On evidence of an active deception operation, see Thomas C Thayer, ed., 'A Systems Analysis View of the Vietnam War 1965–1972' (Washington, DC: Office of Assistant Secretary of Defense, Systems Analysis), Vol. 4, pp. 46–50, and Jones. For an analysis of other deception operations mounted to influence US public opinion on the war, see Chapter 7.
56. Michael Handel, in Godson, p. 85.
57. Primarily, MACV wished to include in the OB only the forces that American troops were tactically engaged against, that is, large unit formations. Such was not an unreasonable position. On the whole, American forces were explicitly directed against the large-unit threat by a national policy, originated in Washington during 1962, namely, the United States Overseas Internal Defense Plan, see, Charles Maechling Jr. 'Our Internal Defense Policy—A Reappraisal', *The Foreign Service Journal* (January 1969), pp. 19–21, 27, and 'Insurgency and Counterinsurgency: The Role of Strategic Theory', *Parameters*, Autumn, 1984, pp. 32–41. As well, there were also bureaucratic incentives for the Army to fight the type of war it was designed to fight rather than the subversive insurgency presented by the 'irregulars', see Andrew Krepenivich, *The Army and Vietnam* (Baltimore, MD: Johns Hopkins, 1986), passim, and Russell F Weigley, *The American Way of War* (Bloomington: Indiana University Press, 1977), p. 464. Nevertheless, as Walter Laqueur has previously pointed out, neither MACV's intelligence organs, nor the CIA, ever 'accurately assessed the relationship between the US force structure and the nature of the war'. Laquerer, p. 182.
58. CIA review of pre-Tet intelligence, 15 February 1968, pp. 4–5 [DDRS 78:38A]; see also, Vo Nguyen Giap, *Big Victory, Great Task* (New York: Praeger, 1968), pp. 33–86, which presaged the offensive. Giap's treatise had been translated and debated within the US defence establishment before the end of 1967. On the internal debate see, Thayer, 'A Systems Analysis View of the Vietnam War 1965–1972', Vol. 4, pp. 46–50.

59. See, Douglas Pike, *PAVN: People's Army of Vietnam* (Novato, CA.: Presidio Press, 1986), pp. 213–53.
60. Quoted from captured documents cited in 'Missions and Activities of the VC Irregular Forces', 24 April 1968 [CIARR supp. reel 6, 00075]. Cf. *Vietnam: The Anti-US Resistance War For National Salvation 1954–1975: Military Events*, trans. 3 June 1982 (Washington, DC: Foreign Broadcast Information Service), pp. 93–4, and Fleet Marine Force Pacific, 'FMFPac: Historical Summary, March 1965–September 1967', Vol. 1, pp. 3–20, [USMC Historical Center].
61. See Testimony of the former chief of the CIA's Indochina Division, George Allen, quoted in Moise, p. 672, and CIA situation report, 21 February 1968 [DDRS-CIA 1983, 1594].
62. See, for example, Anthony Cave Brown, *Bodyguard of Lies* (New York: Harper and Row, 1975).

Conclusions

O, what a tangled web we weave
When first we practise to deceive!
But when we've practised quite a while
How vastly we improve our style!

J R Pope, expanding on Sir Walter Scott[1]

This book set out to improve knowledge in four areas: the circumstances in which deception has been used; its results; the proclivity of East and West to use deception in international relations under conditions short of war; and the systems' relative vulnerability to such deception. This concluding section addresses these issues and offers observations that are inevitably tentative and incomplete, but which may nevertheless be useful.

The deceptions discussed in the case studies are summarised in tabular form on subsequent pages. The deceptions all fulfilled one of three broad purposes: to mobilise; to achieve concealment or surprise; or to protect legitimacy. Some chapters provide more than one example of deception. These are listed either as separate instances, where there are major differences in purpose or style (for example, between Khrushchev's inflated claims and his Cuban adventure), or as a collective group, where they belong to a series of deception operations (the inflated claims themselves; the Soviet interventions; Hanoi's deceptions, and the CIA's operations in Western Europe). In one case (the Soviet interventions) a collection appears twice, because it gave rise to two quite different types of deception. Categorisation in this manner does not facilitate drawing conclusions based on statistical analysis. In any case, the number of operations examined in this volume is probably too small to constitute a representative sample. Nonetheless, the case studies do permit a more general analysis within the context of the four areas in which improved knowledge is sought.

Summary of Deceptions

a Chapter	b Case Study	c Purpose	d Type of Target	e Results of Deception	f Impact on Sponsor's Wider Policy
			EASTERN		
1	Reichstag Fire	Anti-fascist Mobilisation	Adversary Publics	Successful	Positive
2	Klugmann	Anti-Mihailovich Pro-Tito Mobilisation	Adversary Leaders and Publics	Successful	Positive
3	Orthodox Church	Mobilisation of Christians	Adversary Publics	Partly Successful	Generally Positive
4A	Khrushchev's Inflated Claims	Concealment of Strategic Inferiority	Adversary Leaders and Publics	Successful	Negative
4B	Soviet Missiles for Cuba	Concealment of Deployment	Adversary Leaders	Failure of dissimulation exposed operation	Negative
5A	Four Soviet Interventions	Concealment and Surpirse	Adversary Leaders	Successful	Positive
5B	Four Soviet Interventions	Legitimacy	Domestic and Adversary Publics	Partly Successful	Positive but Created Problems
6	Six-Day War	Anti-Israel Mobilisation	Client Leaders	Successful but ran out of control	Negative
7	Vietnamese Deceptions	Anti-US, Pro-Hanoi Mobilisation	Adversary Publics	Successful	Positive
8	Aldo Moro	Anti-US Mobilisation	Adversary Publics	Partly Successful	Positive but Very Weak
9	KAL 007	Legitimacy	Domestic and Adversary Publics	Partly Successful	Positive
10	Peace Offensive	Anti-Defence Mobilisation	Adversary Publics	Partly Successful	Positive

a	b	c	d	e	f
			WESTERN		
11	Protocol M	Anti-Soviet Mobilisation	Domestic Publics	Successful Until Exposed	Positive
12A	CIA in West Europe	Anti-Communist Mobilisation	Allied Publics	Successful Until Exposed	Positive
12B	CIA and the Munich Radios	Anti-Soviet Mobilisation	Allied and Domestic Leaders and Publics	Successful Until Exposed	Initially Positive; Negative when Exposed
13	Suez, 1956	Legitimacy	Domestic and Allied Leaders and Publics	Partly Successful	Negative; Distorted Policy & Undermined Trust
14A	U-2 Operations	Concealment	Adversary Leaders	Failure of Dissimulation Exposed Operation	Negative
14B	Post U-2 Cover-up	Legitimacy	Domestic Leaders and Publics	Failed: Simulations not credible	Negative; Undermined Trust
15	Bay of Pigs	Legitimacy	Domestic Leaders and Publics	Failure of Dissimulation Exposed Truth	Negative
16	Enemy Strength in Vietnam	Mobilise US Confidence in War Effort	Domestic Publics	Successful until Tet Offensive Exposed True Enemy Strength	Negative

Circumstances of Use

Mobilisation deceptions have the goal of persuading the target to commit itself in support of a cause. Sometimes, deception is used to break an existing commitment; in other cases deception provides the illusion that old and new causes are compatible. Since allegiance lies at the heart of ideological conflict, and as ideology is a key factor in East–West relations, mobilising operations may be particularly important. Moreover, they tend to be strategic in scope.

Willi Münzenberg's deceptions were classic mobilising operations, laying the groundwork for a popular front against fascism and for the recruiting of Soviet spies. SOE-Cairo played a key role in discrediting General Mihailovich so that British support could be mobilised behind Tito. Through the Foreign Office's use of Protocol M, and a multitude of covert CIA operations, British and West European publics were mobilised to resist Communism. The spirit of resistance was also fostered among Soviet and East European publics by Radio Liberty and Radio Free Europe, whose provenance was deceptively presented.

The Vietnam War provided several examples of mobilising deceptions. By creating false pictures of: a democratic, non-communist, indigenous Southern insurrection; an American military employing criminal deeds and genocide as a matter of policy; and a benevolent North Vietnamese regime with no ambitions towards the domination of the South, the Vietnamese communists succeeded in mobilising many Americans and much Western and Third World opinion against the war and the United States. Conversely, by presenting to the American public a Viet Cong 'Order of Battle' that omitted irregular but potent enemy resources, the United States Command in Vietnam deceived itself as well as its wider audience in a way that may have temporarily sustained optimism and support for the war.

Moscow's role in bringing about the Arab–Israeli War of 1967, which rested entirely on deception, was also a mobilising operation. It prepared Egypt and Syria for confrontation and, whether intentionally or by error, set events in motion which culminated in war. In a very different setting, Moscow also tried to mobilise Italian and world opinion against the United States by the deception operation blaming Aldo Moro's death on the CIA. The remaining two examples in this category are the Soviet peace deceptions and the attempt to mobilise Christianity on the side of 'scientific atheism'.

Examination of the targets of these mobilisation operations

throws some interesting light on the shape of this ideological struggle. Six out of the seven Soviet-initiated deceptions were aimed at the West, and all but two were directed exclusively at Western publics rather than leadership elites. The only Soviet operation aimed at her client states, the deception of Arab nations prior to the 1967 War, was directed at leaders. In the discrediting of Mihailovich, British leaders and publics were both targeted. The emphasis on publics as targets is carried over to Western operations, and this is clearly a characteristic of mobilising deceptions. But unlike Soviet stratagems, which concentrated on adversary audiences, all the Western operations were aimed primarily at domestic audiences, friends and allies.[2]

The use of deception to achieve surprise or to conceal is typical of wartime operations, and the inclusion of several cases reflects the "Cold War" nature of East–West relations, with its emphasis on covert and other 'low-intensity' operations. Surprise was important in the Soviet interventions in Hungary, Czechoslovakia and Afghanistan and in the internal crackdown in Poland. In Cuba in 1962 the Soviet Union tried unsuccessfully to deceive the United States over the installation of missiles. This category also contains an example that illustrates the classic military formula of masking weakness with shows of strength. Khrushchev tried to conceal the Soviet Union's strategic inferiority by false presentations of bomber and missile strengths. Concealment, however, stimulates curiosity. The United States, anxious to penetrate the wall of secrecy guarding the true levels of Soviet strategic strength, used the U-2 reconnaissance aircraft. The deception plan designed to conceal its mission provides the last example in this category.

The targets for surprise and concealment deceptions were in all cases the adversary. In the Soviet intervention operations this included 'fraternal' parties that had stepped out of line. As in full-scale war, the targets were political and military leaders, although in the case of the bomber and missile deceptions, Khrushchev seemed concerned to reach American publics too.

Presumably deception operations in this category accompany all intelligence or covert actions, and many small-scale military operations. Indeed, it is virtually impossible to undertake such activities without the utmost secrecy as a minimum. Wherever a cover story or legend is necessary to disguise some person or deployment, secrecy (or dissimulation) is joined by simulation and a deception begins.

The final category consists of deceptions to achieve, maintain or restore legitimacy. Typically, these take place prior to, during, or after some political or military action; they are defensive, even apologetic, in their style. The interventions in Hungary, Suez, Czechoslovakia, the Bay of Pigs, Afghanistan and Poland were accompanied by deceptions to protect legitimacy. The American cover-up that followed the U-2 incident and the Soviet attempts to shift the blame after the shootdown of the Korean airliner were 'after the event' improvisations.

All legitimacy deceptions were directed at publics, with domestic audiences as first priority. Instigators may have hoped to influence adversary publics as well, especially in the KAL 007 case, but legitimacy starts at home and it is this base that must be protected at all costs. Leaders were also targeted, at least in the Anglo-French and the two American examples. Since the West is inclined to dismiss Russian 'public opinion' as a factor in the formulation and execution of Soviet foreign policy, the attention the Soviets gave to the domestic audience might appear surprising. But it is important not to overlook the ideological imperative; all events must be explained, so that the Soviet people will understand the 'correct' interpretation of those events. Since such understanding cannot be left to chance, the deception must be run at home as well as abroad.

Results

Deception is a matter of perception. All deceptions succeed or fail in the minds of intended victims. As noted in the Introduction, a deception succeeds if the presentation attracts the intended victim's attention, holds interest, forms in his mind the meaning intended by the planners, and does not alert him to the fact that he is being deceived. The deception achieves its goal as the victim 'buys' the false presentation, believing it to be true. It fails if the target takes no notice of the presentation, notices but judges what he sees irrelevant, misconstrues the intended meaning, or detects the method of deception. These are the criteria governing the assessments of success or failure of the *deceptions* in this book.

In addition, however, the purpose of 'selling' a false presentation is to create in the mind of the target a motive for *acting* in a way that seems logical to the target but which is really to the initiator's advantage—what the Soviets call 'reflexive control'.[3] So although the deception is successful when the victim 'buys' the presentation, the *operation* succeeds only when the victim acts in the manner intended. Operations succeed or fail according to reactions, but these may result from collective decisions on how to deal with a perceived new situation. It is quite possible that a decision-making group may be divided in its acceptance of a presentation but nevertheless agree to react on majority opinion. Sometimes the target may be a public—an electorate or powerful faction. In such cases, the likelihood of a divided response to the presentation is high. Some will 'buy', some will smell a rat. In the type of 'peacetime' operations studied in this volume, where publics are frequently targeted, partial acceptance of presentations is quite common, unlike wartime when a decision is often reserved to key decision-makers, and the deception either succeeds or fails completely.

Finally, there is the effect of the operation on the policy it was designed to serve. Did the victim's manipulated reactions help or hinder? Was the exposure of a flawed operation embarrassing? And what unplanned secondary effects arose later? The judgement here is between a positive or negative impact on policy. Clearly, while assessments of success rates in deceptions and operations are interesting in a technical sense, it is the impact on policy that really matters and is the major factor in analysing deception in the East–West context.

Of the 20 deceptions presented in the case studies, the false presentation was accepted completely on 10 occasions, partly on six and was rejected four times. Although the success rate appears reasonably high, the injunction against drawing firm conclusions from these figures remains.

The rejections occurred when the intended victims detected the attempted deception. In three cases, dissimulation was so weak that the targets could see the true facts that ought to have been hidden: the real operational purpose of the U-2; American support for the invasion of Cuba; the missile equipment and its base facilities in Cuba. The fourth example—the U-2 cover-up—failed because the simulation, the cover story launched after the downing of the plane, was blown by physical evidence in Soviet hands.

Partial acceptance was usually the result where deceptions were aimed at publics. By no means did this imply failure. The deception of an influential minority, or even the casting of doubt, can bring advantage. The examples of the Soviet peace offensive, the use of the Russian Orthodox Church, and the Korean airliner are testimony to this. For the vital domestic audience, Soviet legitimacy deceptions over their interventions were probably adequate. And even where the proportion of the target audience that 'bought' the simulation was quite small, as with the French and British over Suez and the Soviets in respect of Aldo Moro, at least the operations provided the faithful with a standard argument.

In all the other case studies the targets 'bought' the deceptions in their entirety. Had the Vietnamese deceptions against the American war effort been listed separately, it probably would have been accurate to describe each as a partial success, since there was always a measure of resistance. However, piled on one another, to form what may be termed a 'layered deception', the three operations cumulatively seem to have achieved success. In complex operations of this kind the conclusion of one operation becomes the assumption of the next.

In trying to assess the impact which these operations had on the wider policy they were designed to serve, it is necessary to consider any unwanted secondary effects. Obviously, where operations failed at the deception stage, adverse effects occurred. Khrushchev was humiliated over the Cuban missile debacle. Eisenhower was trapped by the failure of the U-2 deception and humiliated internationally by the bungled cover-up. American credibility was perhaps damaged more by the mendacity surrounding the Bay of Pigs incident than by the actual event.

There were three cases where the deception worked well but turned out to be counter-productive to the wider mission. Khrushchev's inflated claims were taken seriously by the Americans; but by the time they were exposed, the United States had already overtaken the Soviets in the strategic arms race. The Soviet deception of client Arab states worked beautifully, but the operation ran out of control and dragged these allies into war and defeat. The following year, whatever benefit the United States military in Vietnam could claim from their deceptively optimistic reports on enemy strengths turned to ruin as the Tet Offensive exposed the truth.

In addition to these spectacular reverses of fortune, several other operations that were completely or partly successful at the deception stage suffered from unwanted side effects immediately or later. Soviet legitimising deceptions over Czechoslovakia and Afghanistan raised morale problems when troops came face to face with reality. Insofar as the Anglo-French Suez operation worked at all, its undermining of allied trust and its skewering of strategy marked it as a policy disaster. In three other cases, Protocol M, the CIA's operations in Europe, and the establishment of the Munich radios, later exposure caused varying degrees of political embarrassment.

This analysis has made no distinction between minor tactical deceptions and those which influenced or might influence the course of East–West relations. An operation might have a totally positive impact within its own terms of reference and yet be a drop in the bucket of world affairs. Deceptions providing cover or surprise for clandestine operations are such. It is now necessary to stand back from the detail and attempt to assess the real significance of deception in the East–West context.

Operations which seem to have made important political impacts begin with Münzenberg's anti-fascist mobilisation, which influenced the West European climate of opinion in the years leading up to the Second World War. The decision to abandon Mihailovich in favour of Tito was certainly a very important contribution to the eventual communist take-over in Yugoslavia. The CIA's role in mobilising intellectual and political power to implement America's decision to help West Europe resist communist subversion was of historic importance. It is unlikely that North Vietnam would have prevailed had America's commitment to the defence of the South remained unbroken. The break probably came from several causes: war weariness, casualties, and the elusiveness of victory being among them. But it is difficult to dismiss the notion that the three deceptive themes propagated by North Vietnam were instrumental in converting American doubts into opposition, thus releasing the passion that forced a political decision to abandon the South. Nevertheless, it seems fair to conclude the task was made easier by the existence of a 'credibility gap' and the 'shock' of the Tet Offensive. The 'gap ' and the 'shock' were unwanted byproducts of American deception.

In any limited span of time, the Soviet operations involving the Orthodox Church and the World Peace Council might be dismissed as window-dressing. Over the longer term, their combined impact as layered deceptions could have a significant impact on Western public opinion, in the manner demonstrated in Vietnam.

Finally, two of the operations which failed miserably nevertheless had important results for East–West relations. The U-2 incident provided Khrushchev with an excuse to break off his summit meeting with Eisenhower; Khrushchev's subsequent attempt to introduce missiles into Cuba brought the world to the brink of war. In these two cases it was policy that affected East–West relations. Deception's failure to hide policy brought matters to a head, and its exposure aggravated matters by demonstrating deceitfulness.

Proclivity to Use Deception

There is no doubt that deception has always been a part of human affairs and in the period covered by this book it has remained a constant ingredient of diplomacy and extra-diplomatic activity in the East–West context. It is not the intention here to establish whether the East or the West were the worst liars. The ethical shortcomings of a deception are in the eye of the beholder, and objectivity is impossible. The point of this section is to discuss the proclivity of East and West to undertake deception operations, from an operational viewpoint. The only ethical constraint among deceivers is shared by both sides. It is an 'Eleventh Commandment': 'Thou shalt not be found out.'

At the time of the Crusades, the French and English had no difficulty legitimising intervention in the Near East. Christianity, as then interpreted, provided a 'world view' and any questioning would have been regarded as heretical. Even 50 years before the Suez expedition, the imperial myth would have armed these nations with a self-confidence almost immune to criticism. By 1956, however, attitudes at home and abroad had changed; actions of this sort could be justified by Western powers only within the Charter of the United Nations and international law. Hence, presumably, the Anglo-French pretence at 'peace-keeping'. Hence, too, the American desire not to be associated with the Cuban interventions, or the deliberate overflying of foreign territory.

For the Soviets, the Marxist–Leninist world view provides the modern equivalent, in operational if not in theological terms, of medieval religion and nineteenth century imperialism combined. Legitimising deception therefore raises no theoretical problems: the incident is simply subordinated to the world view and interpreted accordingly. In practice, problems may arise when what the people see with their own eyes contradicts all they have been told by their superiors, a difficulty which in earlier times also troubled crusaders and imperialists.

Such problems notwithstanding, the East sustains its legitimising deceptions as part of the Higher Truth. Official versions take their place in 'history', for at least as long as they are deemed to be useful. When the Party's requirements change, so does history, as can be seen from various rewritings of the Stalin period under Khrushchev and Gorbachev.

The policy called *glasnost* seems to recognise the dangers inherent in applying the ideological world view in total disregard of the facts. The gap between 'objective truth' and reality that had affected troop morale in Afghanistan appears to have had similar effects among the Soviet population, so that official pronouncements lost credibility. *glasnost*, it would seem, seeks to protect the Higher Truth by separating it from the day-to-day reporting of world events, to which has been restored a far greater measure of honesty. But the AIDS deception and other current Soviet operations indicate clearly that *glasnost* does not inhibit deception.[4] At the strategic level, the appearance of openness and honesty that *glasnost* has created among non-communist audiences might become a potent deception asset.

All the Western deception cases in this volume occurred before or during the 1960s. However, the Libyan example discussed in the Introduction demonstrates that deception remains in the West's inventory. Clearly, Western leaders can be expected to deceive whenever it is necessary to preserve surprise or concealment of covert or military operations, and to protect their own legitimacy under circumstances where the truth will not do. However, public and especially media reaction to the Libya deception[5] suggests that the spectres of the U-2 and the Bay of Pigs continue to haunt American foreign policy. Perhaps this tempers enthusiasm for major deception operations.

The major difference in proclivity is likely to arise out of operational capability. In organisational terms, the West has a variety of means for promulgating messages to audiences in the East, the West, and the Third World. If it wishes, it can insert deceptive material into such messages. Within intelligence services, it has the means to conduct sophisticated deception. It is difficult to determine how effective these means might be in reaching adversary leaders, but obviously they can be augmented by diplomatic and other channels.

Absent from the West's organisation is the monolithic control and the long-term commitment to policy goals that seem to be characteristic of the East. Also absent from the West's deception inventory is any equivalent of the permanent transmission belts, political organisations, and front organisations that serve the Soviet Union in the non-communist world. Thus the opportunity to conduct *effective* mobilising deceptions against adversary publics is virtually non-existent. The case studies have uncovered four Western mobilising deceptions—all targeted against friends and allies; other than the Libya operations, none were directed against an adversary.

The failure of that effort points to another organisational disparity that greatly restricts the West's operational capability to deceive. The open society and a free news media are likely to detect and publicise any evidence of deception operations, regardless of their target, and thus destroy them. Without sustained dissimulation, there cannot be deception. With or without *glasnost*, the East can preserve most of its secrets: the West, it would seem, cannot. The high failure rate among the Western cases in this book testifies to the power of the media in this respect.

The example of the Libya deception also points to another constraint that an unfettered media imposes on Western deception operations. This is the problem of 'blowback', in which disinformation planted overseas is picked up as reliable information both by the deceiver's news media, and by the intelligence service that planted it. Not only is such self-induced deception embarrassing; if believed and acted upon, it could skew estimates and resulting actions.[6]

Historical, cultural and ideological differences divide East from West and have created two quite different governing structures and styles. John Dziak has described the Soviet Union as the perfect 'counter-intelligence state'.[7] As such, those features—institutionalised secrecy, centralised control and strategic planning, and a large apparatus with a diverse range of assets—clearly gave the Soviet Union an advantage in conducting deception operations. At the very least, it provided a head start in seizing upon and exploiting opportunities. Hardly flawless of course; Khrushchev, apparently acting on impulse, was not a good deceiver. But the apparatus, when properly co-ordinated, has an enviable track record that bears out Pope's observation quoted at the head of this chapter.

Relative Vulnerability

On this matter, assessment necessarily rests on more speculative ground. This being the case, the fact that the evidence points to a Soviet Bloc advantage calls for caution in drawing hard and fast conclusions on the basis of the small sample that constitutes this study.

No one is ever invulnerable to deception. This is one common denominator between East and West. At the leadership level, information is much desired, particularly information that may support a favoured proposal or perception, and an adversary might provide this in order to deceive. In the West, competition in ideas close to top leadership levels may make targeting difficult for an opponent, since there is no way of being sure which faction will prevail. At the same time, such competition heightens the need for information, and hence the opportunities for deception.

Until recently in the Soviet Bloc, the prevalent party line provided a relatively predictable framework within which a Western deceiver could operate. From the Soviet perspective, moreover, the very unpredictability of Western leadership style probably lent, and still lends, credibility to fears of quite outlandish deception schemes. Yet, even where there is factional infighting, it has been more difficult for Western operators to know enough about the true situation to exploit it.

On the face of it, then, vulnerability at this level is much the same in both camps. The main difference is structural, and lies below this level. The Soviet Bloc is no longer impervious to Western influence, but until recently that influence, whether overt or covert, faced the formidable obstacles of the 'counter-intelligence state'. Many of those obstacles remain, and only time will tell whether *perestroika* will mean greater Western access and influence.

The West remains, by preference, relatively open in terms of access and influence, both for its citizens and for leaders. Western governments are oriented to responding to the demands of the electorate, pressure groups, and even individuals, and a free and sceptical media usually ensures that the voices of dissent and criticism are heard. The media itself is sensitive to 'issues' and controversy, and plays a role in determining which are to be amplified and to be given significance, and which are to be played down. This is the nature of the Western political arena. It is, without question, open and vulnerable to Soviet Bloc deception operations, through agents, fronts, and disinformation. And it cannot be denied that these efforts have yielded some successes; the Münzenberg, National Liberation Front, and World Peace Council cases attest to that fact. But it is also a resilient, dynamic arena, capable of absorbing, diluting, or rejecting influence. What is significant today may be irrelevant or forgotten a year from now. While there are risks in this approach to politics, there are benefits as well. The Soviet Bloc cannot plan on long-term consistency in its influence on the West.

Epilogue

Throughout the "Cold War", both sides have used deception. The East, able to plan and sustain deceptions over the long term and to exploit the open nature of Western societies, has been more successful. There is nothing in recent developments which suggests that either side is about to forsake this weapon. And even if *glasnost* is sustained and expanded in the Soviet Union, the asymmetries of the two operational theatres are likely to persist for the foreseeable future. This is hardly a foundation for détente, and attests to deception's inherently destructive nature. For if truth is the first casualty of war, trust is the first casualty of deception. Without trust, how can relations between blocs, states, or individuals be anything other than a state of undeclared war?

Endnotes

1. J R Pope, 'A Word of Encouragement', in Kingsley Amis, ed., *The New Oxford Book of English Light Verse* (New York: Oxford University Press, 1978).
2. As early as the 1960s, Western writers were speculating that the USSR confined ideological warfare to the non-communist world. See, for eg, A Wiseman, 'Peaceful Coexistence—Myth or Menace', *NATO Letter*, November 1960, pp. 6–8.
3. Clifford Reid, 'Reflexive Control in Soviet Military Planning', in Brian D Dailey and Patrick J Parker, eds., *Soviet Strategic Deception* (Lexington, Mass: Lexington Books, 1987), pp. 293–311.
4. For examples see, *Active Measures: A Report on the Substance and Process of Anti-US Disinformation and Propaganda Campaigns* (Washington, DC: US Department of State, August 1986—Publication 9630); *Soviet Influence Activities: A Report on Active Measures and Propaganda, 1986–87* (Washington, DC: US Department of State, August 1987—Publication 9627); *Soviet Active Measures in the Era of Glasnost*, A Report to Congress by the United States Information Agency, March 1988.
5. *Washington Post*, 2 October 1986; *New York Times*, *Los Angeles Times*, 3 October 1986, 2 November 1986; 'A Bodyguard of Lies', *Newsweek*, 13 October 1986.
6. Several studies, among them Gregory Treverton, *Covert Action: the Limits of Intervention in the Postwar World* (New York: Basic Books, 1987), p. 165; and James Adams, *The Financing of Terror* (London: New English Library, 1986), p. 3, have stated that CIA analysts—directed to revise their estimates of Soviet involvement in international terrorism on the basis of information in Claire Sterling's 1981 book, *The Terror Network*—discovered that her assertions were based on CIA disinformation planted overseas. Neither book elaborates on the matter, nor do the authors cite an authority (Treverton claims personal knowledge; Adams acknowledges that the story may be apocryphal). It is noteworthy that the Soviets made the same charge in their overt propaganda several years earlier. See Andrei Grachev, *In the Grip of Terror* (Moscow: Progress Publishers, 1982), p. 139.
7. John J Dziak, *Chekisty: A History of the KGB* (Lexington, Mass: Lexington Books, 1988), pp. 16–17, 39, 103, 167–9.

Select Bibliography

This bibliography comprises a selected list of sources on the theory and practice of deception in war, diplomacy, psychological warfare and intelligence activities, and on deception operations in the East-West context. It consists of bibliographies, documents, books and monographs, articles, chapters and periodicals. It does not include unpublished papers or items from the print or broadcast media. Readers seeking a more comprehensive listing should consult the bibliography compiled by Stanley Zell (see below).

I. Bibliographies

Blackstock, Paul W, and Frank L Schaf. *Intelligence, Espionage, Counter-espionage and Covert Operations*. Detroit: Gale Research, 1978.

Bryant, Melrose M, comp. *Deception in Warfare*. Air University Library, Maxwell Air Force Base, Alabama, July 1985.

Cline, Marjorie W, Carla E Christiansen, and Judith M Fontaine, eds. *Scholar's Guide to Intelligence Literature: Bibliography of the Russell J Bowen Collection, Georgetown University*. Frederick, Maryland: University Publications of America, 1973.

Constantinides, George C. *Intelligence and Espionage: An Analytical Bibliography*. Boulder, Colorado: Westview Press, 1983.

Harris, William R. *Intelligence and National Security: A Bibliography with Selected Annotations*. Revised ed., Cambridge, Massachusetts: Center for International Affairs, Harvard University, 1968. 3 vols.

Pforzheimer, Walter, ed. *Bibliography of Intelligence Literature, Eighth Edition, 1985*. Washington DC: Defense Intelligence College, 1985.

Rocca, Raymond G, and John J Dziak. *Bibliography on Soviet Intelligence and Security Services*. Boulder, Colorado: Westview Press, 1985.

Smith, Myron J Jr. *The Secret Wars: a Guide to Sources in English*. War-Peace Bibliography Series. Santa Barbara, California: ABC-CLio, 1980-1981. 3 vols. (Nos. 12–14).

Wilcox, Laird. *Master Bibliography on Political Psychology and Propaganda: Espionage and Intelligence Operations; and Terrorism Assassination*. Kansas City: Editorial Research Service, 1985.

Zell, Stanley. *An Annotated Bibliography of the Open Literature on Deception*. N–2332-NA. Santa Monica, California: Rand Corporation, December 1985.

II. Documents

Lambert, D R. *A Cognitive Model for Exposition of Human Deception and Counterdeception*. Technical Report 1076. San Diego, California: US Navy, Naval Ocean Systems Center, October 1987.

Reitz, James T. *Lexicon of Selected Soviet Terms Relating to Maskirovka (Deception)*. Washington, DC: Dept of Defense, Defense Intelligence Agency, International Applications Office, October, 1983, 52 pp.

United States Arms Control and Disarmament Agency. *Soviet Propaganda Campaign Against NATO*. Washington, DC, 1983.

United States Arms Control and Disarmament Agency. *The Soviet Propaganda Campaign Against the US Strategic Defense Initiative*. Washington, DC, 1986.

United States. Congress. House. Select Committee on Intelligence, Hearing Before the Sub-Committee on Oversight. *Soviet Covert Action: The Forgery Offensive*. 90th Congress 2nd Session, 1980.

United States. Senate, Committee on the Judiciary, Hearing Before the Subcommittee to Investigate the Administration of the Internal Security Act and Other Internal Security Laws. *The Techniques of Soviet Propaganda*. Report prepared by Suzanne Labin. 90th Congress, 1st Session, 1967.

United States. Senate. Select Committee to Study Government Operations with Respect to Intelligence Activities. *Book I: Foreign and Military Intelligence*. Washington, DC: US Government Printing Office, 1976.

United States. Senate. *Interim Report: Alleged Assassination Plots Involving Foreign Leaders*. New York: Norton, 1976.

United States. Department of Justice, Federal Bureau of Investigation. *Soviet Active Measures in the United States*. Washington, DC, June 1987.

United States. Department of State. *Active Measures: A Report on the Substance and Process of Anti-US Disinformation and Propaganda Campaigns*. Publication 9630. Washington, DC, 1986.

United States. Department of State. *Soviet Active Measures*. Special Report No. 110. Washington, DC, September 1983.

United States. Department of State. *Soviet Active Measures: Focus on Forgeries*. Foreign Affairs Note. Washington, DC, April 1983.

United States. Department of State. *Soviet Active Measures: Forgery, Disinformation, Political Operations*. Special Report No. 88. Washington, DC, October, 1981.

United States. Department of State. *Soviet Influence Activities: a Report on Active Measures and Propaganda, 1986-87*. Publication 9627. Washington, DC: DOS, 1987.

United States. Department of State, and Central Intelligence Agency. *Contemporary Soviet Propaganda and Disinformation: a Conference Report*. Publication 9536. Washington, DC: DOS, 1987.

III. Books and Monographs

Agee, Philip. *Inside the Company: CIA Diary*. Harmondsworth, Middlesex, UK: Penguin Books, 1975.

Andrew, Christopher. *Secret Service: The Making of the British Intelligence Community*. London: Heinemann, 1985.

App, Austin J. *Power and Propaganda in American Politics and Foreign Policy*. New York: Revisionist Press, 1984.

Arbatov, Georgi. *The War of Ideas in Contemporary International Relations*. Moscow: Progress Publishers, 1973.

Atkinson, James. *The Politics of Struggle: The Communist Front and Political Warfare*. Chicago, Illinois: Regnery, 1966.

Bailey, Geoffrey. *The Conspirators*. New York: Harper, 1960.

Balfour, Michael. *Propaganda in War 1939–1945*. London: Routledge and Kegan Paul, 1979.

Barghoorn, Frederick. *Soviet Foreign Propaganda*. Princeton, New Jersey: Princeton University Press. 1964.

Barron, John, *KGB: The Secret Work of Soviet Secret Agents*. New York: Reader's Digest, 1974.

Barron, John. *KGB Today: The Hidden Hand*. New York: Reader's Digest, 1983.

Beaumont, Roger. *Maskirovka: Soviet Camouflage, Concealment and Deception*. College Station, Texas: Texas A & M, 1983.

Betts, Richard K. *Surprise Attack*. Washington, DC: Brookings, 1982.

Bittman, Ladislav. *The Deception Game*. Syracuse, NY: Syracuse University Research Corporation, 1972.

Bittman, Ladislav. *The KGB and Soviet Disinformation*. New York and Oxford: Pergamon-Brassey's, 1985.

Bittman, Ladislav, ed. *The New Image-Makers: Soviet Propaganda and Disinformation Today*. McLean, Virginia: Pergamon-Brassey's 1989.

Blackstock, Paul W. *Agents of Deceit: Frauds, Forgeries and Political Intrigue Among Nations*. Chicago, Illinois: Quadrangle, 1966.

Blackstock, Paul W. *The Secret Road to World War Two: Soviet Versus Western Intelligence 1921-1939*. Chicago, Illinois: Quadrangle, 1969.

Bloch, Jonathan and Patrick Fitzgerald. *British Intelligence and Covert Action*. Dingle, Ireland: Brandon Book Publishers, 1983.

Bok, Sissela. *Lying: Moral Choice in Public and Private Life*. New York: Pantheon, 1978.

Bok, Sissela. *Secrets: On the Ethics of Concealment and Revelation*. New York: Pantheon Books, 1982.

Bowyer, J Barton. *Cheating: Deception in War and Magic, etc*. New York: St Martin's Press, 1982.

Breckenridge, Scott D. *The CIA and the US Intelligence System*. Boulder, Colorado: Westview Press, 1986.

Cave Brown, Anthony. *Bodyguard of Lies*. New York: Harper and Row, 1975.

Chakotin, Serge. *Rape of the Masses: The Psychology of Totalitarian Political Propaganda*. Studies in Philosophy, No. 40. New York: Haskell, 1971.

Clews, John. *Communist Propaganda Techniques*. New York: Praeger, 1964.

Cline, Ray S. *Secrets, Spies and Scholars: Blueprint of the Essential CIA*. Washington, DC: Acropolis, 1976.

Corson, William R. *The Armies of Ignorance: The Rise of the American Intelligence Empire*. New York: Dial, 1977.

Crozier, Brian. *The Conflict of Information: Conflict Studies No. 56*. London: Institute for the Study of Conflict, 1975.

Cruickshank, Charles C. *Deception in World War II*. New York: Oxford, 1979.

Dailey, Brian D and Patrick J Parker, eds. *Soviet Strategic Deception*. Lexington, Massachusetts: Lexington Books, 1987.

Dallin, David. *Soviet Espionage*. New Haven, Connecticut: Yale University Press, 1955.

Daniel, Donald C, and Katherine L Herbig, eds., *Strategic Military Deception*. New York: Pergamon Press, 1982.

Daugherty, William E. *A Psychological Warfare Casebook*. [1958] Reprint. New York: Arno Press, 1979.

Deacon, Richard. *A History of the Russian Secret Service*. London: Frederick Muller, 1972.

Deacon, Richard. *The Truth Twisters*. London: Futura Publications, 1988.

DeLeon, Peter. *Soviet Views of Strategic Deception*. P-6685. Santa Monica, California: Rand Corporation, 1981.

Delmer, Sefton. *Black Boomerang: an Autobiography*. London: Secker and Warburg, 1961.

Delmer, Sefton. *The Counterfeit Spy*. New York: Harper and Row, 1971.

Dulles, Allen Welsh. *The Craft of Intelligence*. New York: Harper and Row, 1971.

Dziak, John J. *Soviet Perceptions of Military Power: The Interaction of Theory and Practice*. New York: Crane Russak & Co., 1981.

Dziak, John J. *Chekisty: a History of the KGB*. Lexington, Massachusetts: Lexington Books, 1988.

Ebon, Martin. *The Soviet Propaganda Machine*. New York: McGraw-Hill, 1979.

Ellul, Jacques. *Propaganda: The Formation of Men's Attitudes*. New York: Alfred Knopf, 1965.

Epstein, Edward Jay. *Deception: The Invisible War Between the KGB and the CIA*. New York: Simon and Schuster, 1989.

Fain, Tyrus G, Katharine C Plant and Ross Milloy, eds. *The Intelligence Community: History, Organization, and Issues*. New York: Bowker, 1977.

Findlay, P T *Protest, Politics and Psychological Warfare*. Melbourne: Hawthorne Press, 1968.

Freedman, Lawrence. *US Intelligence and the Soviet Strategic Threat*. Boulder, Colorado: Westview, 1977.

Garwood, Darrell. *Under Cover: Thirty-five Years of CIA Deception*. Stafford, Virginia: Dan River Press, 1980.

Gerson, Leonard D. *The Secret Police in Lenin's Russia*. Philadelphia: Temple University Press, 1976.

Godson, Roy, ed. *Intelligence Requirements for the 1980's*. 7 vols. New York: National Strategy Information Center, 1979–1982; Lexington, Massachusetts: Lexington Books, 1986.

Golitsyn, Anatoliy. *New Lies for Old*. New York: Dodd, Mead & Co., 1984.

Gooch, John and Amos Perlmutter, eds. *Military Deception and Strategic Surprise*. London: Frank Cass, 1982.

Gordon, Joseph S, ed. *Psychological Operations: The Soviet Challenge*. Boulder, Colorado: Westview, 1988.

Grant, Natalie. *Deception: a Tool of Soviet Foreign Policy*. Washington, DC: Nathan Hale Institute, 1987.

Griffith, Samuel B. *Sun Tzu: The Art of War*. Oxford: Clarendon Press, 1963.

Handel, Michael. *The Diplomacy of Surprise: Hitler, Nixon, Sadat*. Cambridge, Massachusetts: Harvard Center for International Affairs, 1981.

Hannah, Gayle Durham. *Soviet Information Networks*. Washington, DC: Center for Strategic and International Studies, 1977.

Haswell, Jock. *The Tangled Web: The Art of Tactical and Strategic Deception*. Northampton: Defence Publishers, 1986.

Hazan, Baruch A. *Soviet Propaganda: A Case Study of the Middle East Conflict*. Jerusalem: Keter Publishing, 1976.

Heilbrunn, Otto. *The Soviet Secret Services*. London: Allen and Unwin, 1956.

Hingley, Ronald. *The Russian Secret Police: Muscovite, Imperial Russian, and Soviet Political Security Operations*. New York: Simon and Schuster, 1970.

Hinsley, F H, E E Thomas, C F G Ransom, and R C Knight. *British Intelligence in the Second World War: Its Influence on Strategy and Operations*. 4 Vols. London: Her Majesty's Stationery Office, 1979–1988.

Hollander, Paul. *Political Pilgrims*. Oxford: Oxford University Press, 1981.

Horelick, Arnold L, and Myron Rush. *Strategic Power and Soviet Foreign Policy*. Chicago and London: University of Chicago Press, 1965.

Hutchings, Raymond. *Soviet Secrecy and Non-Secrecy*. London: Macmillan, 1987.

Jervis, Robert. *Perception and Misperception in International Politics*. Princeton, New Jersey: Princeton University Press, 1976.

Jones, R V. *Most Secret War*. London: Hamish Hamilton, 1978.

Kam, Ephraim. *Surprise Attack: the Victim's Perspective*. Cambridge, Massachusetts: Harvard University Press, 1988.

Karalekas, Anne. *History of the Central Intelligence Agency*. Laguna Hills, California: Aegean Park Press, 1977.

Kenez, Peter. *The Birth of the Propaganda State: Soviet Methods of Mass Mobilization, 1917-1929*. New York: Cambridge University Press, 1985.

Kirkpatrick, Jeanne, ed. *The Strategy of Deception*. New York: Farrar, Straus and Co., 1963.

Kirkpatrick, Lyman B. *The US Intelligence Community: Foreign Policy and Domestic Activities*. Boulder, Colorado: Westview, 1985.

Kirkpatrick, Lyman B. and Howland H Sargent. *Soviet Political Warfare Techniques: Espionage and Propaganda in the 1970s*. New York: National Strategy Information Center, 1972.

Knightley, Phillip. *Philby: KGB Masterspy*. London: André Deutsch, 1988.

Knorr, Klaus and Patrick Morgan, eds. *Strategic Military Surprise*. New Brunswick, New Jersey: Transaction Books, 1983.

Krivitsky, W G. *In Stalin's Secret Service*. New York: Harper and Bros., 1939.

Laqueur, Walter. *A World of Secrets: The Uses and Limits of Intelligence*. New York: Basic Books, 1985.

Lefebvre, Vladimir A, and Victorina D Lefebvre. *Reflexive Control: The Soviet Concept of Influencing an Adversary's Decisionmaking Process*. Denver, Colorado: Science Applications, Inc., February 1984.

Leites, Nathan. *A Study of Bolshevism*. Glencoe, Illinois: Free Press, 1953.

Leggett, George. *The Cheka: Lenin's Political Police*. New York: Oxford University Press, 1981.

Lendvai, Paul. *The Bureaucracy of Truth: How Communist Governments Manage the News*. London: Burnett Books; Boulder, Colorado: Westview, 1981.

Levchenko, Stanislav. *On the Wrong Side*. New York and London: Pergamon-Brassey's, 1988.
Machiavelli, Niccolo. *The Prince, The Discourses and Other Writings*. Edited by John Plamenatz. London: Fontana, 1972.
Marchetti, Victor, and John D Marks. *The CIA and the Cult of Intelligence*. New York: Laurel, 1974.
Margulies, Sylvia. *The Pilgrimage to Russia: The Soviet Union and the Treatment of Foreigners, 1924–1937*. Madison, Wisconsin: University of Wisconsin Press, 1968.
Martin, David C. *Wilderness of Mirrors*. New York: Harper and Row, 1980.
Masterman, J C. *The Double Cross System in the War of 1939 to 1945*. New Haven, Connecticut: Yale University Press, 1972.
Matsulenko, Maj. Gen. Viktor Antonovich, *Operativnaia Maskirovka Voisk (Po Opymn vedikoy otechostvennoy voyny)* [Operational Military Camouflage and Deception: Based on the Experience of the Great Patriotic War]. Moscow: Voenizdat, 1975.
Maurer, Alfred C, Marion D. Tunstall, and James M. Keagle, eds. *Intelligence Policy and Process*. Boulder, Colorado: Westview, 1985.
McLaurin, Ron D, ed. *Military Propaganda: Psychological Warfare and Operations*. New York: Praeger, 1982.
Mickelson, Sig. *America's Other Voice: The Story of Radio Free Europe and Radio Liberty*. New York: Praeger, 1983.
Mickiewicz, Ellen P. *Media and the Russian Public*. New York: Praeger, 1981.
Montagu, Ewen. *The Man Who Never Was*. Philadelphia: J B Lippincott, 1954, rev. ed., 1967.
Mure, David. *Master of Deception: Tangled Webs in London and the Middle East*. London: William Kimber, 1980.
Mure, David. *Practice to Deceive*. London: William Kimber, 1977.
Orman, John M. *Presidential Secrecy and Deception: Beyond the Power to Persuade*. Westport, Connecticut: Greenwood, 1980.
Owen, David. *Battle of Wits: A History of Psychology: Deception in Modern Warfare*. London: Leo Cooper, 1978.
Pacepa, Ion Mihai. *Red Horizons: Chronicles of a Communist Spy Chief*. Washington, DC: Regnery Gateway, 1987.
Page, Bruce, David Leitch and Phillip Knightley. *The Philby Conspiracy*. Toronto: Fontana Books, 1968.
Pfaltzgraff, Robert L, Uri Ra'anan, and Warren H Milberg, eds. *Intelligence Policy and National Security*. London: Macmillan, 1981.
Philby, Kim. *My Silent War*. [1968] repr. New York: Ballantine Books, 1983.
Phillips, David Atlee. *The Night Watch*. New York: Ballantine, 1977.
Pincher, Chapman. *The Secret Offensive: Active Measures: A Saga of Deception, Disinformation, Subversion, Terrorism, Sabotage and Assassination*. London: Sidgwick and Jackson, 1985.
Potter, William C, ed.*Verification and SALT: The Challenge of Strategic Deception*. Boulder, Colorado: Westview, 1980.
Powell, David E. *Anti-religious Propaganda in the Soviet Union: A Study of Mass Persuasion*. Cambridge, Massachusetts: MIT Press, 1975.
Powers, Thomas. *The Man Who Kept the Secrets: Richard Helms and the CIA*. New York: Alfred Knopf, 1979.
Prados, John. *The Soviet Estimate: US Intelligence Analysis and Russian Military Strength*. New York: Dial, 1982.
Ramaswamy, T N. *Essentials of Indian Statecraft: Kautilya's Arthasatra for Contemporary Readers*. London: Asia Publishing House, 1962.
Ranelagh, John. *The Agency: the Rise and Decline of the CIA*. New York: Simon and Schuster, 1986.
Rasberry, Robert W. *The Technique of Political Lying*. New York: Universal Press of America, 1981.
Richelson, Jeffrey T. *The U.S. Intelligence Community*. Cambridge, Massachusetts: Ballinger, 1985. 2nd. ed., 1989.
Richelson, Jeffrey T. *Sword and Shield: Soviet Intelligence and Security Apparatus*. Cambridge, Massachusetts: Ballinger, 1986.
Robertson, K G, ed. *British and American Approaches to Intelligence*. London: Macmillan/RUSI, 1987.

Roetter, Charles. *Psychological Warfare*. London: Batsford, 1974.
Scott, Andrew. *The Revolution in Statecraft: Informal Penetration*. New York: Random House, 1965.
Sejna, Jan. *We Will Bury You*. London: Sidgwick and Jackson, 1982.
Selznick, Philip. *The Organizational Weapon: A Study of Bolshevik Strategy and Tactics*. New York: McGraw-Hill, 1952.
Shultz, Richard H, and Roy Godson. *Dezinformatsia: Active Measures in Soviet Strategy*. New York: Oxford: Pergamon-Brassey's, 1984.
Sleeper, Raymond S, ed. *Mesmerized by the Bear: The Soviet Strategy of Deception*. New York: Dodd Mead and Company, 1987.
Soley, Lawrence, and John S Nichols. *Clandestine Radio Broadcasting: a Study of Revolutionary and Counter-Revolutionary Electronic Communication*. New York: Praeger, 1987.
Soviet Disinformation. Columbia School of Journalism and University of Missouri at Columbia, 1981.
Stockwell, John. *In Search of Enemies: a CIA Story*. New York: W W Norton, 1978.
Suvorov, Viktor, [pseud.]. *Inside Soviet Military Intelligence*. New York: Macmillan, 1984.
Thompson, James W, and Saul K Padover. *Secret Diplomacy: Espionage and Cryptography, 1500–1815*. New York: Frederick Unger, 1963.
Thompson, Oliver. *Mass Persuasion in History: a Historical Analysis of the Development of Propaganda*. Edinburgh: Paul Harris, 1977.
Treverton, Gregory F. *Covert Action: the Limits of Intervention in the Postwar World*. New York: Basic Books, 1987.
Turner, Stansfield. *Secrecy and Democracy: the CIA in Transition*. Boston: Houghton Mifflin, 1985.
Tyson, James L *Target America*. Chicago: Regnery Gateway, 1981.
Vigor, Peter H. *Soviet Blitzkrieg Theory*. New York: St Martin's Press, 1983.
Wettig, Gerhard. *Broadcasting and Detente*. New York: St Martin's, 1977.
Whaley, Barton. 'Stratagem: Deception and Surprise in War'. Master's thesis, MIT Centre for International Studies, 1969.
Wheatley, Dennis. *The Deception Planners*. London: Hutchinson, 1980.
Wise, David. *The Politics of Lying, Government Deception, Secrecy and Power*. New York: Random House, 1973.
Wohlstetter, Roberta. *Pearl Harbour: Warning and Decision*. Stanford, California: Stanford University Press, 1962.

IV. Articles, Chapters, and Periodicals

Beaumont, Roger. 'Soviet Psychological Warfare and Propaganda.' *Signal: Journal of the Armed Forces Communications and Electronics Association*, Vol. 42, No. 3 (1987).
Beichman, Arnold. 'Soviet Active Measures and Democratic Culture.' *World Media Report*, Vol. II, No. 1 (Spring 1987).
Berezkin, A. 'On Controlling the Actions of an Opponent.' *Voyennaia Mysl'* [Military Thought], No. 11 (1972), pp. 91–4.
Betts, Richard K. 'Surprise Despite Warning: Why Sudden Attacks Succeed.' *Political Science Quarterly* (Winter 1980/81), pp. 551–72.
Bok, Sissela. 'Secrets and Deception: Implications for the Military.' *Naval War College Review*, Vol. 38, No. 2 (March–April 1985), pp. 73–80.
Boorman, Scott A. 'Deception in Chinese Strategy: Some Theoretical Notes on the *Sun-Tzu* and Game Theory.' In Colonel William W Whitson, USA, ed. *The PLA in the 1970s*. New York: Praeger, 1972.
Burhans, William A. 'Radiodezinformatsiya.' *Journal of Electronic Defense*. (May 1986), pp. 22–3.
'Chronicle: Greece's Disinformation Daily?' *Columbia Journalism Review* (November/December 1983), pp. 5–7.
Covert Action Information Bulletin. Washington, DC: Covert Action Publications (1978–date).
Crossman, R H S. 'Psychological Warfare.' *Journal of the Royal United Services Institute for Defence Studies*, Vol. 97 (August 1952), pp. 319–32.

Daniel, Donald C, and Katherine Herbig. 'Propositions on Military Deception.' *Journal of Strategic Studies*, Vol. 5, No. 1 (March 1982), pp. 155–77.

Debo, Richard K. 'Lockhart Plot or Dzerzhinskiy Plot?' *Journal of Modern History*, No. 42 (September 1971), pp. 413–39.

DeMowbray, Stephen. 'Soviet Deception and the Onset of the Cold War.' *Encounter* (July/August 1984).

Dick, C J. 'Catching NATO Unawares: Soviet Army Surprise and Deception Techniques.' *International Defense Review* (January 1986), pp. 21-6.

Disinformation: Soviet Active Measures and Disinformation Forecast. Washington, DC: Institute for International Studies (1985–date).

Douglass, Joseph D Jr. 'Soviet Disinformation.' *Strategic Review* (Winter 1981), pp. 16–25.

Douglass, Joseph D Jr. and Samuel T Cohen. 'Selective Targeting and Soviet Deception.' *Armed Forces Journal International* (September 1983), pp. 95–101.

Douglass, Joseph D Jr. 'Soviet Strategic Deception.' *Defense Science 2002+*. (August 1984), pp. 87–99.

Eagleburger, Lawrence S. 'Unacceptable Intervention: Soviet Active Meaasures.' *NATO Review* (April 1983).

Epstein, Edward Jay. 'Disinformation: Or, Why the CIA Cannot Verify an Arms Control Agreement.' *Commentary* (July 1982), pp. 21–8.

Godson, Roy and Richard Shultz. 'Soviet Active Measures: Distinctions and Definitions.' *Defense Analysis*, Vol. 1, No. 2 (1985), pp. 101–110.

Grant, Natalie. 'Deception on a Grand Scale.' *International Journal of Intelligence and Counter-Intelligence*, Vol. 1, No. 4 (1986), pp. 51–77.

Grant, Natalie. 'Forgery in International Affairs.' *Foreign Service Journal*, No. 47 (May, 1970), pp. 31–2, 46.

Handel, Michael. 'Intelligence and Deception.' *Journal of Strategic Studies*, Vol. 5, No. 1 (March 1982), pp. 122-54.

Handel, Michael. 'Military Deception in Peace and War.' *Jerusalem Papers on Peace Problems*, No. 38 (1985).

Handel, Michael. 'Perception, Deception and Surprise: The Case of the Yom Kippur War.' *Jerusalem Papers on Peace Problems*, No. 19 (1976).

Handel, Michael. 'Surprise and Change in Diplomacy.' *International Security*, Vol. 4, No. 4 (1980), pp. 57-85.

Heuer, Richards J Jr. 'Strategic Deception and Counter-Deception: A Cognitive Process Approach.' *International Studies Quarterly* 25, No. 2 (1981), pp. 294–327.

Hoffmann, Erik P. 'Nuclear Deception: Soviet Information Policy.' *Bulletin of the Atomic Scientists* (August/September 1986), pp. 32–7.

Hotz, Robert. 'Lockheed U-2 Over Sverdlovsk: A Study in Fabrication.' *Aviation Week*, Vol. 72 (26 May 1960), pp. 20–21.

Hant, David. 'Remarks on "A German Perspective on Allied Deception Operations".' *Intelligence and National Security*, Vol. 3, No. 1 (January 1988), pp. 190–94.

Joshua, Wynfred. 'Soviet Manipulation of the European Peace Movement.' *Strategic Review* (Winter 1983), pp. 9–18.

Kahler, Hans. 'Soviet Psychological Warfare.' *International Defense Review* (February 1986), pp. 157-9.

Kobrin, Col N. 'Operational Deception: Examples from WW II.' *Soviet Military Review* (4 April 1981), pp. 42–4.

Kux, Dennis. 'Soviet Active Measures and Disinformation: Overview and Assessment.' *Parameters*, Vol. 155, No. 4 (Autumn 1985), pp. 29–8.

Marro, Anthony. 'When the Government Tells Lies.' *Columbia Journalism Review* (25 March 1985), p. 29.

Martin, L John. 'Disinformation: An Instrumentality in the Propaganda Arsenal.' *Political Communication and Persuasion*, 2, No. 1 (1982).

Mel'nikov, Col Gen P. 'Wartime Experience in Camouflage, Concealment and Deception.' *Voyenna-Istoricheskii zhurnal* (1982).

Mescheryakov, Col Gen V. 'Strategic Disinformation in the Achievement of Surprise in the World War II Experience' [in Russian]. *Voyenna-Istoricheskii zhurnal*, No. 2, February, 1985, pp. 74–80.

Mihalka, Michael. 'Soviet Strategic Deception, 1955–1981.' *Journal of Strategic Studies* 5, No. 1 (1982), pp. 40–93.
Oberg, James E. 'The Sky's No Limit on Disinformation.' *Air Force* (March 1986), pp. 52–55.
Possony, Stefan T. 'Kremlin Ideology and Psychological Warfare.' *Defense and Foreign Affairs* (June, 1983), pp. 24–6.
Quester, George H. 'On The Identification of Real and Pretended Communist Military Doctrine.' *Journal of Conflict Resolution*, Vol. 10 (June, 1966), pp. 172–79.
Remington, Thomas. 'Policy Innovation and Soviet Media Campaigns.' *Journal of Politics* (February 1983), pp. 220–27.
Satter, David. 'The Foreign Correspondent in Moscow—On Manipulation and Self Deception.' *Encounter* (May 1987), pp. 58-64.
'Special Issue on Strategic and Operational Deception in the Second World War.' *Intelligence and National Security*, Vol. 2, No. 3 (July, 1987).
Stein, M L. 'The Soviet KGB and the Press.' *Editor and Publisher* (26 October, 1985).
Stevens, Jennie A, and Henry S Marsh. 'Surprise and Deception in Soviet Military Thought', [Parts 1, 2]. *Military Review*, Vol. 62 (June and July 1982), pp. 2–11, 24–35.
Sullivan, David S. 'The Legacy of SALT I: Soviet Deception and US Retreat.' *Strategic Review*, Vol. 7, Winter, 1979, pp. 26–41.
Suvorov, Viktor, [pseud.] 'GUSM: The Soviet Service of Strategic Deception.' *International Defense Review*, Vol. 18, No. 8 (September 1985), pp. 1235–40.
Ulsamer, Edgar. 'The Fog of War.' *Air Force* (October 1985), pp. 74–6.
Valenta, Jiri. 'Soviet Use of Surprise and Deception.' *Survival*, 24, No. 2 (March–April 1982), pp. 50–61.
Vermaat, J S Emerson. 'Moscow Fronts and the European Peace Movement.' *Problems of Communism* (November–December 1982), pp. 43–56.
Whaley, Barton. 'Toward a General Theory of Deception.' *Journal of Strategic Studies*, 5, No. 1 (1982), pp. 178–92.
Wohlstetter, Roberta. 'Cuba and Pearl Harbor: Hindsight and Foresight.' *Foreign Affairs*, Vol. 43, No. 4 (July 1965), pp. 691–707.
Whohlstetter, Roberta. 'The Pleasures of Self-Deception.' *The Washington Quarterly*, 2 (Autumn, 1979), pp. 54–63.
Yefimov, V A, and S G Chermashchentsev. 'Maskirovka.' *Sovetskaia Voyennaia Entsiklopediia* [Soviet Military Encyclopedia]. Moscow: Voyenizdat, Vol 5, 1978, pp. 175–77. Translated in USAF, *Soviet Military Concepts*, February 1979.

Index

PACE UNIVERSITY LIBRARIES
BIRNBAUM STACKS
D842.D43 1990
Deception operations
3 5061 00520 7460